TRANSPORTATION ENGINEERING

Revised Edition

DR. L. R. KADIYALI
B.E. (Hons.), (Bombay),
P.G. Diploma in Highway and
Traffic Engineering (Newcastle-Upon-Tyne),
Ph.D. (Kakatiya University)
Formerly Chief Engineer, (Roads Wig),
Ministry of Road Transport & Highways, New Delhi
&
Study Director, Road User Cost Study,
Central Road Research Institute, New Delhi

Edited by:

DR. A. I. DHATRAK
B. E. (Civil), M. E. (Geotech. Engg.), Ph.D (Civil)
Associate Professor and Dean Academic,
Government College of Engineering, Amravati, Maharashtra.

KHANNA BOOK PUBLISHING CO. (P) LTD.

Publisher of Engineering and Computer Books

4C/4344, Ansari Road, Darya Ganj, New Delhi-110002
Phone : 011-23244447-48 **Mobile:** +91-9910909320
E-mail : contact@khannabooks.com
Website : www.khannabooks.com

Dear Readers,

To prevent the piracy, this book is secured with HIGH SECURITY HOLOGRAM on the front title cover. In case you don't find the hologram on the front cover title, please write us to at contact@khannabooks.com or whatsapp us at +91-99109 09320 and avail special gift voucher for yourself.

Specimen of Hologram on front Cover title:

Moreover, there is a SPECIAL DISCOUNT COUPON for you with EVERY HOLOGRAM.

How to avail this SPECIAL DISCOUNT:

Step 1: Scratch the hologram
Step 2: Under the scratch area, your "coupon code" is available
Step 3: Logon to www.khannabooks.com
Step 4: Use your "coupon code" in the shopping cart and get your copy at a special discount
Step 5: Enjoy your reading!

Price: ₹ 450.00

ISBN: 978-93-82609-85-8

Transportation Engineering
by L.R. Kadiyali

First Edition: 2016
Revised Edition: 2019
Reprint Edition: 2022, **2026**

To view complete list of books, Please scan the QR Code:

Published by:
Khanna Book Publishing Co. (P) Ltd.
CIN: U22110DL1998PTC095547

Visit us at: **www.khannabooks.com**
Write us at: **contact@khannabooks.com**

Printed in Bharat.

Preface

India's transport system has several deficiencies such as inadequate capacity, poor safety record, emission of pollutants and outmoded technology. But as the economy is poised for a big growth in the coming years, transportation engineers will have to come up with innovative ideas.

The book addresses these issues and it is hoped that the engineering students studying transportation engineering will have a clear idea of the problems involved and how they can be overcome in their professional career. This book has been designed keeping in mind the latest syllabi of all major Indian Universities.

Efforts have been made to include latest developments in different areas of transportation engineering and also minimise errors. The authors sincerely welcome constructive criticisms and suggestions for further improvement.

DR. L. R. KADIYALI

Contents

SECTION—1: TRANSPORTATION SYSTEMS AND PLANNING

Section—2: HIGHWAY ENGINEERING

Section—3: RAILWAY ENGINEERING

Section—4: AIRPORT ENGINEERING

Section—5: CONTINUOUS FLOW SYSTEMS

Section—6: WATER TRANSPORT

Section—7: MULTI-MODAL TRANSPORT

SECTION—1

TRANSPORTATION SYSTEMS AND PLANNING

1 Chapter

Introduction

1.1. Common Terms and their Meaning

Transport deals with the movement of people and materials from one place to another. Movement of people is called passenger transport and movement of materials is called goods transport oı freight transport.

The term 'transportation' is generally used as an adjective, for example transportation engineering

The word 'communications' has a different meaning, as it deals with wire-less, radio, television internet and media, and should not be confused with transport.

Urban transport deals with transport in towns and cities and rural transport deals with transport in the rural area. Urban transport is also known by the term 'intra-city', whereas 'inter-city' transport deals with movement between cities.

Public transport deals with movement of people in vehicles other than their own, and includes non-personalised modes like buses, trains, trams and other intermediate modes of transport. Intermediate Public Transport (IPT) means non-personalised passenger transport in hired modes such as taxis, vans, mini-buses, autorickshaws, rickshaws (three-wheeler cycles). Rapid Transport means mass movement of passengers by road or rail in cities. For example, BRT means Bus Rapid Transport, and MRTS means Mass Rapid Transport System (generally railways).

1.2. Historical Evolution

The history of transport dates back to the period before the advent of recorded history. With the desire to hunt for animals for food, the early human began to form pathways and tracks. As civilisation advanced, the growth of agriculture took place and human settlements began to be formed. From one settlement to another, tracks were formed. These tracks might be the skeletal framework of modern highways. Man soon began to domesticate animals and use them as beasts of burden (horses, oxen, camels, mules, ponies and elephants). It is interesting to note that even in the present century, these beasts of burden are still in use, particularly in less developed countries and remote areas.

The invention of the wheel (approximately 3500–5000 BC) revolutionised transport, and even now wheels are the mainstay of many forms of transport. The advantages of an axle joining two wheels was realised, and two-wheeled and four-wheeled carts, chariots and carriages were developed.

Many civilizations have been known for their excellence and attainments in road building. The Roman civilization, Persian civilization, Chinese civilization and the Indus Valley civilization may be mentioned here.

As settlements began to be developed along the banks of rivers and on the sea coast, boats propelled by oars and sails began to be used. The introduction of iron and steel as ship-building materials and the invention of the steam engine gave an impetus to oceanic shipping.

The early attempts in flying by man centred around balloons filled with hot air or hydrogen. Gliders (which are engineless planes) were successfully built and tried in the last decade of the nineteenth century. The first power-driven plane was flown by the Wright brothers at Kitty Hawk (USA). The airplane contributed significantly to the outcome of World War I. Charles Lindbergh's flight from New York to Paris in 1927 was a historic event. The World War II saw a dramatic shift to the use of aircraft for transport of troops and bombing, ending with the atomic bombing of Hiroshima and Nagasaki. Since then, aviation has grown steeply, contributing to the shrinking of world. In India, J.R.D. Tata's solo flight in 1932 from Karachi to Bombay in a single engined Puss Moth heralded the advent of civil aviation.

1.3. Role of Civil Engineers in Transportation Engineering

Transportation engineering, primarily involves planning, design, construction, maintenance, and operation of transportation facilities. The facilities support air, highway, railroad, pipeline, water, and even space transportation are planned, constructed and maintained by civil engineers. The design aspects of transportation engineering include the sizing of transportation facilities, determining the materials and thickness used in pavement designing the geometry (vertical and horizontal alignment) of the roadway. The transportation is the key component of a nation's progress therefore the role of a transportation engineer is to develop efficient transportation systems that fulfill all the needs. Transportation engineers also need to study the mindset and mannerisms of transportation system users, to come up with appropriate solutions.

1.4. Role of the Government in Transportation

Historically, the government had a major role in the promotion of transport. It constructed roads, railway lines, docks and harbours and airports, which facilitated the movement of goods and passengers. Even today, the government in most countries retains the power to plan, fund, construct and maintain the transport infrastructure. Funds are raised through various taxes. Of late, increasing involvement of the private sector is being sought in constructing and maintaining roads (toll roads), airports and sea terminals. But, the government in most countries retains the power to regulate certain aspects of transport in public interest. For examples, the government enforces rules of registration of motor vehicles and aircraft, for licensing of drivers and aircraft pilots and for observance of safety and environmental standards. In India, the government, through the Planning Commission, decides on the allocations of government funds in the various Five Year Plans for the different modes of transport, with a view to ensure a coordinated development.

2
Chapter

Role of Transport in Society

2.1. Transport: A Vital Infrastructure

Transport, along with power and telecommunications, is a vital infrastructure for the overall development of an economy. The great industrial revolution happened because it was possible to transport raw materials to the factories and the finished goods to the markets. Transport and Communications contribute nearly 5-15 per cent to a country's economy. (In India, this contribution was 7 per cent in 1998-Ref 1).

2.2. Transport and Economy

Transportation shapes, sustains and enhances the economy of a country. Transport and economy are closely inter-related, the one depending on the other vide Fig. 2.1.

Fig. 2.1: Interdependence of Transport and Economy

It is often said that America did not build its roads, rather the roads built America.

The rate at which the country's economy grows closely determines the rate at which transport grows. Generally, the rate of growth of transport is higher than the rate of growth of economy. The ratio between the rate of growth transport and the rate of growth of the economy (known as the elasticity) can be marginally above 1.0 in the case of highly developed countries, but can be as high as 2.0 in the case of low income and middle income economies. The trend is seen in Fig. 2.2.

Thus when India's economy was growing at 3.5 per cent, its road transport was growing at about 7 per cent, and now when its economy is growing at 8.5 per cent, road and air transport can grow at 12-15 per cent.

The relationship between transport and economy can be seen from Fig. 2.3, where the gross national product (GNP) of various countries are plotted on the X – axis and the car ownership rates (cars per 1000 persons) are plotted on the Y – axis. (Long form of GNP).

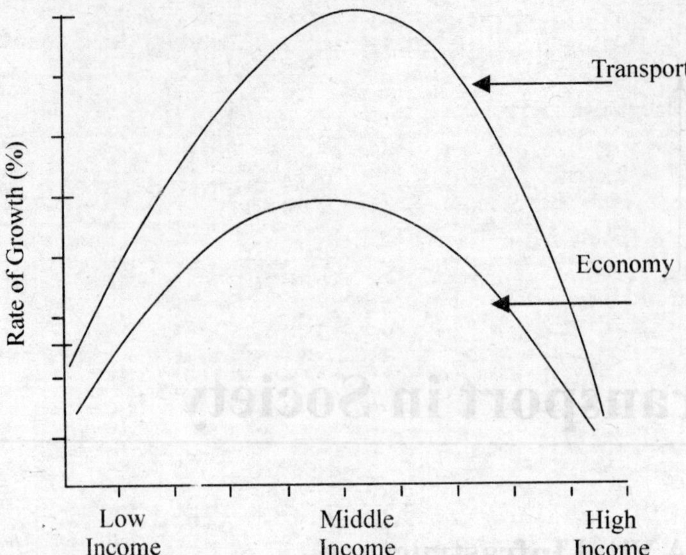

Fig. 2.2: Rate of Growth of Transport and Economy

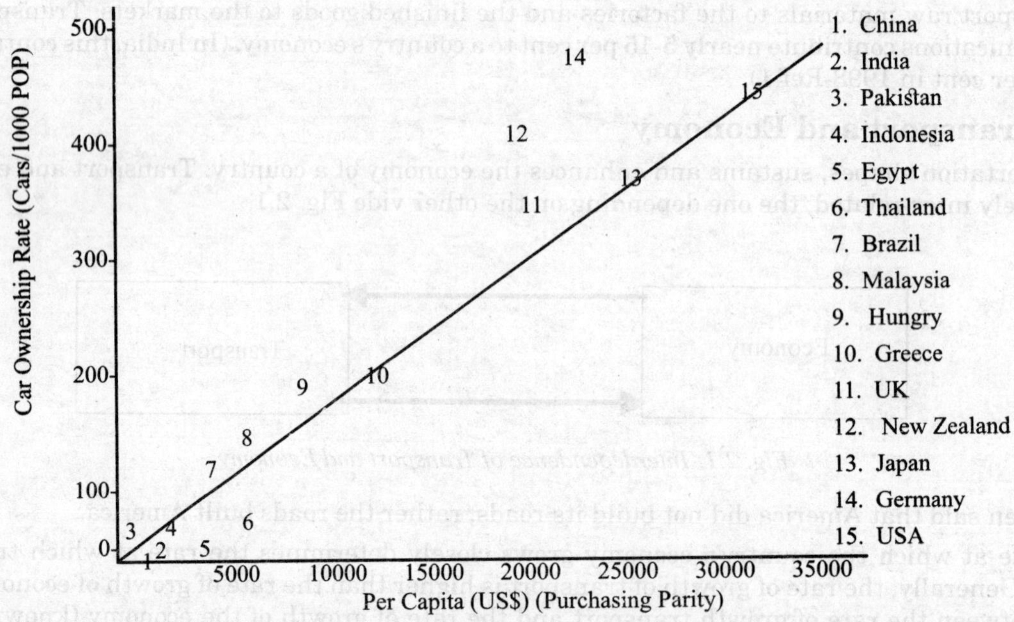

Fig. 2.3: GNP versus Car Ownership Rate

2.3. Transport and Poverty Alleviation

Provision of a good road connectivity to villages can increase the agricultural yield, employment potential, general health of the rural population and per capita income. In India nearly 46 per cent of the villages were not connected by an all weather road in 2000. Some states in India have good connectivity, and in these states the percentage of rural poverty is small. (Refer the Figure 2.4).

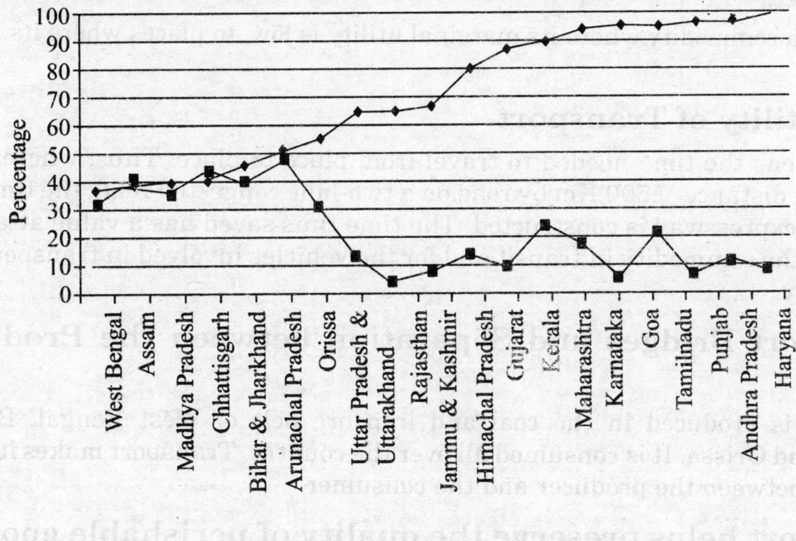

Fig. 2.4: Relationship between Village Connectivity and Rural Poverty in India

2.4. Transport and Urbanisation

In the low income countries, the bulk of the population lives in the villages depending solely on subsistence farming. For example, in Ethiopia, a typical African low income country, the urban population is only 16 per cent. As the transport infrastructure is expanded, the rural population will slowly migrate to towns and cities, and the shifted population will gain employment in manufacturing and services in the towns and cities. In a fully developed country, the urban population is very high: for example, in the USA, the urban population is 79 per cent. In India, the urban population now is 29 per cent, whereas in China it is 43 per cent. China's urban population was only 18 per cent in 1985, and the rapid growth of its economy has been facilitated by urbanisation, which in turn took place because of investments in the transport sector. India's urban population is expected to reach 40 per cent by the year 2020, which can take place if the government's current programme of connecting all villages by an all-weather road by 2021 gets implemented.

2.5. Exploitation of Natural Resources

Most of the low-income and middle income economies depend upon exploitation of natural resources and their export to sustain their economy. India, for example, is able to export its iron ore deposits because of its transport infrastructure (railways, roads and ports). Saudi Arabia, rich in oil, is able to export oil because of a good system of pipe lines and shipping.

2.6. Transport and Place Utility

Raw materials needed for the production of goods are spread over the world and country rather unevenly. These have to be processed and transformed into consumer goods, either at the place of their occurrence or at a distantly located industry, and the goods produced have to be transported to consumer centres. Transport systems thus give "space utility" to the goods, making it possible for transferring a commodity where its marginal utility is low, to places where its marginal utility is high.

2.7. Time Utility of Transport

Transport shortens the time needed to travel from place to place. Thus, whereas it may take 10 hours to travel a distance of 500 Km by road on a two-lane congested road, the time may be reduced to 5 hours if an expressway is constructed. The time thus saved has a value attached to it, for the passengers, for the commodity in transit and for the vehicles involved in transport. Thus transport gives time utility.

2.8. Transport Bridges and Separation between the Producer and the Consumer

In India, steel is produced in the coal and iron ore belt of West Bengal, Bihar, Jharkhand, Chhattisgarh and Orissa. It is consumed all over the country. Transport makes it possible to bridge the separation between the producer and the consumer.

2.9. Transport helps preserve the quality of perishable goods

Perishable goods like milk, fruits, vegetables, fish, poultry products, meat and flowers can be transported quickly from the places of production to the consumer over long distances by means of refrigerated containers travelling by road, rail and air.

2.10. Transport Reduces the Cost of Goods

A good system of transport (for example: good roads) has a low cost of transport. This helps in bringing the goods to the consumer at a reduced cost.

2.11. Transport Accelerates Agricultural Development

Modern agriculture requires inputs like better seeds, pesticides and fertilisers. The inputs can reach the farmer at a low cost if there is a good transport system. Similarly, the farmer is encouraged to produce more only if the price of his produce at the market centre is attractive. A good road system reduces his cost of transportation, and the farmer is induced to produce more to earn a surplus.

2.12. Transport and Tourism Promotion

Tourism can be promoted only if the tourists can reach the destination in a short time and at affordable travel costs. Thus, the key to tourism development is a good transport system.

2.13. Transport Promotes Industrial Development

The Industrial Revolution which took place in Europe owes a lot to the break-through in transport such as in Railways (steam engines) and shipping. Newcastle in U.K. produced coal which could be transported to other cities like Birmingham (to produce steel) and Manchester (to produce textiles). Cotton from India could be shipped overseas to run the Manchester textile mills in U.K. Jute mills got established in Calcutta as raw jute could be transported cheaply by rail and river transport. Textile mills got established in Bombay and Ahmedabad as the hinterland produced cotton which was transported by the newly laid railway lines. It is thus amply clear that primary industries get located where they enjoy the advantages of a good transportation system.

2.14. Transport is Vital for Defence and Strategic Needs

The US Interstate Highway system was conceived to provide a high level of service to interstate travel and to meet defence needs. The Autobahn system in Germany helped the Germans to move their tanks and supplies to the war front. In India, the vast network of border roads has served to guard the country's frontiers. Air Transport is also essential to move troops and supplies to the war front.

2.15. Transport Facilitates International Trade

Singapore port, which handles several millions of containers, has developed as an international hub of trade. India's major ports like Bombay, Calcutta and Madras, have handled the country's international trade, backed by a good system of railways and roads in the hinterland. London, New York and Hamburg are other examples of ports that handle large quantity of international trade, backed up by good port facilities and railways and roads in the hinterland.

2.16. Transport Facilitates Administration of Vast Areas

The Trans-Siberian Railway in Russia is a classic example of how a major transport link has helped in administering the remote area of Siberia. The Qinghai Lhasa railway line recently completed in 2006 has given quick access to Tibet from China. The Jammu-Kashmir valley Railway line, slated for connecting Kashmir Valley to the rest of India's vast rail network will facilitate the administration. The Srinagar-Leh National Highway has facilitated the administration of the remote area of Ladak.

2.17. Transport Facilitates the Specialisation and Economies of Scale

Taking the case of lap-top computers, several countries are associated with the production of components like key board, display screen, mother board, cables, mouse, each country specialising in a particular part and producing it on a mass scale.

This results in economies of scale. In India, the automobile industry has thrived on specialisation and mass production of ancillaries such as clutch assembly, gear box, speedometer, headlight, piston, valves, ignition system, tyres, and so on.

A good system of transport enables the ancillaries to be moved to the central assembly plant.

2.18. Transport Influences Growth of Cities

As motor vehicles became popular in USA, the people found it more convenient to move out from the crowded central part of towns to the suburbs. Good roads made it possible for people to commute to the work place in a reasonable time.

The underground tube railway of London has made it possible for people to reside on the outskirts and travel to work comfortably.

The construction of a new bridge across the Thane creek separating the island of Bombay and the mainland has given birth to a new flourishing township on the mainland (New Bombay). Cities have grown historically on the banks of great rivers which facilitated the movement of people and goods.

Examples are: London (on the Thames River) and Newcastle (upon the Tyne river) in U.K.; Patna (on the Ganges river) and Calcutta (on the Hooghly river) in India.

2.19. Transport Facilitates, Evacuation and Relief Operations during Natural Disasters

The importance of a good system of transport to reach out to people affected by natural disasters is enormous. Examples of the Tsunami in India, Katrina in US, Bhuj earthquake in India amply demonstrate this.

REFERENCES

1. Book of the year 2007, Encyclopaedia Britannica, Chicago, 2007.
2. Rural Road Development Plan: Vision 2015, Ministry of Rural Development, Government of India, New Delhi, 2007.
3. The Challenge of Development, World Development Report, 1991, Published for the World Bank by Oxford University Press, Delhi, 1991.
4. India Vision 2020, Report of Planning Commission's Committee, Academic Foundation, New Delhi, 2003.

3

Chapter

Modes of Transport and Criteria for Choice

3.1. Multiple Modes

To meet any given demand for transport of people and goods within a country and between countries, multiple modes are available, each mode having its own advantages and disadvantages. The changes in technology taking place often renders the choice of a mode a difficult task and the transport planner has to reckon future possibilities. Considerations of sustainability, environmental effects and economy are important. The various modes available, their characteristics and criteria for choice are discussed in this Chapter.

3.2. Important Modes currently available

The important modes of transport currently available are:

1. Roads and Road Transport
2. Raiiways
   - Surface
   - Underground (Tube)
   - Elevated
   - Mass Rapid Transport (popularly known as Metro in India)
   - Light Rail Transit (Tram and other modern system)
3. Air Transport
4. Water Transport
   - Coastal Shipping
   - International Shipping
   - Inland Water Transport
5. Pipelines
6. Aerial Ropeways
7. Combination of various modes such as container movement
8. Conveyor Belts

3.3. Criteria for Choice

The choice of a particular mode is governed by various criteria such as:

1. Speed
2. Adequacy
3. Frequency
4. Regularity
5. Safety
6. Environmental Impact
7. Responsibility
8. Cost
9. Fuel Efficiency
10. Employment Generation
11. Comfort

3.4. Speed

Speed is of prime importance in the movement of people and goods.

Table 3.1 gives some typical speeds of various types of vehicles.

Table 3.1: Speeds of Various Transport Modes and Vehicles

	Mode/ Vehicle	Speed (Range as prevalent in India) (Km/hr)	Remarks
1.	Railway (Broad Gauge)	50–120	High Speed Trains such as the Bullet trains which can travel at 250 Km/hr
2.	Road Transport		
	Cars	30–120	Speeds are low in hills. High speeds are possible on Expressways
	Buses and Trucks	20–80	
	Two-wheelers	25–60	
	Autorickshaws	15–30	
	Cycles	8–15	
3.	Air Transport		
	Concorde	2250	
	Boeing 747/ Air bus	1000	

(Contd...

	Mode/ Vehicle	Speed (Range as prevalent in India) (Km/hr)	Remarks
	Boeing 707	950	
	Propeller Aircraft	300–400	
	Helicopter	200–300	
4.	Water Transport		
	Ocean Liners	35-55	
	Coastal Ships	15-35	
	Barges for Inland Water Transport	10–25	
5.	Pipelines	5–10	
6.	Ropeways	5–10	

3.5. Adequacy

Adequacy is the capacity of a mode to transport people and goods. The capacity of a railway system is governed by the number of trains that can be handled per day. Table 3.2 gives some values.

Table 3.2: Capacity of Railway System

1.	Single Track	20-30 trains per day
2.	Double track with manual track control	60-80 trains per day
3.	Double track with centralised track control	160-200 trains per day

In India, super-fast trains consist of over 20 coaches and carry over 1500 passengers. Metros (Mass Rapid System) can cater to trains with a headway of 2 minutes, and can carry 60,000 persons per hour per direction. In modern Light Rail Transit, capacity upto 20,000 persons/hour/direction is possible. Goods trains in India can haul 50-60 wagons, each wagon having payload of 50-80 tonnes.

As regards roads, traffic is heterogeneous consisting of fast moving vehicles of various types and slow vehicles like cycles, cycle rickshaws and animal-drawn vehicles. Traffic is represented in terms of Passenger Car Units (PCU) and the following design capacity values as follows (Ref 1).

Table 3.3: Capacity of Roads

Carriageway Width	Design Capacity (PCUs/day)
Single Lane (3.75 m)	2000
Intermediate Lane (5.5 m)	6000
Two Lane (7.0 m)	15,000–17,500
Four-Lane (Divided)	35,000–40,000

The following values of passenger occupancy in vehicles have been found from Indian Studies (Ref 2).

Table 3.4: Average Occupancy of Passenger Vehicles in India

Car	4.8
Bus	4.3
Two-wheeler	1.5
Autorickshaws	2.5

An average two-axled single unit truck is permitted to carry 10 tonnes of pay load, tough overloading upto 20 tonnes is quite common. A single unit three truck can carry 15 tonnes of pay load, though overloading upto 30 tonnes is quite common. Truck-trailer combinations can carry pay loads of 25-40 tonnes.

The advantage of road transport is that any number of buses and trucks can be mobilised to cater to the demand. Hence it has great flexibility.

The capacity of barges and ships is given in Table 3.5.

Table 3.5: Capacity of Barges and Ships

Small barge of IWT	100–400 tonnes
Large barge of IWT	1,000–2,000 tonnes
Coastal Ships	3,000–20,000 tonnes (Dead Weight Tonnage DWT)
Ocean Liners	
(a) Crude Oil	150,000 DWT for berths 300,000 DWT for single point moorings
(b) Dry Bulk	150,000 DWT
(c) Containers	
Feeder Vessels	600 Twenty Feet Equivalent Units (TEU)
Main Line Vessels	2,000–3,000 TEUs

The capacity of water transport is also governed by the handling capacity of port berths, whose typical values are given in Table 3.6.

Table 3.6: Handling Capacity of Port Berths

	Type of Berth	**Handling Capacity per Year**
1.	General Cargo	
	Break Bulk	60,000 T

	Type of Berth	Handling Capacity per Year
	Neo-bulk (steel, auto etc.)	200,000–400,000 T
2.	Dry Bulk	300,000–1,000,000 T
3.	Containers	100,000–150,000 TEUs
4.	Petroleum	1.5–6.0 million T
5.	Liquid Bulk other than Petroleum	80,000 T

The seating capacity of modern air-craft is given in Table 3.7.

Table 3.7: Seating Capacity of Air-Craft

	Type	Maximum Passengers
1.	Douglas DC-4	86
2.	Caravelle	139
3.	Boeing 707	90
4.	Boeing 737	130
5.	Boeing 747	390
6.	Airbus A-300	345

The handling capacity of a single runway with modern facilities is in the range of 170,000 to 215,000 landings and take-offs per year.

Pipelines can handle gases, liquids and solids in a slurry form. A 50 cm dia pipeline can handle about 250 tonnes of solids per day.

Capacity of conveyor belts depends upon the length and height to be negotiated, and the speed.

3.6. Frequency and Regularity

Almost all modes of transport offer frequent and regular schedules, which are very important for customers.

3.7. Safety

Road transport is known for its poor record of safety in India. Nearly 140,000 persons get killed every year on India's roads. Engineering, Enforcement and Education are the three Es of safety, and in all of these, India has a lot to improve.

Rail transport is comparatively safe since trains move on a dedicated track and crossing and overtaking take place only in stations under controlled conditions. Unmanned level crossings present a safety hazard. But when accidents do take place, the death toll is very high.

Air Transport has an excellent record of safety because of strict measures of recruitment of pilots, maintenance of aircrafts, control of movement of aircrafts, modern instruments for landing and strict vigilance against terrorists. Accidents are caused by bird hits, lightning, storms and fog. Whenever an accident does take place, the death toll is very high.

Oceanic water transport is very safe. However, Inland Water Transport lacks modern instrumented aids and hence barges are liable to be affected by cyclones and storms. Pipelines are very safe. Aerial Ropeways also have a good safety record.

3.8. Environmental Impact

Roads and Road Transport use liquid fuels which emit considerable amount of pollutants causing health hazards and smog which is dangerous to driving. Noise pollution and visual intrusion are other serious effects. Road construction consumes considerable quantity of stone aggregates and thus raises serious doubts about its sustainability. Rail transport does not create environmental hazards. In India, the toilets in the trains discharge into the track and this needs to be stopped by adopting modern methods such as those used in aircraft. Water transport also does not create adverse environmental impact. So also pipelines and ropeways. Aircrafts produce noise which is of a high order causing discomfort to residents near airports.

3.9. Responsibility

Responsibility is the extent to which the facility operator meets liabilities or compensation for damages, loss or theft. A government owned operator like railways and reputed airlines are better than private truck operators.

3.10. Cost

Air transport is the costliest for passengers and goods movement. For urban travel, passenger fares are almost similar for buses, suburban trains and Metros. For long distance passenger travel, bus fares and rail fares (by ordinary second class sleeper) are almost same. But air-conditioned fares by rail are higher than bus fares.

As regards freight movement, road transport is cheaper for short hauls upto 300-350 Km, but beyond this the advantage lies with railways. High value commodities like tea, cotton textiles, TV sets, electronic goods, vegetables and fruits etc. prefer road transport because of lower costs and other advantages. As regards long hauls, the cost advantages lies with railways. Bulk goods like iron ore, coal, fertilisers prefer railways for long hauls, though road transport may be choice for short hauls upto 200-250 Km.

3.11. Fuel Efficiency

The energy intensity of air transport is obviously very high, followed by rail, road, water transport and pipe-line. Table 3.8 gives some typical values

Table 3.8: Energy Intensity Of Various Transport Modes

	MODE	ENERGY REQUIREMENT
A.	***Passenger***	***Litres Per Passenger-Km***
1.	Modern Aircraft like Boeing 747	0.05
2.	Car (4 seats)	0.025
3.	Bus (52 seats)	0.005

(Contd...)

MODE		ENERGY REQUIREMENT
A.	***Passenger***	***Litres Per Passenger-Km***
4.	AC Railway Berth	0.016
5.	Non-AC Railway Sleeper Berth	0.003
B.	***Freight***	***Litres Per Tonne-Km***
1.	Rail	0.008
2.	Heavy Commercial Truck	0.03–0.06
3.	Light Commercial Vehicle	0.104

3.12. Employment Generation

For a labour-surplus country like India, employment generation is an important consideration. The National Transport Policy Committee of Planning Commission have given the following values of employment potential (Ref 3).

Table 3.9: Employment Potential of Various Transport Modes

	Mode	Employment Potential Per 100,000 Rupees of Investment (1977-78 Price Level) Person-Year
1.	Inland Water Transport	33.59
2.	Development of navigational channels	13.20
3.	Road construction and maintenance	27.50
4.	Road Transport Operation	
	Bullock Cart	27.00
	Three-wheeler	17.15
	Trucks	16.95
	Buses	9.26
	Taxis	3.71
	Vehicle Production	1.78
5.	Coastal Shipping	
	Sailing vessels	20.00
	Other coastal shipping	2.40
6.	Railways	4.30
7.	Air Transport	0.97

(Contd...)

3.13. Comfort

Air travel is the most comfortable. Rail travel, particularly in AC coaches is also comfortable. Bus travel by AC sleeper coaches is not uncomfortable for journeys upto 300 Km. Travel by car can be comfortable for medium distance upto 300 Km.

REFERENCES

1. Guidelines on Capacity of Roads in Rural Areas, IRC:64-1990, Indian Roads Congress, New Delhi, 1990.
2. Manual for Economic Analysis for Highway Projects, IRC:SP:30-2009, Indian Roads Congress, New Delhi, 2009.
3. Report of the National Transport Policy Committee, Planning Commission, New Delhi, 1980.

4

Chapter

Transport Planning

4.1. Introduction

Transport is an important sector of a nation's economy. In fact, transport and economy are inter-related as shown below:

Fig 4.1: Inter-relationship between Economy and Transport

As a country's economy grows, the transport system grows, and similarly as the transport system grows the country's economy also grows. Transport planning is a science that seeks to study the need and manner in which transport facilities have to be provided in the urban, regional or national setting.

4.2. Inter-relationship between Land Use and Traffic

In 1954, Mitchell and Rapkin propounded a theory that traffic was a function of land use (Ref 1). Though the statement is not spectacular, it sums up the ground reality. To give an example, Connaught place in New Delhi is the centre of commercial activity, and thus attracts considerable traffic.

Another example is the head quarters of Infosys in Bangalore where a large number of software engineers work, thus attracting high volume of traffic. If a dedicated freight corridor between Delhi and Mumbai comes up, many satellite townships will develop along the corridor, manufacturing goods which can be easily transported.

4.3. Hierarchy of Transport Plans

Transport plans can be prepared and implemented at various levels such as:

1. National Level
2. Regional Level, such as State or Region
3. Local Level, such as city, town or a district

An example of national level plans is the National Highway Development Programme of the National Highways Authority of India. The recently completed Strategic Road Network Master Plan for Uttar Pradesh (Ref 2) is an example of a regional level plan. The Rural Road Network plan prepared for each district by the National Rural Road Development Agency and the transport plans for major cities like Delhi are examples of local level plans.

4.4. Systems Approach to Transport Planning

The processes forming part of the systems approach to transport planning consist of various inter-connected steps are shown in Fig 4.2.

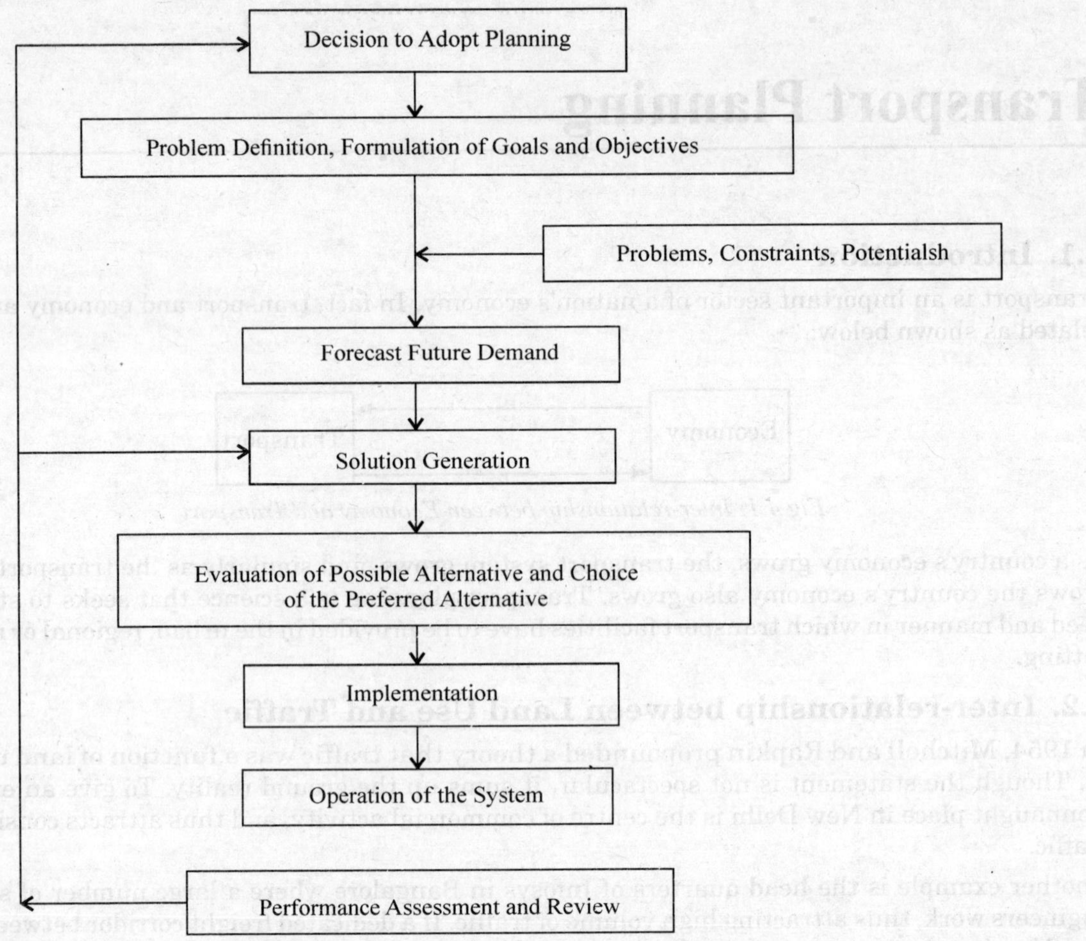

Fig 4.2: Systems Approach to Transport Planning

4.5. Goals and Objectives

To start the transport planning process, the goals and objectives must be clearly understood. Some of these could be:

1. The transport plan should be in conformity with the nation's overall economic development plan. For example, in India the Five Year Plans give broad guidelines on development of the transport infrastructure and these should be respected.
2. The transport plan should be a harmonious mix of various alternatives and modes so as to give the maximum benefits such as economy in cost of operation, conveniences, comfort, safety and speed.

3. Environmental considerations and sustainability should be given importance.
4. Transport development should be in step with the development plans for other sectors such as agriculture, industry, mining, forestry, employment generation and export/ import trade.
5. A balanced development of urban settlements and rural habitations should be aimed at. The PMGSY rural connectivity plan is a good example.
6. Conservation of energy should be given high priority in the plans.
7. Public transport should be given priority over use of personalised vehicles.
8. Pedestrians and cyclists should get adequate attention.

4.6. Time Horizon for Planning

Short-term plans having a time horizon of 5 years or less are intended to give quick relief and remove bottlenecks. Examples are improvement of junctions, signalisation, providing dedicated cycle tracks and "one-way street" operation. Medium term plans have a period of about 10 years and involve substantial outlays on engineering improvements. Examples are provision of pedestrian and cattle subways, construction of grade separators, provision of off-street parking facilities and construction of bypasses. Long-term transport plans have a time horizon of about 20 years. They are also known as Master Plans, Strategic Plans and Comprehensive Plans.

4.7. Stages in the Transport Planning Process

The various stages involved in transport planning process are depicted in Fig. 4.3.

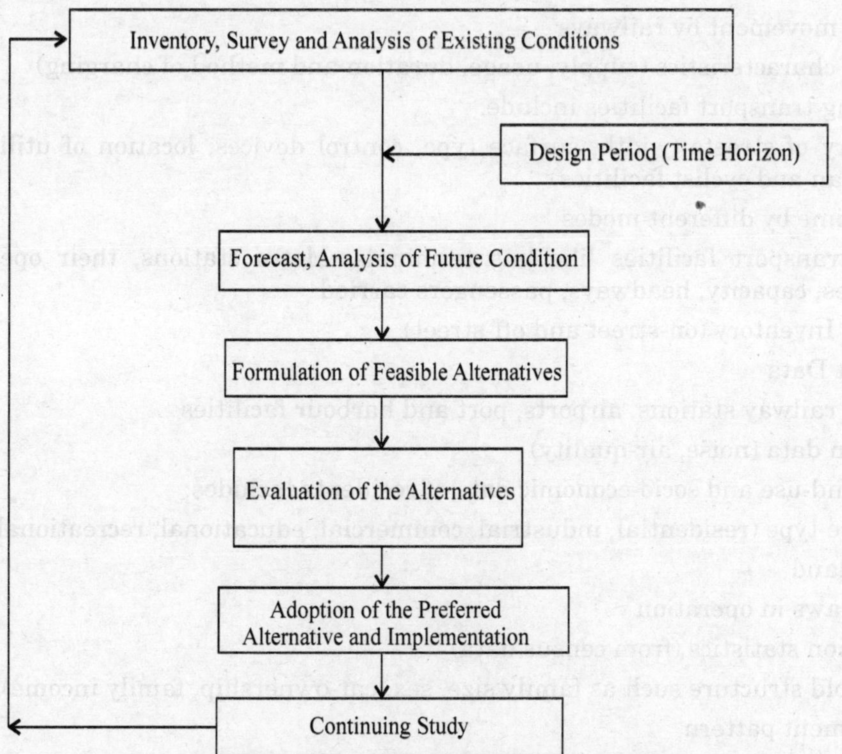

Fig 4.3: Stages in the Transport Planning Process

The stages involved in the Transport Planning Process are:

1. Inventory, survey and analysis of existing conditions
2. Forecast, analysis of future conditions
3. Formulation of feasible alternatives
4. Evaluation of the alternatives
5. Adoption of the preferred alternative and implementation
6. Continuing study

4.8. Inventory, Survey and Analysis of existing Conditions

The above stage consists of three different tasks:

(i) Collection of data on traffic
(ii) Collection of data on existing transport facilities
(iii) Inventory of land-use and socio-economic data of residents

Data on traffic to be collected includes:

(i) Origin and Destination of journeys
(ii) Traffic volume on various links including vehicle class, hourly variation, daily variation and any seasonal variation that can be expected
(iii) Data on movement of goods
(iv) Data on movement of public transport (Buses, Metro, LRT)
(v) Data on movement by railways
(vi) Parking characteristics (supply, usage, duration and method of charging)

Data on existing transport facilities include:

(i) Inventory of streets (width, surface type, control devices, location of utilities, drainage, pedestrian and cyclist facilities)
(ii) Travel time by different modes
(iii) Public transport facilities like bus terminals, Metro stations, their operating speeds, schedules, capacity, headways, passengers carried
(iv) Parking Inventory (on-street and off-street)
(v) Accident Data
(vi) Data on railway stations, airports, port and harbour facilities
(vii) Pollution data (noise, air-quality)

Inventory of land-use and socio-economic data of residents includes:

(i) Land-use type (residential, industrial, commercial, educational, recreational etc.)
(ii) Vacant land
(iii) Zoning laws in operation
(iv) Population statistics (from census data)
(v) Household structure such as family size, sex, car-ownership, family income)
(vi) Employment pattern
(vii) School and college attendance

The data collected is analysed to determine any quantifiable relationships between travel pattern and land-use and other socio-economic characteristics of the household.

4.9. Transportation Survey

4.9.1. This activity is crucial to the transport planning process, as the data forms the basis for formulating plans. Since the survey costs are high, proper organisation of the work and following guidelines evolved by various organisations are very important.

4.9.2. The study begins with the definition of the study area, which can be at the national, regional or local level. For planning at the city level, it is necessary that the area not only covers the existing city limits but should include areas of possible future growth. The boundary of the study area is an imaginary line known as the "external cordon", and should be continuous and should intersect roads where it is safe and convenient to carry out traffic surveys.

4.9.3. The study area is then sub-divided into "zones", so that the data from a zone reflects the same land-use (for example: residential, industrial, commercial, educational, recreational etc) and the average characteristics of individual households. The zones are later used for associating the origins and destinations of travel. Zones within the study area are called "internal zones" and those outside the study area where trips can originate and terminate are called "external zones". Guidance may be taken from the zoning adopted by other bodies, such as Population Census. The boundaries of zones should match with natural barriers such as canals, rivers etc.

4.9.4. The movements that are possible between zones are represented in the Fig. 4.4 below:

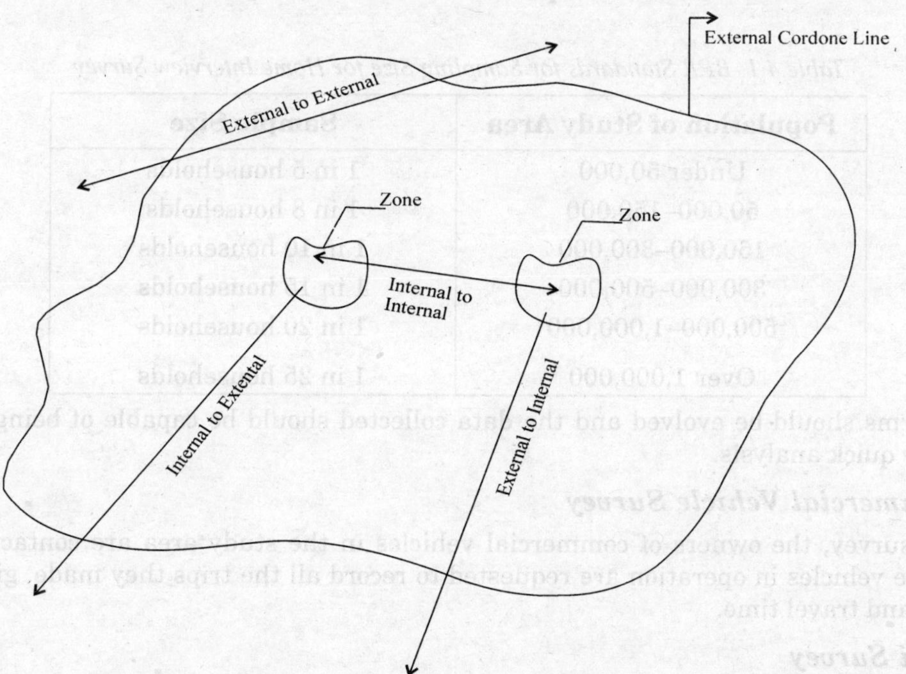

Fig 4.4: Movements possible in a transportation survey

4.9.5. The types of surveys that are usually carried out are:

(i) home interview survey

(ii) commercial vehicle survey

(iii) taxi survey
(iv) public transport survey
(v) road-side interview survey
(vi) post-card questionnaire survey
(vii) registration number plate survey
(viii) tag survey

These are described in detail hereafter.

4.9.6. Home Interview Survey

The home-interview survey is the most reliable survey for collecting the data needed for transport planning. The information collected includes:

- Number of trips made, with their origin, destination, purpose, travel mode, time of departure and arrival
- Household characteristics like type of dwelling unit, number of residents, age, sex, vehicle ownership, number of residents who can drive, family income, school/ college going children, number of employed persons.

Since the number of residents in an area is too large, it is practical to select a few sample households. The Bureau of Public Roads (BPR) in USA have prescribed the following standards for sampling (Ref 3).

Table 4.1: BPR Standards for Sampling Size for Home Interview Survey

Population of Study Area	Sample Size
Under 50,000	1 in 5 households
50,000–150,000	1 in 8 households
150,000–300,000	1 in 10 households
300,000–500,000	1 in 15 households
500,000–1,000,000	1 in 20 households
Over 1,000,000	1 in 25 households

Standard forms should be evolved and the data collected should be capable of being coded in a computer for quick analysis.

4.9.7. Commercial Vehicle Survey

Under this survey, the owners of commercial vehicles in the study area are contacted and the drivers of the vehicles in operation are requested to record all the trips they made, giving origin, destination and travel time.

4.9.8. Taxi Survey

In cities taxis are a convenient mode of travel. Under this survey, taxi owners in the study area are contacted and the drivers of the taxis are requested to record all the trips they made, giving origin, destination, taxi charges and travel time.

4.9.9. Roadside Interview Survey

Roadside Interview Surveys are conducted on the cordon or any other screen line. The vehicles are stopped (with the help of police) on a sampling basis and information is collected on the type of vehicle, origin, destination, trip purpose and travel time. Normally a week's data is collected, and on each day it is customary to collect 16 hour data (with two shifts of enumerators).

4.9.10. Post-card Questionnaire

At the survey point on the cordon line, a reply paid questionnaire is handed over to the drivers with a request to complete the information and send it by return post. The data includes type of vehicle, origin, destination, trip purpose and travel time.

4.9.11. Registration Number Plate Survey

This survey is conducted at roads intersecting the cordon line. The registration number plates of vehicles entering or leaving are noted. By matching the number plates later in the office, it is possible to find out at which point the vehicle entered and at which point the vehicle exited. The time interval gives the travel time taken.

4.9.12. Tags on Vehicles

Under this survey, the vehicles entering the study area at the cordon line are stopped and a tag is affixed to the windscreen. The tags have different colour and shape for different survey stations. At the exit points, the vehicles are stopped and the tags are removed. Thus it is possible to identify at which survey point the vehicle entered and at which survey point it exited. If the time at entry and exit are marked on the tags, the travel time can be determined.

4.9.13. Public Transport Survey

This survey is conducted at roads intersecting a cordon line. The buses are stopped and the passengers are interviewed and information on their journey and socio-economic characteristics is collected. Alternatively, a prepared questionnaire is handed out with a request to complete the form and send it by post. Similar surveys can also be conducted on passengers travelling in a train.

4.10. Inventory of Transport Facilities

The inventory of existing transport facilities should cover:

1. **Inventory of streets :**

   (i) layout (with a map); *(ii)* length; *(iii)* width of street (carriageway, footpath, median, cycle track); *(iv)* capacity; *(v)* traffic signals

2. **Inventory of bus transport :**

   (i) Number of buses; *(ii)* Routes; *(iii)* Location of depots and terminals; *(iv)* Schedules; *(v)* Operating speeds; *(vi)* Passengers carried; *(vii)* Fare structure.

3. **Inventory or fail-based facilities :**

   (i) Sub-urban rail system, with routes, stations; *(ii)* Metro; *(iii)* LRT; *(iv)* number of passengers carried; *(v)* fare structure; *(vi)* schedules; *(vii)* operating speed.

4. **Accident Data for the past 3 years :**

   (i) Location; *(ii)* Number of accidents; *(iii)* Serious Injuries; *(iv)* Minor Injuries

5. **Pollution Data :**

   (i) Noise levels recorded; *(ii)* Air-quality levels recorded

4.11. Urban Transportation Planning Process (UTPP)

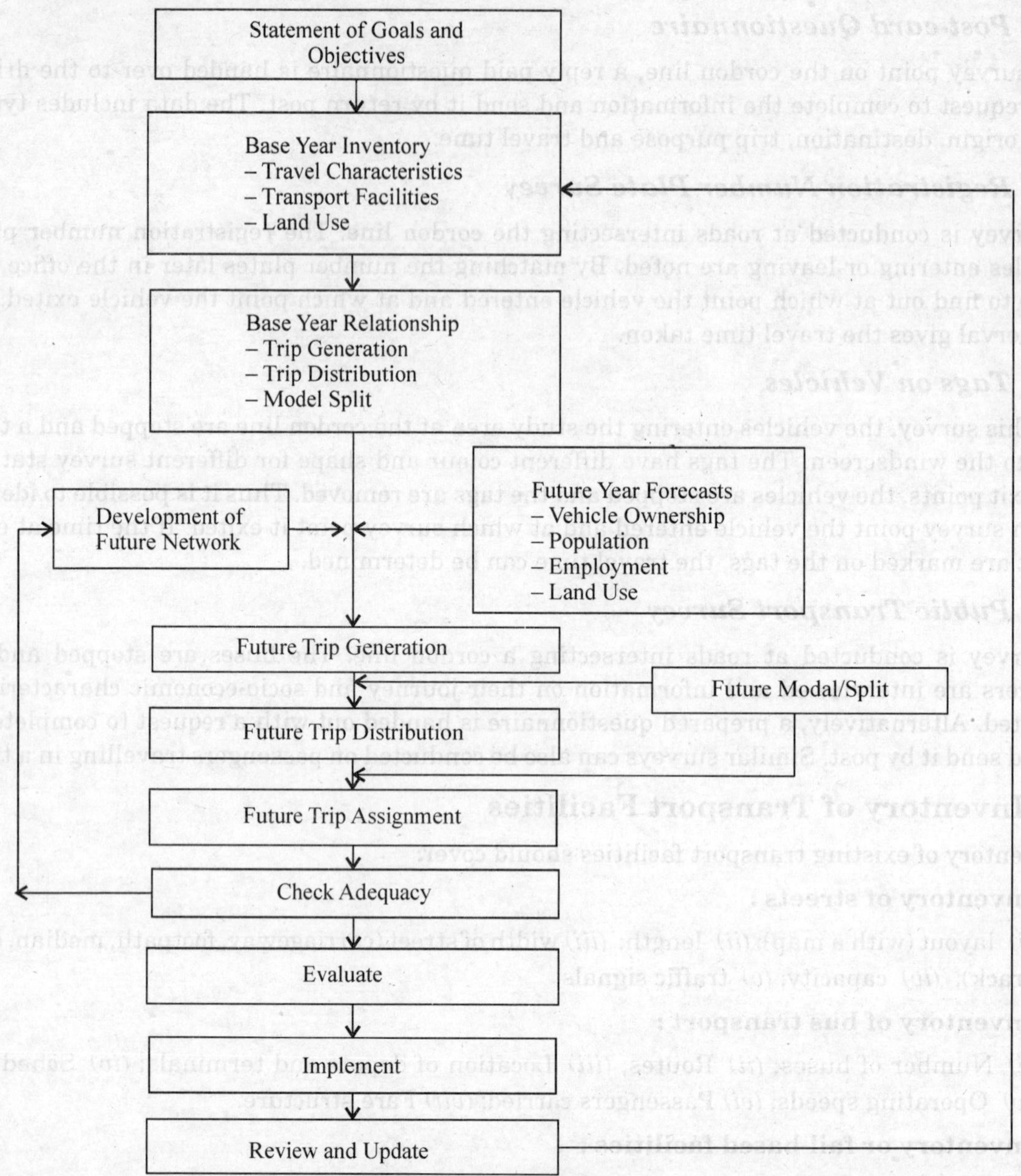

Fig 4.5: Flow Chart of UTPP

As the vehicle population began to grow and streets became increasingly choked, the need for a well-planned strategy to overcome the problem began to be recognised. In the 1950s a new approach to the problem was developed and is well-known now by the term "Urban Transportation Planning Process-UTPP)". The sequential approach in the UTPP is given above (Fig 4.5).

4.12. Trip Generation

After the surveys and data collection are completed, the next step is to analyse the data. In the JTPP, the first step in the analysis is Trip Generation. Trip generation modelling is building up mathematical relationships between the numbers of observed trips and factors influencing the generation of trips. A trip is defined as the one-way person movement by a mechanised mode of transport, having two trip ends, an origin and a destination. Thus if a person travels from his home to his work place, his trip is called home-based. A non-home based trip is one which has neither the start of the trip nor the end of the trip at the home. When a person travels from his home to a work place and returns home, there are two home-based trips. The start of the home-based trip is the generation and the end of the home-based trip is the attraction.

The factors governing trip generation and attraction are:

1. Income
2. Car ownership
3. Family size and composition
4. Land use characteristics
5. Distance of the zone from the town centre
6. Accessibility to public transport system and its efficiency
7. Employment opportunities, floor space in the industrial units and offices, sales figures in shops etc.

A typical model can be:

$$T = a_0 + a_1 x_1 + a_2 x_2 + \ldots\ldots\ldots\ldots a_n x_n$$

Where T = number of trips

a_o = constant term

$x_1, x_2 \ldots x_n$ = independent variables related to land-use and socio-economic characteristics

$a_1, a_2 \ldots a_n$ = coefficient of respective variables $x_1, x_2 \ldots x_n$

As an example, let us consider the following model that has been generated:

T = 30 + 0.318 (population 5 years and older)

+ 0.458 (number of households)

+ 0.890 (number of cars owned in the zone)

+ 0.425 (number of two-wheelers owned in the zone)

If a zone has 1000 households, with a population of above 5 years as 4800, with 300 cars in the zone and 350 two-wheelers in the zone, the number of trips generated on an average working day:

= 30 + 0.318 x 4800 + 0.458 x 1000 + 0.890 x 300 + 0.425 x 350

= 30 + 1526 + 458 + 267 + 149

= 2430

Since the collection of data and modelling is an elaborate process, for small towns a simpler approach is often followed. For example, an average trip rate for each land-use is used. These techniques are known as Quick Response Techniques.

Trips are classified according to the purpose of trips, such as:

(i) work

(ii) educational

(iii) business

(iv) social, sports, recreational

4.13. Trip Distribution

Trips generated in the various zones are attracted to various zones. Trip distribution is the process of distributing the trips between zones.

To understand this process, on O-D Matrix becomes handy. The vertical axis consisting of n zones (1, 2, 3 .... n) represents the trip generations and the horizontal axis represents the trips attracted to the n zones. The term t_{i-j} represents the trip ends generated in a zone i and terminating in zone j. Fig. 4.6 illustrates this.

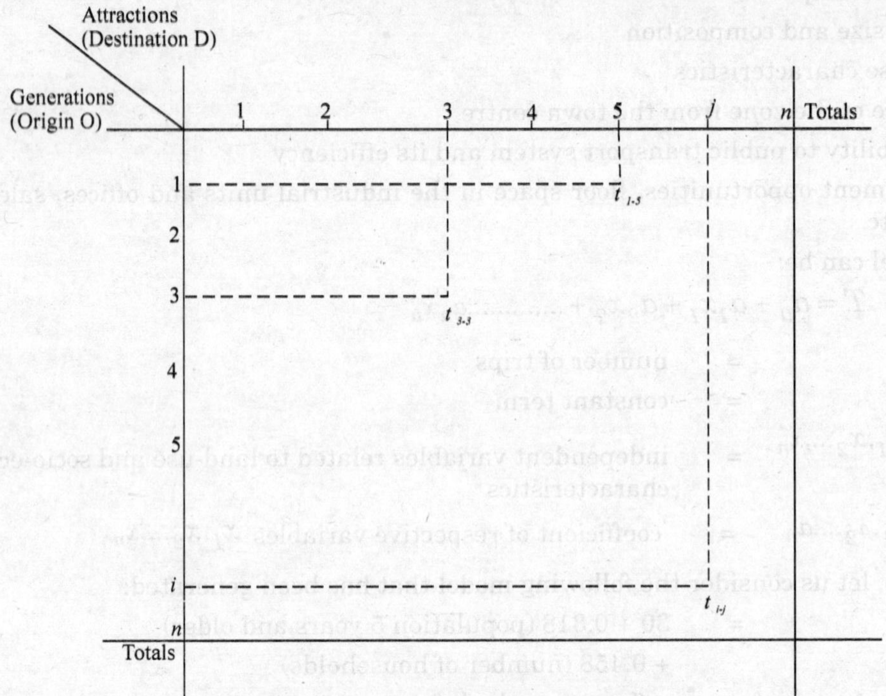

Fig 4.6: Origin - Destination (O - D) Matrix

After the trips generated in each zone in the base year is determined, the distribution of these trips in the horizon year between the various zones is done by either of the following methods:

1. Growth Factor Method
2. Synthetic Method

In the growth factor method, the present year trips are projected to the horizon year by using expansion factors. A simple and convenient method is the Average Factor Method. This is illustrated in the following example.

Given (i) the distribution of trips in the base year (Fig 4.7)

(ii) the expected future trips generated in each zone (Fig 4.8)

O \ D	1	2	3
1	30	150	450
2	150	45	300
3	450	300	75

Fig 4.7: Distribution of Trips in the Base Year

Zone	Future Trips Generated T_i
1	900
2	1200
3	1500

Fig 4.8: Future Trips Expected to be Generated from Zones 1, 2 and 3

Fig 4.9: Gives the growth factors.

O \ D	1	2	3	pi	Pi	$Ei = \frac{Pi}{pi}$
1	30	150	450	630	900	1.43
2	150	45	300	495	1200	2.42
3	450	300	75	825	1500	1.82
aj	630	495	825	1950	3600	
Aj	900	1200	1500			
$Ej = \frac{Aj}{aj}$	1.43	2.42	1.82			

Fig 4.9: Growth Factors

Using the average growth factor , $\frac{Ej + Ei}{2}$

$$T_{1-1} = 30 \times \frac{1.43 + 1.43}{2} = 42.9$$

$$T_{1-2} = 150 \times \frac{1.43 + 2.42}{2} = 288.75$$

$$T_{1-3} = 450 \times \frac{1.43 + 1.82}{2} = 731.25$$

$$T_{2-1} = 150 \times \frac{2.42 + 1.43}{2} = 288.75$$

$$T_{2-2} = 45 \times \frac{2.42 + 2.42}{2} = 108.9$$

$$T_{2-3} = 300 \times \frac{2.42 + 1.82}{2} = 636.0$$

$$T_{3-1} = 450 \times \frac{1.82 + 1.43}{2} = 731.25$$

$$T_{3-2} = 300 \times \frac{1.82 + 2.42}{2} = 636.0$$

$$T_{3-3} = 75 \times \frac{1.82 + 1.82}{2} = 136.5$$

O \ D	**1**	**2**	**3**	**pi**	**Pi**	$Ei = \frac{Pi}{pi}$
1	42.9	288.75	731.25	1062.9	900	0.85
2	288.75	108.9	636.0	1033.65	1200	1.16
3	731.25	636.0	136.5	1503.75	1500	1.00
aj	1062.9	1033.65	1503.75			
Aj	900	1200	1500			
$Ej = \frac{Aj}{aj}$	0.85	1.16	1.00			

Fig 4.10: Average Growth Factor

The newly derived values of Ei and Ej are used for the next iteration. The iteration is carried out till the growth factor approaches unity.

The growth factor methods are approximate. The more accurate method is the Gravity Model which is based on the Newton's gravity law. Thus, the Gravity Model calculates trips between zones on the basis of relative attraction between zones and the spatial separation between the two. Thus,

$$T_{i-j} = \frac{K\, Pi\, Aj}{(di - j)^n}$$

$$T_{i-j} = \frac{Pi \times \frac{Aj}{(di-j)^n}}{\frac{Aj}{(di-j)^n} + \ldots\ldots + \frac{Ak}{(di-k)^n}}$$

Where T_{i-j} = Trips between zones i and j

Pi = Trips produced in zone i

Aj = Trips attracted to zone j

d i - j = distance between zones i and j or the time or cost of travelling between them

K = A constant

n = exponential, usually taken as 2, and can vary between 1 and 3

k = total number of zones

The following example illustrates the above:

There are four residential zones, A, B, C and D which serve too industrial zones X and Y. The matrix below gives the distances between the zones.

Industrial Zones / Residential Zones	X	Y
A	3	5
B	2	1
C	4	5
D	3	2

The trips generated in the 4 zones are:

A = 2000

B = 3000

C = 5000

D = 4000

There are 5000 jobs in X and 9000 in Y. Distribute the trips.

$$T_{i-j} = \frac{Pi\dfrac{A_j}{(di-j)^2}}{\sum\dfrac{Ai}{(di-n)^2}}$$

$$T_{A-X} = \frac{2000 \times \dfrac{5000}{3^2}}{\dfrac{5000}{3^2} + \dfrac{9000}{5^2}} = 1214$$

$$T_{A-Y} = \frac{2000 \times \dfrac{9000}{5^2}}{\dfrac{5000}{3^2} + \dfrac{9000}{5^2}} = 786$$

$$T_{B-X} = \frac{3000 \times \dfrac{5000}{2^2}}{\dfrac{5000}{2^2} + \dfrac{9000}{1^2}} = 366$$

$$T_{B-Y} = \frac{3000 \times \dfrac{9000}{1^2}}{\dfrac{5000}{2^2} + \dfrac{9000}{1^2}} = 2634$$

$$T_{C-X} = \frac{5000 \times \dfrac{5000}{4^2}}{\dfrac{5000}{4^2} + \dfrac{9000}{5^2}} = 2323$$

$$T_{C-Y} = \frac{5000 \times \frac{9000}{5^2}}{\frac{5000}{4^2} + \frac{9000}{5^2}} = 2677 \qquad T_{D-X} = \frac{4000 \times \frac{5000}{3^2}}{\frac{5000}{3^2} + \frac{9000}{2^2}} = 792$$

$$T_{D-Y} = \frac{4000 \times \frac{9000}{2^2}}{\frac{5000}{3^2} + \frac{9000}{2^2}} = 3208$$

The matrix then becomes

O \ D	X	Y	TOTAL CALCULATED GENERATIONS	TOTAL PREDICTED GENERATIONS
A	1214	786	2000	2000
B	366	2634	3000	3000
C	2323	2677	5000	5000
D	792	3208	4000	4000
TOTAL CALCULATED ATTRACTIONS	4695	9305	14000	14000
TOTAL PREDICTED ATTRACTIONS	5000	9000		

There is a slight difference between calculated and predicted attractions. The calculations are redone in the next iteration on the same basis till there is a perfect match.

4.14. Trip Assignment

After the trips are distributed between zones, the next stage is what path these trips will follow since there may be more than one path. The common criteria for the best path is the journey time since it considers both distance and congestion. The most widely used technique is the "Moore's Algorithm" which traces the tree of the minimum path. The Fig 4.11 shows an area having 16 centroids of zones. The time taken to travel is indicated in each link. Starting from centroid 1, the thick lines show the minimum path.

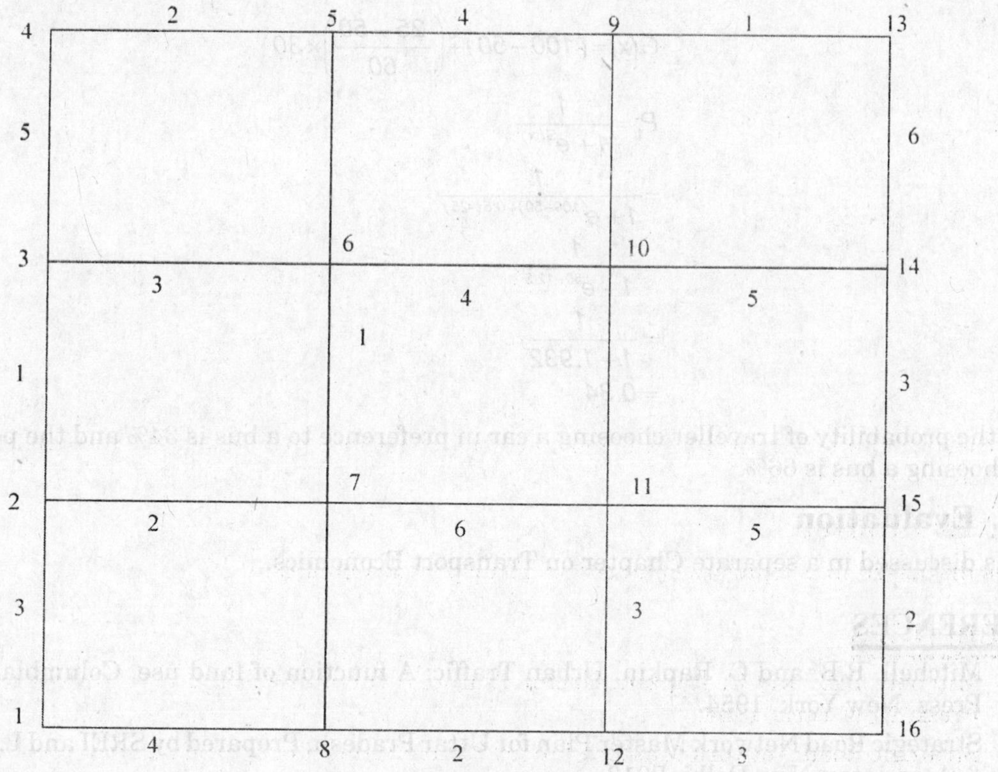

Fig 4.11: Minimum Path Tree

4.15. Model Split

This stage of the UTPP is allocation of trips according to modes preferred. The most widely used model is the Logit Model given below:

$$P_1 = \frac{1}{1+e^{G(x)}}$$

Where P_1	=	Possibility of choosing Mode 1
$1 - P_1$	=	Possibility of choosing Mode 2
e	=	base of Naperian logarithm
$G(x)$	=	$a_1 (C_1 - C_2) + a_2 (t_1 - t_2)$
C_1, C_2	=	Cost of travel by modes 1 and 2
t_1, t_2	=	Time of travel by modes 1 and 2

Other parameters in addition to cost and time can also be used.

The following example illustrates the use of the above formula:

It has been found that the cost of travel between two places is Rs 100 and Rs 50 by car and bus respectively. The time taken for travel by car is 25 minutes and by bus is 50 minutes. The value of time of the traveller is Rs 30 per hour. Find the probability of a traveller choosing a car.

$$G(x) = (100 - 50) + \left(\frac{25 - 50}{60}\right) \times 30$$

$$P_1 = \frac{1}{1 + e^{G(x)}}$$

$$= \frac{1}{1 + e^{(100-50)+0.5(-25)}}$$

$$= \frac{1}{1 + e^{50-12.5}}$$

$$= \frac{1}{1 + 1.932}$$

$$= 0.34$$

Thus the probability of traveller choosing a car in preference to a bus is 34% and the possibility of him choosing a bus is 66%.

4.16. Evaluation

This is discussed in a separate Chapter on Transport Economics.

REFERENCES

1. Mitchell, R.B. and C. Rapkin, Urban Traffic: A function of land use, Columbia University Press, New York, 1954.
2. Strategic Road Network Master Plan for Uttar Pradesh, Prepared by SREI and L.R. Kadiyali & Associates, New Delhi, 2013.
3. Bureau of Public Roads, Procedure Manual: Conducting a Home Interview Origin-Destination Survey, Vol 2-B, Washington, 1956.

SECTION—2

HIGHWAY ENGINEERING

5

Chapter

History of Growth of Highways

5.1 Historical Background

The invention of the wheel around 3500–5000 BC revolutionised transport. Instead of moving on horses and by walking, man began to move in wheeled chariots. Many civilisations, like the Roman, have built a good system of roads. The saying: "All roads lead to Rome" indicates that the Romans built a good road network with Rome as the focal point. The Indus Valley Civilisation dealing back to 5,000 year B.C., flourished well with well-planned towns like Mohenjo Daro having good roads. The Mauryan Kings (14th to 2nd Century BC) built good roads. Chandragupta Maurya built a road connecting Pataliputra (now known as Patna) to Takshashila now in Pakistan) connecting cities like Varanasi, Kausambhi, Mathura, Indraprastha (Delhi) and Kurukshetra. The road is famously known as Grand Trunk Road. Kautilya, the author of Arthashastra, laid down standards for various classes of roads. Rajapathas were the main highways and Banijapathas were trade routes. The Muslim rule consolidated the road system. Sher shah, the Afghan monarch who ruled in the 16th century developed the Grand Trunk Road further. The Mughal rulers, in their attempts to extend their domain, built good roads.

In Europe, the Industrial Revolution created great demand for transport, mainly in wheeled coaches. Pierre Tresaguet of France recognised the need for good drainage of roads and introduced the concept of camber. Tomas Telford in Britain constructed pavements with single-sized boulders as foundation. John Mc Adam, a Scottish engineer, proposed pavement courses with broken stones which interlock with each other and impart great strength. His name has now been associated with various specifications even now (like water-bound macadam, wet mix macadam and bituminous macadam). **A great fillip to road construction was provided by the introduction of steam road rollers by Eveling and Barford.**

The British rule concentrated its attention to maintain roads for military and administrative importance. William Bentinck modernised the Grand Truck Road from Calcutta to Delhi by building arch bridges and providing 'Kankar' (limestone) bed. Lord Dalhousie organised the provincial Public Works Department. Engineering colleges began to be established for training civil engineers. The introduction of railways in the 1850s was a set-back to the growth of roads.

The development of the internal combustion engine and the pneumatic tyres saw the introduction of cars. The Americans exploited this invention fully. In India, the motor car made its appearance in the beginning of the twentieth century. The First World War and the years following it registered rapid growth of road transport.

The need for good roads led the Government of India to appoint a Committee under the chairmanship of M.R. Jayakar to suggest how roads could be improved and funded. The Committee's Report

(Ref 1) is a major land-mark, and emphasised the need to develop a good system of roads in the country, funded by taxation of fuel and vehicles and licence fee for vehicles on hire. The Central Road Fund (CRF) was created, out of which ten per cent was the central government's share and the balance was distributed among the provinces.

Another recommendation of the Jayakar Committee was the need to hold periodic road conferences to discuss all matters relating to roads. As a result, the first conference was held in 1930 and led to the establishment of the Indian Roads Congress (IRC) in 1934. The mandate for the IRC is to promote and to encourage the practice of road building and hold periodic meetings to discuss all technical matters.

The IRC during its 80 years of existence have held annual sessions and have several technical committees to deal with various aspects. A number of specifications, codes of practices and guidelines have been brought out by the IRC, setting uniformity of practices throughout the country. It has also established the Highway Research Board to deal with research needs and disseminate research findings.

The Second World War saw a rapid growth of road traffic which caused serious deterioration of the condition of roads. Recognising this situation which demanded satisfactory solution, the Government convened a Conference of Chief Engineers of the Provinces at Nagpur. The deliberations resulted in the formulation of the now famous Nagpur Plan (Ref 2). Some of the recommendations of the Nagpur Plan are:

1. The lead for the road development should be taken by the Central Government even though "roads" was a provincial subject. A long-term plan for road development should be formulated.
2. Roads should be divided into four classes:
   - *(a)* National Highways (NH) traversing the length and breadth of the country, linking provincial capitals and serving strategic and administrative needs of the country. National Highways should be the charge of the Central Governments.
   - *(b)* State Highways would be the other main roads of the Province.
   - *(c)* District Roads are those which take off from the main roads leading to the interior of the districts. Depending upon the importance, some of them would be classified as Major District Roads (MDRs) and the others as Other District Roads (ODRs).
   - *(d)* Village Roads are those which give connectivity to the villages.
3. All roads should be regarded as important for the nation's economy and one class should not progress at the expense of the other.
4. All NHs, and MDRs should be provided with a durable hard pavement.
5. ODRs and Village Roads should be provided with a properly engineered earth surface, but gravelling, soil stabilisation and macadam surface could be considered wherever necessary.
6. All NHs, SHs and MDRs should be adequately bridged.
7. The following formulae have been suggested for determining the lengths of various classes of roads:

Length of NHs, SHs and MDRs in miles

$$= \frac{A}{5} + \frac{B}{20} + N + 5T + D - R$$

where A = Agricultural area of the State in sq. miles

B	=	Non-agricultural area of the State in sq. miles
N	=	Number of towns of population 2,000–5,000
T	=	Number of towns of population over 5,000
D	=	An allowance for development during the next 20 years (about 15 percent)
R	=	Length of railways in miles in the area under consideration.

The above formula ensures that in a developed agricultural area no village will be more than 5 miles from a main road and the average distance will be 3 miles in many cases. In a non-agricultural area, no village will be more than 20 miles from a main road and the average distance will be 6 or 7 miles.

The length of ODRS and Village Roads in miles.

$$= \frac{V}{5} + \frac{Q}{2} + R + 2S + D$$

where	V	=	number of villages with population 500 or less
	Q	=	number of villages with population 501–1000
	R	=	number of villages with population 1001–2000
	S	=	number of villages with population 2001–5000
	D	=	Allowance for development during the next 20 years (usually 15 per cent)

As per the above formulae, the targeted lengths to be achieved in a 20 year period for post-partition India are given in Table 5.1.

Table 5.1 : Nagpur Plan Targets of Road Lengths for Post-partition India

S. no.	Class of Road	Targeted Road Length (in miles)	Targeted Road Length (in Km)
1.	National Highways	16,600	26,715
	National Trails	4,150	6,680
2.	State Highways	53,950	86,825
3.	Major District Roads	49,800	80,145
4.	Other District Roads	83,000	1,33,580
5.	Village Roads	1,23,000	1,98,755
	Total	**3,30,500**	**5,32,700**

REFERENCES

1. Report of the Road Development Committee, 1927-28, Govt. of India, Central Publications Branch, Calcutta, 1928.
2. Chief Engineers' Conference on Road Development (1943), Post War Road Development in India, (Nagpur Plan), Indian Roads Congress, New Delhi.

6 Chapter

Development of Roads in India After Independence

6.1 Development of Roads in India

After independence, the Central Government took over full financial responsibility for the construction, improvement and maintenance of NHs. Under the Constitution of India, NHs became a central subject. This was enabled by the National Highways Act, 1956 (Ref 1). The development of roads was taken up under the various Five Year Plans.

India prepared the following long-term road development plans in the post-independence period:

1. Bombay Plan 1961–81 (Ref 2)
2. Lucknow Plan 1981–2001 (Ref 3)
3. Vision 2021 (Ref 4)
4. Rural Road Vision 2025 (Ref 5)

The Bombay Plan gave the following formulae for determining the lengths of various classes of roads:

1. National Highways

$$\text{Mileage} = \frac{A}{40} + \frac{B}{50} + \frac{C}{60} + \left[20\,K + \frac{10\,M}{2}\right] + D$$

2. National Highways and State Highways

$$\text{Mileage} = \frac{A}{12.5} + \frac{B}{15} + \frac{C}{20} + (30\,K + 15\,M + 7\,N + P) + D$$

3. National Highways, State Highways and Major District Roads

$$\text{Mileage} = \frac{1}{5}\left[A + \frac{B}{2} + \frac{C}{3}\right] + (30\,K + 15\,M + 7\,N + 6\,P + 4\,Q + 1.5\,R) + D$$

4. National Highways, State Highways, Major District Roads and Other District Roads

$$\text{Mileage} = \frac{1}{5}\left[A + \frac{B}{2} + \frac{C}{3}\right] + (30\,K + 15\,M + 7\,N + 6\,P + 4\,Q + 1.5\,R) + D$$

5. National Highways, State Highways, Major District Roads, Other District Roads and Village Roads

$$\text{Mileage} = \frac{2}{5}\left[A + \frac{B}{2} + \frac{C}{3}\right] + \left(30K + 15M + 7N + 6P + 8Q + 3.75R + S + \frac{T}{2.5} + \frac{V}{8}\right) + D$$

where A = Developed and agricultural area in sq. miles
B = semi-developed area in sq. miles
C = Undeveloped and uncultivable area in sq. miles
K = Number of towns with population over 1,00,000
M = Number of towns with population between 1,00,000 and 50,000
N = Number of towns with population between 50,000 and 20,000
P = Number of towns with population between 20,000 and 10,000
Q = Number of towns with population between 10,000 and 5,000
R = Number of towns with population between 5,000 and 2,000
S = Number of villages with population between 2,000 and 1,000
T = Number of villages with population between 1,000 and 500
V = Number of villages with population below 500
D = Allowance (assumed 5 per cent) of future development and other foreseen factors

Some of the broad objectives and recommendations in the Bombay Plan are:

1. Road length should be increased to give a road density of 32 Km per 100 sq. Km.
2. Every village in a developed area should be brought within 6.5 Km of metalled road and 2.4 Km from any road. In a semi-developed area, every village should be brought within 13 Km of any road. In an undeveloped and uncultivable area, every village should be brought within 20 Km of any road.
3. The targeted road lengths and costs are:

Table 6.1 :Bombay Plan Targets

Class of roads	Length (Km)	Cost (Rs-crores)
National Highways	51,000	980
State Highways	1,12,650	1,580
Major District Roads	2,41,400	1,360
Other District Roads	2,89,680	650
Village Road	3,62,100	630
Total	10,56,830	5,200

4. Stage construction strategy should be adopted in view of limited funds available.
5. Traffic Engineering cell should be established in each state.
6. Maintenance of roads should be given high priority.
7. All arterial roads should be two-lane wide and fully bridged.
8. Research on roads should be given adequate attention.

After the Bombay Plan period expired in 1981, the Indian Roads Congress formulated the Lucknow Plan for a further twenty year period (1981-2001) (Ref 3). Some of the features of this Plan are:

1. Lengths of various classes of roads should be decided on the basis of the following formulae:

$$\text{NH (in Km)} = \frac{\text{Area (in sq Km)}}{50}$$

$$\text{SH (in Km)} = \frac{\text{Area (in sq Km)}}{25}$$

or

$$62.5 \times \text{Number of Towns} - \frac{\text{Area (in sq Km)}}{50}$$

$$\text{MDR (in Km)} = \frac{\text{Area (in sq Km)}}{12.5}$$

or

$$90 \times \text{Number of Towns}$$

Total Road Length (in Km) = 4.74 x Number of villages and towns

Rural Road length (ODR + Village Roads) = Total road length – lengths of NH, SH and MDR

[Note: Towns are defined as places having population above 5,000]

2. The above formulae give a square grid of 100 Km for NH, a road length of 27,66,089 Km, a road density of 0.82 Km per sq. Km.
3. The outlay suggested is Rs 64,250 crores
4. Expressways should be constructed on major traffic corridors.
5. Toll roads and private sector participation should be explored.
6. R & D should receive adequate attention
7. Management Information System (MIS) should be set up to handle large volume of data on roads.
8. Training of staff should receive attention

In 2001, the Indian Roads Congress published the next twenty year road plan covering the period 2001-2021 (Ref 4).

Some of major recommendations of the above Plan are:-

1. Targets of road length for 2021 (Km)

Expressways	**15,766**
NH	80,000
SH	1,60,000
MDR	3,20,00
ODR and Village Roads	No target suggested

2. Half the length of NH should have carriageway of four/ six lanes and the remaining half should have two-lane carriageway with hard shoulders.
3. 10,000 Km of SH should have four-lanes and the balance should have two lanes
4. Forty per cent of MDRs should have two lane carriageway
5. The targets for basic road access to village should be:

Population above 1000	Year 2003
Population 500–1000	Year 2007
Population below 500	Year 2010

6. Maintenance of roads should be given adequate priority.
7. R & D in the sector should receive good attention.
8. Toll financing and creation of a dedicated Road Fund should be explored as alternate source of financing.
9. Road safety should receive attention.
10. Training of engineers should receive attention.
11. Upgradation of technology through adoption of innovative procedures and specifications should be favoured.

The development of Rural Roads got a major thrust with the launching of the Pradhan Mantri Gram Sadak Yojana (PMGSY) in 2000. This aimed at connecting all villages with a population of 500 and above in plain areas and 250 and above in hill states with an all-weather road. Under the next phase of PMGSY, upgrading and constructing nearly 50,000 Km of rural roads is planned.

In the year 2005, the Indian Roads Congress brought out the Vision 2025 Plan for Rural Roads (Ref 5). The main recommendations are:

1. All habitations with population above 100 will be connected by all-weather roads and Master Plans should be prepared accordingly.
2. Upgradation of 1,237,000 Km of rural roads will be taken up.
3. Cost should be brought down by the use of locally available materials, soil stabilisation, use of low cost drainage structures, stage construction and tractor-bound technology for construction and maintenance.

Table 6.2 gives a comparison of road lengths in various countries (Ref 6). The following facts emerge:

1. USA has the target network (6.3 mil Km) and India comes next with 4.1 mil Km. It is followed by China (4.0 mil Km).
2. The density of roads in India is 129.8 Km/ sq. Km as against 68.7 of US and 41.9 of China.
3. Only 50 per cent of India's roads is paved, as against 67 per cent of USA and 44 per cent of China.
4. It is in the Expressway that India lags far behind (with only 353 Km) as against USA's 96,597 Km and China's 97,355 Km.

Table 6.2 :Road Lengths in Selected Countries (Ref 6)

S. No.	Country	Road Length (Km)	Percent Paved Road	Area (Sq Km)	Road Density Km/ 100 sq Km	Expressway Length (Km)
1.	U.S.A	6,545,326	67	9,526,488	68.7	95,597
2.	U.K	394,428	100	242,495	162.7	6,016
3.	Germany	643,969	NA	357,104	180.3	15,249
4.	France	951,260	100	543,965	174.9	14,685
5.	Spain	681,298	100	505,991	134.6	16,205
6.	Russia	903,000	79	17,098,200	5.3	39,143
7.	Japan	1,213,000	81	377,873	321	11,739
8.	China	4,008,200	44	9,572,900	41.9	97,355
9.	Australia	810,641	42	7,692,202	10.5	1,700
10.	South Africa	362,099	20	1,220,813	29.7	1,927
11.	Egypt	160,472	89	996,603	10.1	988
12.	Brazil	1,751,868	6	8,515,767	20.6	11,000
13.	Mexico	374,262	38	1,964,375	19.1	6,144
14.	Pakistan	259,463	70	796,096	32.6	2,225
15.	India	4,109,592	50	3,166,414	129.8	353
16.	Indonesia	487,314	57	1,910,931	25.5	722
17.	Sri Lanka	114,093	15	65,610	412.4	180
18.	Bangladesh	270,565	30	147,570	183.3	707

REFERENCES

1. The National Highway Act, 1956.
2. Report of the Chief Engineers on Road Development Plan for India, 1961-1981, presented to the Ministry of Transport, 1958.
3. Road Development Plan for India, 1981-2001, Published by the Indian Roads Congress, New Delhi, 1984.
4. Road Development Plan, Vision 2021, Indian Roads Congress, New Delhi, 2001.
5. Rural Road Development Plan, Vision 2025, Indian Roads Congress, New Delhi, 2005.
6. From the Internet, Wikipedia.

7 Chapter

Geometric Design

7.1. Introduction

Geometric design of a highway deals with design of features such as the horizontal alignment, vertical profile, cross-sectional elements, access control, junctions, grade separated interchanges and median openings. The design should result in an efficient, safe and economic operation of transport. When considering cost, the initial cost plus the maintenance and user costs should be given due attention. In India, the standards are laid down by the Indian Roads Congress and these should govern.

7.2. Highway Classification

The geometric design is related to the highway type. In India, the highways in the rural (non-urban) areas are classified as under:

1. Expressways, which are superior facilities along arterial routes where only cars, buses, trucks and two-wheelers with engine capacity greater than 50 c.c. are allowed, all at-grade crossings are not allowed, separation of opposing traffic by a median is mandatory and rest-areas at convenient intervals are provided.
2. National Highways are the main highways traversing the length and breadth of the country, connecting state capitals, ports, large industrial and tourist centres and foreign highways and serve strategic and defence needs.
3. Stage Highways are the main trunk routes of a State, connecting up with National Highways or highways of adjacent States, and linking the district headquarters and important cities within the State.
4. District Roads are the roads traversing each district, serving areas of production and markets and connecting these with each other or with National Highways, State Highways, Railways or important navigational routes. They should be capable of taking road traffic into the heart of rural areas throughout the year with only minor interruptions. District Roads are divided into to classes on the basis of traffic as under:
   (i) Major District Roads (MDR) for the higher order of traffic
   (ii) Other District Roads (ODR) for the lower order of roads
5. Village roads linking villages to the nearest main road.

Village roads and ODRs are often grouped together as "Rural Roads" in India.

As regards the urban streets, the following classification is followed:

1. Expressways which are divided arterial highways for motorized traffic with full or partial control of access and generally provided with grade separation at intersections. Parking loading and unloading of goods and pedestrian traffic are not permitted on them.
2. Arterial Streets are those which form the principal network for though traffic, and serve intra-urban travel between the Central Business District (CBD) and outlying suburban centres. They are generally spaced at less than 1.5 Km in highly developed areas and the CBD at 8 Km or move in the suburban fringe areas. They are generally divided carriageways facilities with full or partial control of access. Parking and loading and unloading of goods are usually restricted and regulated. Pedestrians are allowed to cross only at designated locations.
3. Sub-arterial Streets are of somewhat lower level of travel mobility than arterial streets. Spacing varies from about 0.5 Km in the CBD to 3 to 5 Km in the suburban fringe. Parking and loading and unloading of goods are restricted and regulated. Pedestrians are allowed to cross only at designated locations.
4. Collector Streets are intended for collecting and distributing traffic to and from local streets and providing access to arterial streets. Full access is allowed on these streets from abutting properties. There are few parking restrictions.
5. Local streets are those intended for access to residences, business premises or abutting property. Parking is unrestricted. Pedestrian movements are not controlled.

7.3. Terrain

The topography and the physical features of the area through which the road passes governs the design elements. The terrain is classified as under (Ref 1, 2).

Terrain Classification	Percent of Cross-slope of the Country
Level	0–10
Rolling	10–25
Mountainous	25–60
Steep	Greater than 60

7.4. Design Speed

The speed for which a road is designed influences the geometric features of the road. The design speed should be appropriate for the class of the road and the terrain. Usually, the 95th or the 98th percentile speed (from the speed distribution curve) is selected as the design speed. The design speeds recommended by the Indian Roads Congress for Rural Highways (Ref 1) and Urban Streets (Ref 2) are given in Tables 7.1 and 7.2 respectively.

Table 7.1: Suggested Design Speeds in India for Rural Highways (Ref 1)

Classification	Plain Terrain		Rolling Terrain		Mountainous Terrain		Steep Terrain	
	Ruling	**Minimum**	**Ruling**	**Minimum**	**Ruling**	**Minimum**	**Ruling**	**Minimum**
National Highways and State Highways	100	80	80	65	50	40	40	30
Major District Roads	80	65	65	50	40	30	30	20
Other District Roads	65	50	50	40	30	25	25	20
Village Roads	50	40	40	35	25	20	25	20

Table 7.2 : Suggested Design Speeds in India for Urban Streets (Ref 2)

Classification of Roads	Speeds in Kmph
Arterial	80
Sub-arterial	60
Collector Streets	50
Local Streets	30

A design speed of 120 Kmph is being adopted for expressways in India.

7. 5. Design Vehicle

Different types of vehicles use the road. The geometric design has to cater to the critical vehicle which governs the radii, width of pavements, clearances and parking geometrics. In India, the Indian Roads Congress standards are followed (Ref 3). For important roads catering to high volume of traffic, the largest vehicle such as a single unit truck, semi-trailer and semi-trailer/ full trailer combination will govern. For the design of junctions in residential areas, passenger car will govern. The turning radii of design vehicles suggested by AASHTO (Ref 4) are given in Table 7.3 below.

Table 7.3 : Turning Radii of AASHTO Design Vehicle (Meters) Design Vehicle Type

	Passenger Car	Single Unit Truck	Semi-trailer	Semi-Trailer/ Full-trailer
Minimum Turning Radius (m)	7.32	12.8	13.72	13.75
Minimum Inside Radius (m)	4.65	8.61	6.00	6.85

These are illustrated in Fig. 7.1

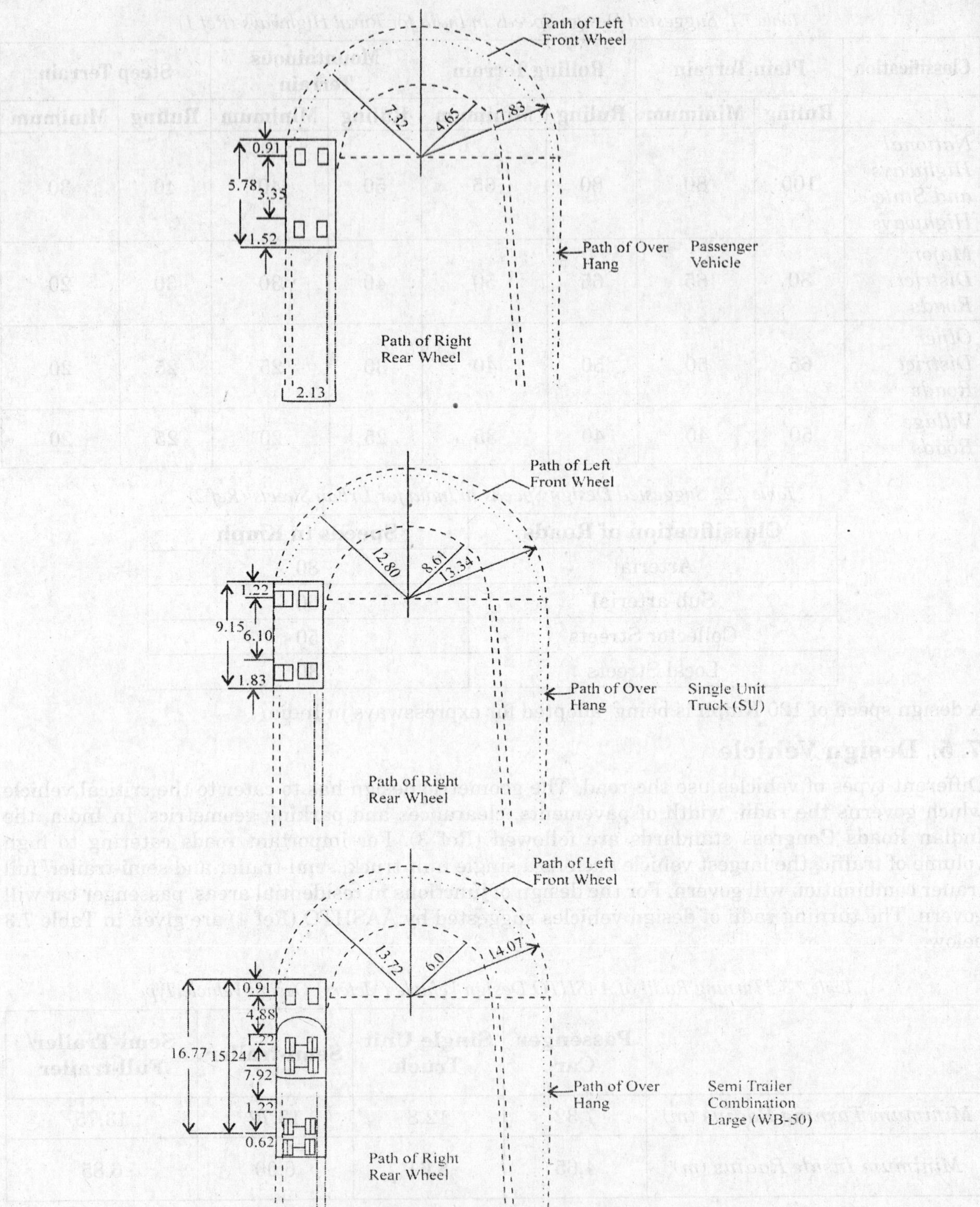

Fig. 7.1: Turning Radii of some AASHTO-Design Vehicles (Ref 4)

The relevant details from the IRC Standard (Ref 3) are given below:

1. ***Maximum Width*** : 2.5 m
2. ***Height***

Buses and Trucks for normal application	:	3.8 m
Trucks carrying containers	:	4.2 m
Double Decker Bus	:	4.75 m

3. ***Length***

Single Unit Truck (2 or more axles)	:	11 m (exclusive of front and rear bumpers)
Single Unit Bus (2 or more axles)	:	12 m (exclusive of front and rear bumpers)
Truck-tractor semi-trailer combination	:	16 m (exclusive of front and rear bumpers)
Truck-trailer combination	:	18 m (exclusive of front and rear bumpers)

4. ***Weight***

Single axle with dual wheels	:	10.2 Tonnes
Tandem Axle (spacing between axles not less than 1.2 m and more than 2.5 m)	:	18 Tonnes

7. 6. Road User and the Vehicle

For designing a highway facility, it is essentiai to know the characteristics of the road user and the vehicles. The road users include the drivers of vehicles, pedestrians and cyclists. The human behaviour can be studied under the following groups:

1. Physiological
   - *(i)* Vision
   - *(ii)* Hearing
2. Psychological
   - *(i)* Perception
   - *(ii)* Intellection
   - *(iii)* Emotion
   - *(iv)* Volition

The retina perceives objects very sharply within a zone formed by a cone whose angle is 3 degrees about the centre of the retina. This zone is known as the zone of acute vision or visual acuity. Vision is still satisfactory if the angle of the cone is 10-12 degrees. Peripheral vision deals with the visual zone for the two eyes. The angle is about 160 degrees in the horizontal field and 115 degrees in the vertical field. The field of peripheral vision falls within speed. Colour vision is very important for discerning the signals. Glare from opposing vehicles and street lights causes discomfort to drivers.

Hearing is important to a driver to alert him when a horn is blown.

The psychological trait of perception, intellection, emotion and volition, collectively known as PIEV is a very important factor in highway design. It governs the perceiving of information (perception), analysing it (intellection), starting the decision making process (emotion) and the will to react to a situation (volition). The AASHTO (Ref 4) recommends a time of 2.5 secs. The IRC recommends a time of 2.5 secs and 3 secs for plain and hilly terrain respectively for the combined perception and brake reaction time.

The speed of walk of pedestrians (needed for design and signals) varies from 0.75-1.8 m/sec (Ref 5). For the design of signals, the recommended speed of pedestrians is 1.2 m per sec in USA (Ref 6).

The space occupied by a pedestrian is considered as an ellipse with a major axis of 0.6 m and a minor axis of 0.45 m. The spacing between pedestrians while walking is generally assumed to be 2.5 m.

The sizes of vehicles commonly seen on Indian roads are given in Table 7.4.

Table 7.4 : Sizes of Vehicles Commonly Found in India

S. No.	Vehicle Type	Breadth (m)	Length (m)	Height (m)	Wheelbase (m)
1.	Maruti 800 Car	1.34	3.30	1.41	2.18
2.	Large Car	1.68	4.31	1.60	2.50
3.	Jeep	1.75	3.73	1.70	2.31
4.	Single Unit Truck	2.4-2.5	6.97	3.10	4.23
5.	Bus	2.5	7.0-9.5	3.22	4.0-5.2
6.	Three-wheeler Autorickshaw	1.30	2.49	1.30	2.0

Para 7.5 gives the maximum permitted dimensions.

7. 7. Power Performance of Vehicles

The power developed by the engine of a vehicle moving at a speed has to overcome all resistance to motion such as:

(*a*) Rolling resistance

(*b*) Air resistance

(*c*) Grade resistance

(*d*) Inertia forces during acceleration and deceleration

(*e*) Transmission losses

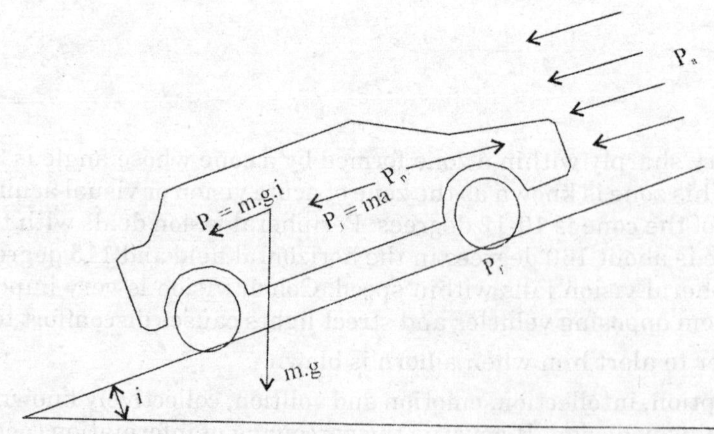

Fig. 7.2: Forces Acting on a Vehicle

(a) Rolling Resistance

When vehicle wheels roll over the road surface, the tyres get deformed due to the irregularities and the roughness of the surface. The road itself may deflect, if the wheel load is high. When in speed, shocks and impact are caused due to motion on an irregular surface. All these factors cause hindrance to motion. The smoother the surface the less is the resistance caused. A study conducted in India has yielded the results given in Table 7.5 (Ref 7).

Table 7.5 : Values of Rolling Resistance Coefficient from Indian Studies (Ref 7)

S. No.	Type of Surface	Coefficient of Rolling Resistance
1.	Asphaltic Concrete	0.01
2.	Premixed carpet in good condition	0.016
3.	Premixed carpet in bad condition	0.022
4.	Water-bound macadam in good condition	0.025
5.	Water-bound macadam in bad condition	0.037
6.	Gravel/ Earth	0.046

The rolling resistance is given by

$$P_f = m.f.g. \tag{7.1}$$

where m = mass of the vehicle in kg

f = coefficient of rolling resistance

g = acceleration due to gravity in m/sec²

P_f = rolling resistance in N

The coefficient of rolling resistance also varies with speed, increasing in value at higher speeds. The following approximate equation accounts for this increase (Ref 8):

$$f_u = f_o\,[1 + 0.01(\text{V} - 50)] \tag{7.2}$$

where f_u = coefficient of rolling resistance at speed V (Kmph)

V = Speed (Kmph)

f_o = coefficient of rolling resistance, assumed constant upto a speed of 50 Kmph, and having values as given in Table 7.5.

(b) Air Resistance

The motion of a vehicle at speed is resisted in the following ways by air:

(i) Since air has density, it exerts a reaction pressure against the front of the vehicle when it moves at speed.

(ii) The friction of air against the sides of the vehicle body causes resistance.

(iii) The eddying of the air stream behind the vehicle, under the body and around the wheels causes power loss.

(iv) The flow of air through the vehicle for ventilating the cooling causes resistance to motion.

The following formula can be used to determine the air resistance:

$$P_a = C_a A v^2 \quad (7.3)$$

where P_a = air resistance in N

A = projected front Area of the vehicle in sq. m. on a plane at right angles to the direction of motion; for Indian vehicles, the value can be taken from Table 7.6.

v = speed of the vehicle relative to air in m/sec

C_a = coefficient of air resistance, having value set forth in Table 7.6. Kg/m³

Table 7.6 : Values of Frontal area and Coefficient Air Resistance (Ref 7, 9)

S. No.	Type of Vehicle	Frontal Area (m²)	Mass (Kg)	Coefficient of air resistance C_a (Kg/m³)
1.	Maruti 800 Car	1.54	880	0.40
2.	Premier Padmini Car	1.63	1065	0.42
3.	Ambassador Car	2.15	1365	0.39
4.	Jeep	2.38	1200	0.37
5.	Tata Truck	5.37	6120 (Empty)	0.48
6.	Ashok Leyland Beaver Haulage Truck	5.37	8125 (Empty)	0.48

(c) Grade Resistance

When a vehicle which was moving on level stretch at a particular speed has to move up an incline, additional work has to be done in keeping the vehicle at the same speed as on the level stretch. The additional work is equal to the work that will be needed to lift the vehicle through a height represented by the inclination. If the horizontal distance is unity (1 metre) and if the inclination is i per cent, the rise will be $\frac{i}{100}$ metres. If the mass of the vehicle is m kg, the additional force to move the vehicle up the incline, P_i is given by:

$$P_i = \frac{m.i.g}{100} \quad (7.4)$$

(d) Inertia Forces During Acceleration

When a vehicle moves on a road, it has to frequently change its speed by accelerating or decelerating, which involves additional force. Similarly, if a vehicle has to start from a stopped position, additional force is needed to accelerate. The additional force P_j is given by:

$$P_j = \text{Mass x Acceleration} \quad (7.5)$$

Hence $(\pm)\ P_j$ = m.a.

where P_j = Force to overcome inertia while accelerating or decelerating, N

m = mass, kg

a = acceleration or deceleration is m/sec²

The value of P_j will be +ve if the vehicle has to accelerate and –ve if the vehicle has to decelerate.

(e) Transmission Losses

Power losses occur in a vehicle in a number of ways. Some loss occurs in the transmission system, in the gear system itself. At the start of the vehicle, high power is needed. Similarly, a high engine power is needed when climbing uphill, which is accomplished by driving at a low speed. For movement along a good road, where the resistance to motion will be small, a high gear is generally used. The highest forward gear will be usually 1:1, representing direct drive. A further gear reduction is made at the rear axle. The total effect of the above is to consume about 10-15 per cent of the engine horse-power which may be as high as 25 per cent in case of trucks in their lower gear.

7. 8. Power Requirements of the Vehicle

The mechanical power developed by the engine is transmitted to the driving wheels by the transmission system. The torque developed at the flywheel is converted to a torque at the rear axle. The following equation holds good:

Rear-axle torque, $T_a = K.T_e.G_t.G_a$ (7.6)

where T_a = rear axle torque

K = efficiency of the transmission system, representing the losses between the engine and the driving wheels, and can have a value of 0.80 to 0.90

T_e = engine torque at the flywheel

G_t = transmission gear ratio

G_a = rear-axle gear ratio

The rear axle torque imparts a tractive force P_p at the contact of the tyre and the road (Fig. 7.3). This tractive force, known as the tyre-rim pull, is given by the following equation:

$$P_p = \frac{Rear\ axle\ torque}{Radius\ of\ the\ rolling\ drive} = \frac{K.T_e G_t G_a}{r_w} \quad (7.7)$$

r_w is related to the radius of the tyre r_o by the following formula:

$$r_w = r_o \lambda \quad (7.8)$$

where λ = tyre deformation factor (will have a value 0.945–0.950 for the high pressure air tyres and 0.930–0.935 for low pressure tyres, on hard surfaces

The power output corresponding to the tractive effort P_p when the vehicle moves at a speed of v m/sec is:

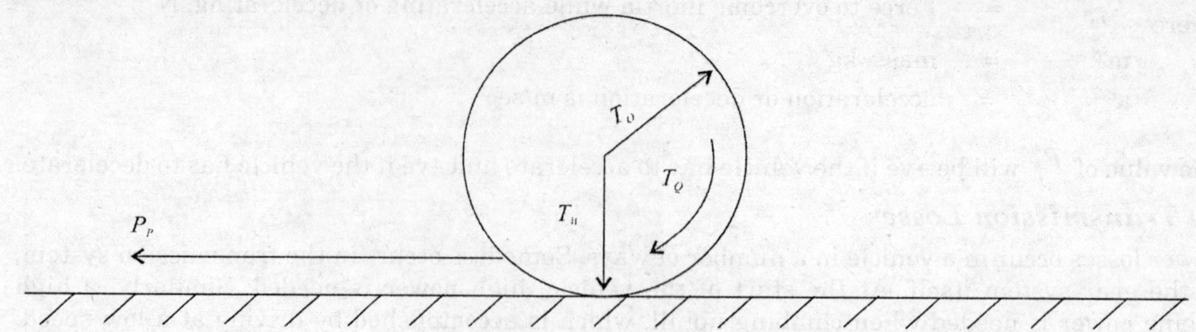

Fig. 7.3: Tractive effort at the driving wheel

$$\text{Power output} = P_p v \tag{7.9}$$

$$v = \frac{V \times 1000}{3600} = \frac{V}{3.6} \tag{7.10}$$

where V = speed in K.P.H.

$$\therefore \text{Power output} = P_p \cdot \frac{V}{3.6} \tag{7.11}$$

$$\text{Also} \quad v = \frac{2.\pi.r_w n}{60.G_t.G_a} \tag{7.12}$$

where n = Engine speed in R.P.M.

$$v = \frac{V}{3.6} = \frac{2.\pi.r_w.n}{60.G_t.G_a}$$

$$\therefore \quad V = \frac{0.377\, r_w.n}{G_t.G_a} \tag{7.13}$$

Substituting (7. 13) in (7.11)

$$\text{Power out put} = \frac{P_p \times 0.377 \times r_w \times n}{G_t.G_a}$$

$$\text{Engine power} = \frac{P_p \times 0.377 \times r_w \times n}{G_t.G_a.K} \tag{7.14}$$

where K = transmission efficiency

Equation (7.14) gives engine power in watts. To convert into Horse-Power,

$$\text{Engine Horse Power (Metric)} = \frac{\text{Engine Power in watts}}{735} \tag{7.15}$$

It may be noted that the power developed at the wheels should be able to overcome the various resistances.

Thus: P_p = Rolling resistance + Air resistance + Grade resistance + inertia forces due to acceleration and deceleration

$$= P_f + P_a \pm P_i \pm P_j \tag{7.16}$$

Problem 7.1. A passenger car weighing 2.25 tonnes is required to accelerate at a rate of 2 m/sec^2 in the first gear from a speed of 8 Kmph The gradient is +3 per cent and the road has a blacktopped surface with the coefficient of rolling resistance equal to 0.02. The frontal projection area of the car is 2.15 m^2. The coefficient of air resistance is 0.39 kg/m^3. The car tyres have a radius of 0.33 m. The inflation pressure reduces this by a factor 0.935. The rear axle gear ratio is 3.82:1 and the first gear ratio is 2.78:1. Calculate the engine power (in Watts and h.p.) needed and the speed of the engine in R.P.M. Assume a transmission efficiency of 0.85.

Solution

Tractive force needed, $P_p = P_f + P_a \pm P_i \pm P_j$

From Equation 7.1

$$P_f = \text{m.g.f} = 2250 \text{ x } 9.81 \text{ x } 0.02 = 441.45 \text{ N}$$

Average speed = 8 Kmph

$$P_a = 0.39 \text{ x } 2.15 \text{ x } \frac{8}{3.6} \text{ x } \frac{8}{3.6} = 4.14 \text{ N}$$

$$P_i = \text{m.g.i} = 2250 \text{ x } 9.81 \text{x} \frac{3}{100} = 662.18 \text{ N}$$

$$P_j = \text{m.a} = 2250 \text{ x } 2 = 4500 \text{ N}$$

$$P_p = 441.45 + 4.14 + 662.18 + 4500 = 5607.77\text{t N}$$

Power Output $= 5607.77 \text{ x } \frac{8}{3.6} = 12461.71 \text{ W}$

$$= \frac{12461.71}{735} \text{hp} = 16.95 \text{ (Metric)}$$

Since transmission efficiency is 0.85,

engine horse power $= \frac{16.95}{0.85} = 19.94$

$$V = \frac{0.377 \text{ x } r_w . n}{G_t . G_a} \tag{7.13}$$

$$r_w = t_{i-j} \tag{7.8}$$

$$= 0.935 \text{ x } 0.33 = 0.308$$

$$n = \frac{8 \text{ x } 3.82 \text{ x } 2.78}{0.308 \text{ x } 0.377} = 732 \text{ R.P.M.}$$

7.9. Horizontal Alignment

7.9.1. Basic Formula for Movement of Vehicles on Curves

When a vehicle is moving on a curved path, it is subjected to an outward force, commonly known as the centrifugal force. In order to resist this force, it is the usual practice to super-elevate the roadway cross-section. Fig. 7.4 shows the forces acting on the vehicle at a super-elevated section.

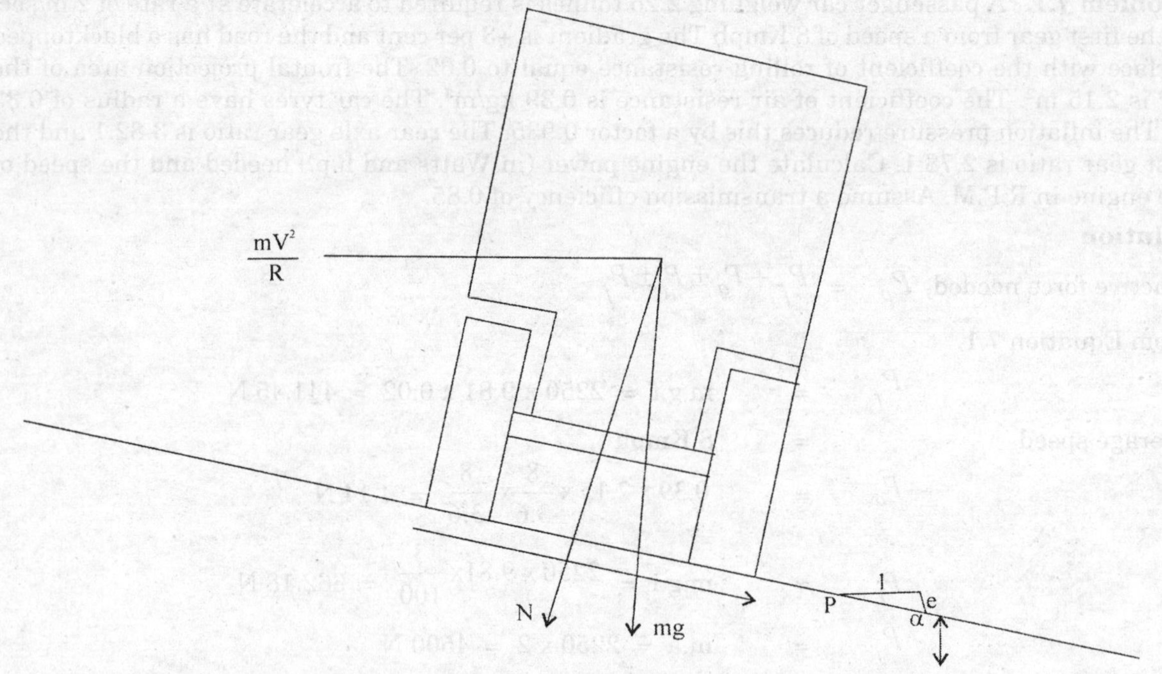

Fig. 7.4: Forces Acting on a Vehicle at a Super-Elevated Section

Let	m	=	mass of the vehicle,
	v	=	speed of the vehicle, m/second
	R	=	radius of the curve, m
	g	=	acceleration due to gravity = 9.81 m/second^2
	p	=	side friction force resisting the centrifugal force = Nμ
	N	=	normal force
	μ	=	coefficient of lateral friction
	α	=	angle of super-elevation
	e	=	super elevation, tan α

The centrifugal force $= \dfrac{mv^2}{R}$

For equilibrium $\dfrac{mv^2}{R} \cos \alpha = m\,g \sin \alpha + P = m\,g \sin \alpha + N\mu$

$$= \quad mg\ \sin\alpha + \mu\left(mg\ \cos\ \alpha + \frac{m\,g\,v^2}{R}\sin\alpha\right)$$

Therefore, $$\frac{v^2}{gR} \quad = \quad tan\ \alpha + \mu + \frac{\mu\,v^2}{gR}\tan\alpha$$

The term $\frac{\mu\,v^2}{gR}\tan\alpha$ is small and can be neglected. Thus,

$$\frac{v^2}{gR} \quad = \quad e + \mu$$

Expressing speed as V in K.P.H., and putting g = 9.81,

$$\frac{v^2}{127R} \quad = \quad e + \mu \qquad (7.17)$$

Equation (7.17) above is the basic equation relating to the speed of vehicles, the radius of the curve, the super-elevation and the coefficient of lateral friction. The equation forms the basis of design horizontal curves.

The value of the coefficient of lateral friction depends upon a number of factors, chief among them being the vehicle speed, type and condition of roadway surfaces, and type and condition of the tyre. The IRC recommends a constant value of 0.15 (Ref 9, 16, 17).

7.9.2. Maximum super-elevation value

If equation (7.17) is to be used for design, it is desirable to know the maximum super-elevation that can be permitted. Practice in this regard varies from country to country. In India the IRC recommendations for hill roads are a maximum value of 0.07 for snowbound areas and 0.10 for area not affected by snow (Ref 9). In other cases, a value of 0.07 is considered maximum.

7.9.3. Maximum radii of curves

Equation (7.17) can be rewritten as:

$$R \quad = \quad \frac{V^2}{127\,(e+\mu)} \qquad (7.18)$$

Adopting the value of e = 0.07 or 0.10 and μ = 0.15 as per current Indian practice the formula becomes:

$$R \quad = \quad 0.0357\ V^2 \text{ (for plain and rolling terrain)} \qquad (7.19)$$

$$R \quad = \quad 0.0315\ V^2 \text{ (for hill roads which are not snow bound)} \qquad (7.20)$$

$$R \quad = \quad 0.0357\ V^2 \text{ (for hill roads which are snow bound)} \qquad (7.21)$$

Since the current Indian Standards give two values for Design Speed (i.e. Ruling and Minimum), the above formulae yield two values for radius, the ruling radius and the absolute minimum radius. These values are indicated in Table 7.7.

7.9.4. Super-elevation Rates

Super-elevating on curves is intended to counteract a part of the centrifugal force, the remaining part being resisted by the lateral friction. Also, super-elevating results in economies in maintenance. This is because skidding and unequal pressure on the wheels of vehicles, which result from high value of sideway force between the tyres and the roadway surface, give rise to pot-holes and necessitate frequent attention to the surface.

Super-elevation (e) can be provided either fully to counteract the centrifugal force or to counteract a fixed proportion of the centrifugal force.

As per Indian practice, super-elevation is calculated on the assumption that it should counteract the centrifugal force developed at three-fourth the design speed. Thus:

$$e = \frac{(0.75V^2)}{127R} = \frac{V^2}{225R} \tag{7.21}$$

The super-elevation calculated from Equation (7.22) is restricted to 0.07 and 0.10 as given in 7.7.

7.9.5. Radii of curves beyond which no super-elevation is required

The normal cambered section of a highway can itself be continued on a curve where the super-elevation calculated is less than the camber. From formula (7.21).

$$R = \frac{V^2}{225e} \tag{7.22}$$

Substituting the value of the camber for e in Equation (7.22) the minimum radius beyond which no super-elevation is required is obtained. Table 7.8 gives the values for various design speeds and cambers.

Table 7.7 : Minimum Radii of Horizontal Curves for Different Terrain Conditions (Source: ref. 1)

Classification of Road	Plain Terrain		Rolling Terrain		Mountainous terrain				Steep terrain			
					Areas not affected by snow		Snow based areas		Areas not affected by snow		Snow based areas	
	Ruling Minimum	Absolute Minimum	Ruling Minimum	Absolute Minimum	Ruling Minimum	Absolute Minimum	Ruling Minimum	Absolute Minimum	Ruling Minimum	Absolute Minimum	Ruling Minimum	Absolute Minimum
1. *National Highways and State Highways*	360	230	230	155	80	50	90	60	50	30	60	33
2. *Major District Roads*	230	155	155	90	50	30	60	33	30	14	33	15
3. *Other District Roads*	155	90	90	60	30	20	33	23	20	14	23	15
4. *Village Roads*	90	60	60	45	20	14	23	15	20	14	23	15

Table 7.8 : Radii Beyond Which Super-Elevation is not Essential for Different Cambers and Speeds (Indian Practice) (Source: Ref. 1)

Design Speed (Kmph)	Radius in Metres Beyond which no Super-Elevation is Needed for Values of Camber of				
	0.04	**0.03**	**0.025**	**0.02**	**0.017**
20	50	60	70	90	100
25	70	90	110	140	150
30	100	130	160	200	240
35	140	180	220	270	320
40	180	240	280	350	420
50	280	370	450	550	650
65	470	620	750	950	1100
80	700	950	1100	1400	1700
100	1100	1500	1800	2200	2600

In such cases where the radii is greater than those given in Table 7.8, it is however, desirable to remove the adverse crown in the outer half of the carriageway and super-elevate at the normal crown slope.

7.9.6. *Method of Attainment of Super Elevation*

The normal cambered surface on a straight reach of road is changed into a super-elevated surface in two stages. In the first stage, outer half of the camber is gradually raised until it is level led, vide, Fig. 7.5(b).

It is desirable to accomplish the raising of the outer-half till it is level led before the starting point of the transition curve. The raising of the outer edge should be done in a slope not exceeding 1 in 150 for plain and rolling terrain and 1 in 60 for hilly terrain.

In the second stage, three methods may be adopted to attain the full super-elevation:

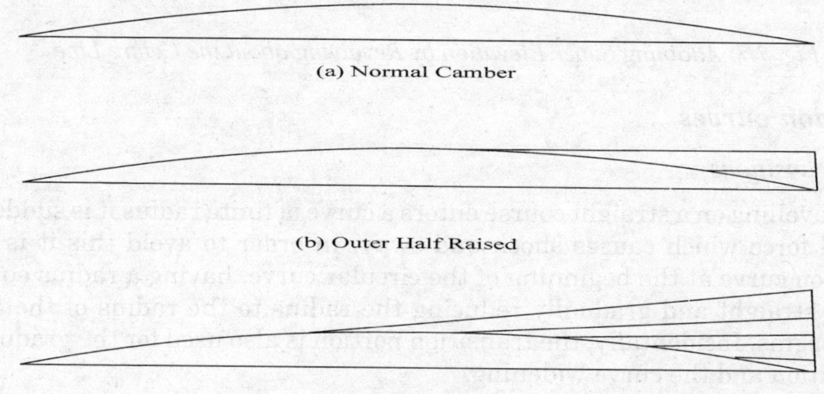

(a) Normal Camber

(b) Outer Half Raised

© Elevation Equal to CamberSuper

Fig. 7.5: First stage in Attaining Super Elevation by Bringing Super-Elevation Equal to Camber

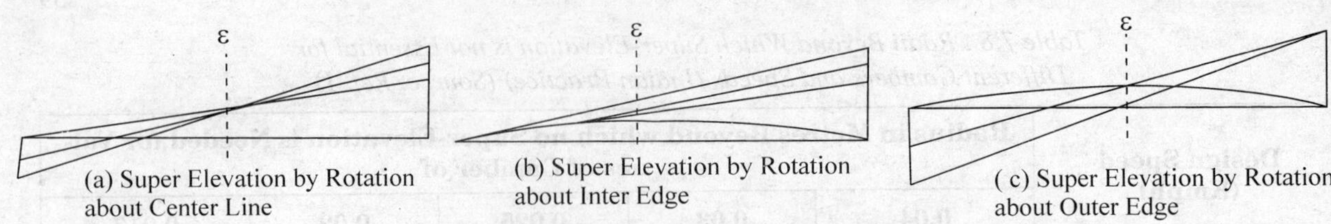

Fig. 7.6: Second Stage in Attaining Super Elevation

(i) The surface of the road is rotated about the centre-line of the carriageway, gradually lowering the inner edge and raising the upper edge, keeping the level of the centre line constant [Fig. 7.6(a)].

(ii) The surface of the road is rotated about the inner edge, raising the centre and the outer edge [Fig. 7.6(b)].

(iii) The surface of the road is rotated about the outer edge, depressing the centre and the inner edge [Fig. 7.6(c)]

Method (i) is most generally used as it results in the least distortion of the pavement. Fig. 7.7 shows the method of attaining super-elevation as per method (i).

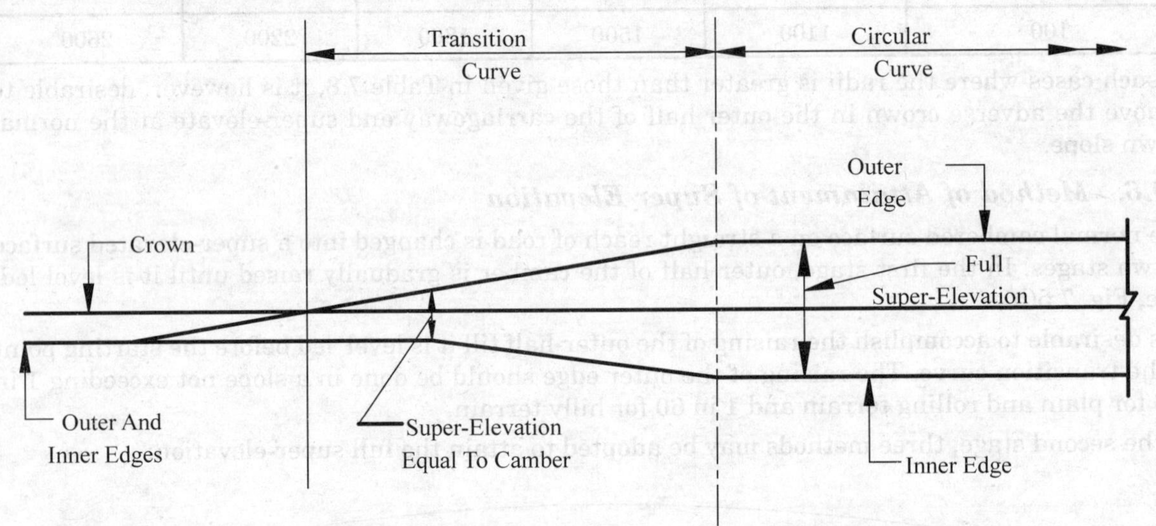

Fig. 7.7: Attaining Super Elevation by Revolving about the Centre Line

7.9.7. Transition curves

7.9.7.1. Need for transitions

When a vehicle travelling on a straight course enters a curve of finite radius it is suddenly subjected to the centrifugal force which causes shock and sway. In order to avoid this it is customary to provide a transition curve at the beginning of the circular curve, having a radius equal to infinity at the end of the straight and gradually reducing the radius to the radius of the circular curve where the curve begins. Incidentally, the transition portion is also used for the gradual application of the super-elevation and the curve widening.

7.9.7.2. The Spiral

Various forms of curves are suitable for highway transitions, but the one most popular and recommended for use in this country is the Spiral. It is easy, to set out in the field and the rate of change of acceleration is uniform through the length of the transition.

Fig. 7.8 indicates the main elements of a circular curve provided with spiral for transitions at its two ends.

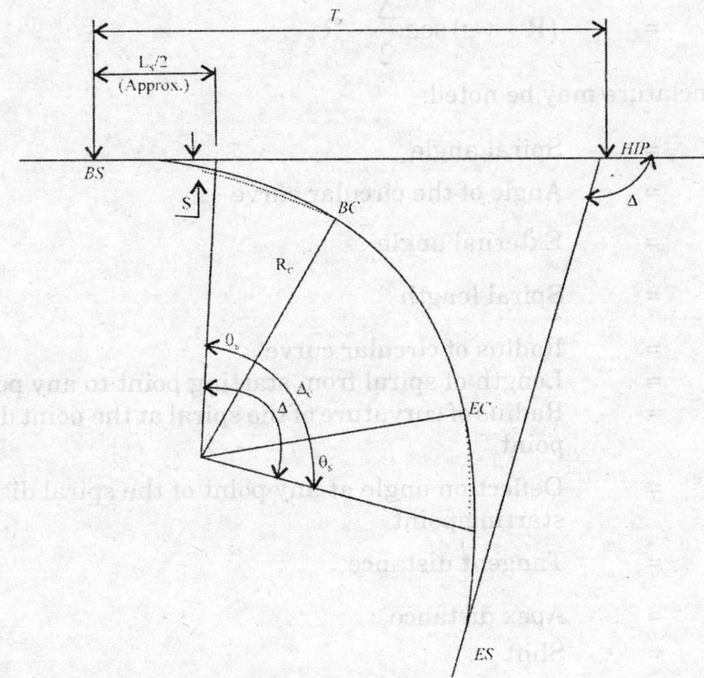

Fig. 7.8: Main Elements of a Circular Curve Provided with Transitions

Some of the important properties of the spirals are given below:

(i) $L_s R_c = LR = \text{constant}$ (7.23)

(ii) $L = M\sqrt{\theta}$ (7.24)

Where M is a constant

$M = \sqrt{2RL}$ (7.25)

Also $L = 2R\theta$ (7.26)

(iii) $\theta = \left(\dfrac{L}{L_s}\right)^2 \theta_s$ (7.27)

(iv) $\theta_s = \dfrac{L}{2R_c}$ radians

$= \dfrac{28.65\,L_s}{R_c}$ degrees (7.28)

(v) $$T_s = \frac{L_s}{2} + (R_c + s)\tan\frac{\Delta}{2} \quad (7.29)$$

(vi) $$S = \frac{L^2_s}{24 R_c} \quad (7.30)$$

(vii) $$E_s = (R_c + s)\sec\frac{\Delta}{2} - R_c \quad (7.31)$$

The following nomenclature may be noted:

θ_s	=	Spiral angle
Δ_c	=	Angle of the circular curve
Δ	=	External angle
L_s	=	Spiral length
R_c	=	Radius of circular curve
L	=	Length of spiral from starting point to any point
R	=	Radius of curvature of the spiral at the point distant L from starting point
è	=	Deflection angle at any point of the spiral distant L from the starting point
T_s	=	Tangent distance
E_s	=	Apex distance
s	=	Shift
HIP	=	Horizontal intersection point
BS	=	Beginning of spiral
BC	=	Beginning of circular curve
EC	=	End of circular curve
ES	=	End of spiral

7.9.7.3. Length of transition

The length of the transition should be determined from the following two considerations:

(i) The rate of change of centrifugal acceleration adopted in the design should not cause discomfort to the drivers.

If C is the rate of change of acceleration

$$C = \frac{v^2}{R_c} \div t$$

$$= \frac{v^2}{R_c} \div \frac{L_s}{v} = \frac{v^3}{R_c L_s} \quad (7.32)$$

where R_c and L_s have the same meaning as in 7.9.7.2,

t = time in second to travel a distance of L_s

v = speed in m/second

Thus,

$$L_s = \frac{V^3}{(3.6)^3\, CR_c} = \frac{0.0215\, V^3}{R_c} \qquad (7.33)$$

where V = Speed in K.P.H.

the value of C is given by the following equation (Ref 11):

$$C = \frac{80}{75 + V} \qquad (7.34)$$

subject to a maximum of 0.8 and minimum of 0.5.

(ii) The rate of change of super elevation should be such as not to cause higher gradients and unsightly appearances. This could be kept 1 in 150 for roads in plain and rolling terrain and 1 in 60 in hilly terrain (Ref 1).

The higher of the values given in considerations (i) or (ii) above may be used. Suggested values are given in Table 7.9.

Table 7.9 : Minimum Transition Lengths for Different Speeds and Curve Radii

Curve radius R (metres)	Plain and Rolling Terrain Design Speed (KPH)						Curve radius (metres)	Mountainous and Steep Terrain Design Speed (KPH)				
	100	80	65	50	40	35		50	40	30	25	20
	Transition length-metres							Transition length-metres				
45					NA	70	14				NA	30
60				NA	75	55	20				35	20
90				75	50	40	25			NA	25	20
100			NA	70	45	35	30			30	25	15
150			80	45	30	25	40		NA	25	20	15
170			70	40	25	20	50		40	20	15	15
200		NA	60	35	25	20	55		40	20	15	15
240		90	50	30	20	NR	70	NR	30	15	15	15
300	NA	75	40	25	NR		80	55	25	15	15	NR
360	130	60	35	20			90	45	25	15	15	
400	115	55	30	20			100	45	20	15	15	
500	95	45	25	NR			125	35	15	15	NR	
600	80	35	20				150	30	15	15		
700	70	35	20				170	25	15	NR		
800	60	30	NR				200	20	15			
900	55	30					250	15	15			
1000	50	30					300	15	NR			
1200	40	NR					400	15				
1500	35						500	NR				
1800	30											
2000	NR											

NA-Not applicable

NR Transition not required

7.9.8. Widening on curves

Widening of pavements is needed on curves for the following reasons:

(i) On curves the vehicles occupy a greater width because the rear wheels track inside the front wheels (Fig. 7.9).

On curves drivers have difficulty in steering their vehicles to keep to the centre line of the line.

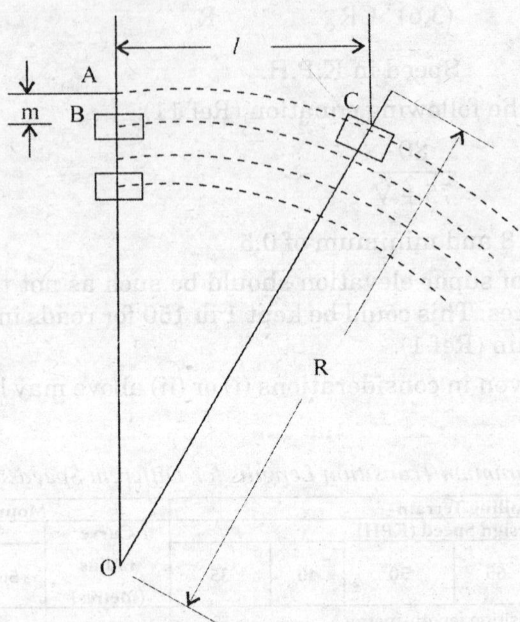

Fig. 7.9: Widening on Curves

From Fig. 7.9, l^2 considering the right angled triangle OBC,

$$l^2 + (R-m)^2 = R^2$$

$$l^2 = 2\,Rm - m^2$$

Neglecting m^2, $l^2 = 2\,Rm$

$$m = \frac{l^2}{2R} \qquad (7.35)$$

Assuming a wheel base of 6 m for a vehicle corresponding to AASHTO Single Unit (SU). Widening in metres

$$m = \frac{18}{R} \qquad (7.36)$$

where R = radius in meters

The widening due to psychological reasons is function of speed and can be assumed to be given b the empirical formula:

$$\frac{0.1\,V}{\sqrt{R}} \qquad (7.37)$$

where V = speed in K.P.H

R = radius in meters

The total widening for pavements of those width which are in excess of two lanes is given by the formula:

$$W_e \text{ (in m)} = n\frac{18}{R_c} + 0.1\frac{V}{\sqrt{R_c}} \qquad (7.38)$$

where n = number of lanes

Table 7.10 gives the widening necessary for single lane and 2-lane pavements (Ref 1, 11).

Table 7.10 : Extra Width of Pavement at Horizontal Curve

Radius of Curve (m)	upto 20	21 to 40	41 to 60	61 to 100	101 to 300	above 300
Extra width (m)						
Two-lane	1.5	1.5	1.2	0.9	0.6	Nil
Single-lane	0.9	0.6	0.6	Nil	Nil	Nil

7.9.9. *General Controls for Horizontal Alignment*

The following general controls for horizontal alignment should be kept in view in a sound design practice.

(i) Alignment should be as directional as possible.

(ii) The alignment should be consistent with the topography and should confirm generally to the natural contours.

A line cutting across the contours involves high fills and large cuts, mars the landscape and is difficult for maintenance.

(iii) The number of curves should, in general, be kept to a minimum.

(iv) The alignment should be fluent and avoid abrupt turns. Widening alignment consisting of short curves should be avoided, since it is the cause of erratic operation.

(v) A sharp curve at the end of a long tangent is extremely hazardous and should be avoided. If sharp curvature is unavoidable over a portion of the route selected, it is preferable that this portion of the road be preceded by successfully sharper curves.

(vi) Short curves giving the appearance of kinks should be avoided, especially for small deflection angles. The curves should be sufficiently long to provide pleasing appearance and smooth driving on important highways. They should be at least 150 metres long for a deflection angle of 5 degrees, and the minimum length should be increased by 30 metres for each 1 degree decrease in the deflection angle.

(vii) For a particular design speed, as large a radius as possible should be adopted. The minimum radii should be reserved only for the critical locations.

(viii) The use of sharp curves should be avoided on high fills. In the absence of cut slopes, shrubs, trees, etc., above the roadway, the drivers may have difficulty in estimating the extent of curvature and fail to adjust to the conditions.

(ix) While abrupt reversal in curvature is to be avoided, the use of reverse curves becomes unavoidable in hilly terrain. When they are provided, adequately long transitional curves should be inserted for super-elevation run-off.

(x) Curves in the same direction separated by short tangents, say 300-500 metres long are called broken-back curves.

They should be avoided as they are not pleasing in appearance and are hazardous.

(xi) Compound curves may be used in difficult topography in preference to a broken-back arrangement, but they should be used only if it is impossible to fit in a single circular curve. To ensure safe and smooth transition from curve to curve, the radius of the flatter curve should not be disproportional to the radius of the sharper curve. A ratio of 2:1 or preferably 1.5:1 should be adopted.

(xii) The horizontal alignment should blend with the vertical harmoniously. General controls for the combination of horizontal and vertical alignments are indicated in section 7.11.

7.10. Vertical Alignment

7.10.1. Gradients

7.10.1.1. Maximum Grades

One of the important considerations in designing a highway is the gradient. The cost of operation of vehicles and the capacity of a highway are profoundly affected by the grades provided. On mountainous and steep terrain, the grades are not only influenced by the ability of vehicles to negotiate them, but also by the altitude of a road above the sea level since rarefaction of air causes loss of engine power. A peculiar consideration for Indian conditions is the presence of animal drawn traffic.

The grades should be selected so as to encourage uniform operation as far as possible for a given terrain and the vertical profile should be as smooth as economically possible. Roads once constructed to a given vertical profile are very difficult to be upgraded to flatter gradients at a subsequent stage without great loss to the initial investment.

Table 7.11 gives the Indian practice as regards maximum gradients (Ref 1, 10). In Table 7.11 gradients up to the ruling may be used as a matter of course in design. The limiting gradients may be used where the topography of the area compels this or where the adoption of gentler gradients would enormously add to the cost. Exceptional gradients are meant to be used only in very exceptional situations and for short lengths prescribed in the Table 7.11.

Table 7.11 : Gradients for Different Terrain Types-Indian Practice

S. No.	Terrain Type	Ruling Gradient (%)	Limiting Gradient %	Exceptional Gradient %
1.	Plain	3.3 (1 in 30)	5 (1 in 20)	6.7 (1 in 15) (for short distances not exceeding 100 m at a stretch)
2.	Rolling	3.3 (1 in 30)	5 (1 in 20)	6.7 (1 in 15) (for short distances not exceeding 100 m at a stretch)
3.	Mountainous	5 (1 in 20)	5 (1 in 16.7)	7 (1 in 14.3) (for a distance not more than 400 m)

(Contd...)

4.	Steep			
(i)	Upto 3000 m height above mean sea level	5 (1 in 20)	6 (1 in 16.7)	7 (1 in 14.3) (for short distances not exceeding 100 m)
(ii)	Above 3000 m height above mean sea level	6 (1 in 16.7)	7 (1 in 14.3)	8 (1 in 12.5) (for short distances not exceeding 100 m)

7.10.1.2. Minimum Gradients for Drainage

Where the drainage of a road is to be effected by means of longitudinal ditches, as in cut sections or hill roads, or along the curbed edge of pavements, a certain minimum gradient is necessary for efficient drainage. The values given in Table 7.12 are commonly adopted.

Table 7.12 :Minimum Gradients for Drainage (Source: Ref 2)

S. No.	Site Condition	Minimum gradients (%) (Desirable)
1.	Longitudinal drainage along curbed edge of pavements, (lined)	0.5
2.	Longitudinal drainage along road side ditches (unlined)	1.0

7.10.2. *Vertical Curves*

7.10.2.1. General Considerations

Vertical curves should be provided at points of change of grade so as to achieve gradual transition. If the point of vertical intersection (termed PVI) is above the road surface, the curve is a summit or crest curve and if it is below the road surface, the curve is a sag curve. If the total change in gradient from one tangent to another does not exceed 0.5 per cent, vertical curves can be dispensed with. The parabola is a vertical curve most frequently used because of the ease with which it can be laid out.

7.10.2.2. Length of Summit Curves

For summit curves, the most important consideration in determining the length of the curve is the sight distance requirement. The following basic formulae can be used to determine the length:

Case I. When the sight distance (S) is less than the length of the curve (L):

$$L = \frac{AS^2}{\left(\sqrt{2h_1} + \sqrt{2h_2}\right)^2} \tag{7.39}$$

Case II. When the sight distance (S) is greater than the length of the curve (L):

$$L = 2S - \frac{2\left(\sqrt{h_1} + \sqrt{h_2}\right)^2}{A} \tag{7.40}$$

In the above formulae,

S = sight distance in metres
L = length of vertical curve, in metres
h_1 = height of eye above the roadway surface

h_2 = height of object above the roadway surface

A = algebraic difference in slopes expressed in absolute values in a decimal form

Adopting h = 1.2 m

h_2 = 0.15 m

as per current Indian practice (Ref 12, 13) Eqs. (7. 39) and (7.40) transform as under for safe stopping sight distance:

When S is less than L,

$$L = \frac{AS^2}{4.4} \tag{7.41}$$

When S is greater than L,

$$L = 2S - \frac{4.4}{A} \tag{7.42}$$

If the gradient is expressed in per cent, the above formulae become:

$$L = \frac{NS^2}{440}$$

$$L = 2S - \frac{440}{N}$$

For intermediate or overtaking sight distance, $h_1 = h_2 = 1.2$ m.

This gives the following relationships:

When L is greater than S,

$$L = \frac{AS^2}{9.6}$$

When L is less than S,

$$L = 2S - \frac{9.6}{A}$$

7.10.2.3. Sag Curves

For sag curves, one of the criteria for determining the length is rider comfort. The other criteria are vehicle headlight, sight distance, drainage control and general appearance. As per rider comfort as the basis of design of sag curves (Ref 13), α, the vertical radial acceleration, is given by,

$$\alpha = \frac{v^2}{R}$$

where α is in m/sec²

v = speed in m/sec

$$= \frac{V^2}{13R} \tag{7.43}$$

where V = speed in K.P.H.

Since the radial acceleration changes from zero to α in a length of L/2, where L = length of the sag curve, the rate of change of radial acceleration C is given by:

$$C = \frac{V^2}{13R} \div \frac{L \times 3.6}{2\,V} \tag{7.44}$$

$$= \frac{V^3}{23.4\,LR}\,m/\sec^3 \tag{7.45}$$

The recommended value of C for design is 0.6 m/sec (Ref 4, 13). Substituting this in (7.44), we get

$$L = \frac{V^2}{14R} \tag{7.46}$$

$$\therefore \quad LR = \frac{V^3}{14} \tag{7.47}$$

For a transition curve of the form of spiral or a cubic parabola, it can be shown that

$$L = (2RLA)^{1/2} \tag{7.48}$$

where A = algebraic difference in grades of the straights, expressed as a decimal

Substituting Eq. (7.47), in (7.48)

$$L = \left(2\frac{V^3}{14}A\right)^{1/2} = \frac{V^{3/2}\,A^{1/2}}{2.65} \tag{7.49}$$

Another criterion for design of valley curves is that for night travel, the design must ensure that the roadway ahead is illuminated by vehicle headlights for a sufficient length, which enables the vehicle to brake to a stop, if necessary. This is known as the headlight sight distance and is equal to the safe stopping distance (Table 7.13). For designing valley curves for the above needs, the following points should be kept in view (Ref 12):

(i) height of headlight above the road surface is 0.75 m,

(ii) the useful beam of headlight is one degree upwards from the grade of the road; and

(iii) the height of the object is nil.

7.10.3. General Controls for Vertical Alignment

The following general controls for vertical alignment should be kept in view in designing the vertical profile of a highway:

(i) The grade line selected should be smooth with gradual changes, consistent with the class of highway and terrain. Numerous brakes and short lengths of grades should be avoided.

(ii) The roller-coaster or hidden type of profile should be avoided as it is hazardous and aesthetically unpleasant.

(iii) Undulating grade lines, involving substantial lengths of momentum grades, should be appraised for their effect upon traffic operation. Such profiles permit heavy trucks to operate at higher overall speeds than when an upgrade is not preceded by a down-grade but may encourage excessive speeds of trucks with consequent hazard to traffic.

(iv) A broken-back grade line (two vertical curves in the same direction separated by short section of tangent grade) should generally be avoided.

(v) On long continuous grades, it may be preferable to place the steepest grades at the bottom

and flatten the grades near the top. Alternatively, the long grades may be broken by short intervals of flatter grades.

(vi) Intersections on grades should be avoided as far as possible. Where unavoidable, the approach gradients and the gradient through the intersections should be flattened to the maximum possible extent.

7.11. Combination of Horizontal and Vertical Alignment

The following general controls should be kept in view in designing a highway for combination of horizontal and vertical alignment:

(i) The overall appearance of highway is enhanced with a judicious selection of horizontal and vertical elements. Best results are obtained when there is a proper balance between curvature and grades. Straight alignment or flat horizontal curves imposed on a road with steep or long grades or excessive curvature imposed on a road with a flat grade are both poor designs. A compromise between the two offers the best in safety, capacity, ease and uniformity of operation.

(ii) Vertical curvature superimposed upon horizontal curvature and vice-versa, generally results in a pleasing effect.

(iii) From the safety point of view, sharp horizontal curvature should not be introduced at or near the top of a pronounced summit vertical curve. In such situations, the horizontal curve is made longer than the vertical curve.

(iv) Sharp horizontal curves should be avoided at or near the low point of a pronounced sag curve.

(v) On two-lane highways, the need for safe passing sections at frequent intervals and for an appreciable percentage of the length of the highway often supersedes the general desirability for combination of horizontal and vertical alignment. In these situations, it is desirable to have long straight sections to secure, sufficient passing sight distance in design.

(vi) Horizontal curvature and vertical profile should be made as flat as feasible at highway intersections when sight distance is important and vehicles may have to slow down or stop.

(vii) On divided highways, variation in the width of the median and the use of separate profiles and horizontal alignments should be considered to derive design and operational advantages of one-way roadways.

7.12. Sight Distance

7.12.1. General

The design of a highway with adequate sight ahead of a travelling vehicle results in safe operation. The distance along the road surface at which a driver has visibility of objects, stationary or moving, at a specified height above the carriageway is known as the sight distance. A knowledge of the sight distance requirement is needed in designing vertical curves. It also governs the set-back of buildings, slopes, fences and other obstructions adjacent to a carriageway on a horizontal curve. Sight distance requirement can be considered under two heads:

(i) Stopping sight distance and

(ii) overtaking (passing) sight distance.

7.12.2. Stopping sight distance (SSD)

Stopping sight distance is the distance required by a driver of a vehicle travelling at a given speed to bring his vehicle to a stop after an object on the roadway becomes visible. Because of its importance to safety, all highways must be designed for the minimum stopping sight distance. It is made up of two components:

(i) the distance travelled during perception and brake reaction time; and

(ii) the distance travelled during the time the brakes are under application till the vehicle comes to a stop.

When sensations received through the eyes, ears or body are strong enough to be recognised and interpreted, they become perceptions. In the case of a motorist, it is the time which elapses between the instant the driver perceives the object on the carriageway and the instant that he realizes that braking of the vehicle is needed. The time lag or the brief interval between the perception of danger and the effective application of the brakes is called the brake reaction time. The perception time and the brake reaction time depend upon a variety of factors, viz., age, sex, alertness and visual acuity of the driver, visibility, vehicle design, the size and type of the object etc.

As per current Indian Practice (Ref 12), a value of 2.5 second is considered reasonable for the perception time and brake reaction time taken together. The distance travelled during this interval is given by

$$d_1 = vt$$

$$d_1 = 0.278\ Vt = 0.278\ V \times (2.5)$$

$$= 0.695\ V \qquad (7.50)$$

where d_1 = distance travelled in meters

v = speed in metres per second

V = speed in Kmph

t = perception and brake reaction time

= 2.5 seconds

The braking distance is the distance in which a moving vehicle comes to a stop after the application of brakes. On a level roads, the distance is given by:

$$d_2 = \frac{V^2}{254f} \qquad (7.51)$$

where d_2 = braking distance in meters

V = Speed in Kmph

f = coefficient of longitudinal friction between the tyres and the pavement.

The coefficient of friction is assumed to vary from 0.40 at 20 Kmph to 0.35 at 100 Kmph (Ref 18). The above considerations yield the value given in Table 7.13 recommended by IRC (Ref 1, 12).

Table 7.13 : Stopping Sight Distance as per Indian Practice

Design Speed Kmph	Safe Stopping Sight Distance (Metres)
20	20
25	25
30	30

Design Speed Kmph	Safe Stopping Sight Distance (Metres)
40	45
50	60
60	80
65	90
80	120
100	180

Some slight adjustments are needed in the values of the braking sight distance to take into account the effect of grades. The following amended formula may be used to calculate d_2:

$$d_2 = \frac{V^2}{254(f \pm 0.01G)} \quad (7.52)$$

where G = longitudinal grade in per cent

The positive sign may be used inside the bracket if the gradient is upward and the negative sign may be used if the gradient is downward. Correction for grade should not be applied on undivided roads with two-way traffic, but must invariably be considered for divided highway which have independently designed profiles.

The safe stopping sight distance is measured between two points, one 1.2 m above the carriageway representing the driver's eye and the other at 0.15 m height representing an object (Ref 12).

7.12.3. Overtaking Sight Distance (OSD)

Overtaking becomes necessary because some vehicles travel fast while some travel slow. The highway should be designed so that the vehicles travelling at the design speed are afforded chance to overtake the slower vehicles. Overtaking manoeuvre is possible only if the drivers have sufficient sight distance available to them so that the whole operation can be accomplished without danger. On undivided roads, overtaking involves the occupation of the road space normally used by opposing traffic. The overtaking sight distance is that distance which should be available to enable the driver to overtake another vehicles safely and comfortably without interfering with the speed of an oncoming vehicle travelling at the design speed should it come into view after the overtaking manoeuvre is started.

For computing the minimum overtaking sight distance on two-lane two-way highways, the following assumptions are made (Ref 12):

(i) The overtaken vehicle travels at a uniform speed which is 16 Kmph less than the design speed of the road.

(ii) The overtaking vehicle has reduced its speed and trails the overtaken vehicle as it enters the overtaking section.

(iii) The overtaking vehicle follows the vehicle ahead for a short while to perceive the clear road ahead before beginning the overtaking movement.

(iv) The overtaking is done by acceleration to the design speed and hurriedly returning to its own side of the road.

(v) When the overtaking vehicle returns to its lane, it arrives alongside an oncoming vehicle in the opposite lane.

[Some authorities such as AASHTO (Ref 4) lay down that a certain clear distance should be available between the two].

Fig. 7.10 below indicates the elements that go to make up the overtaking sight distance.

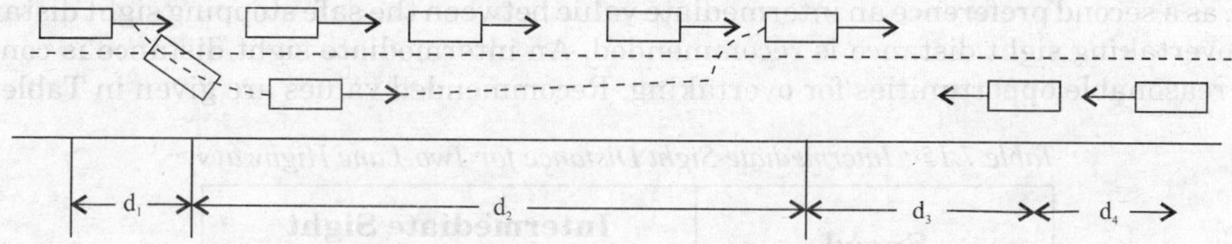

Fig. 7.10: Elements of Overtaking Sight Distance

The dimension d_1 represents the distance travelled during the perception and reaction time and during the initial acceleration to the point of encroachment on the right lane. AASHTO practice (Ref 4) gives a value of 3.6 to 4.5 seconds for this initial manoeuvre. The dimension d_2 represents the actual distance covered by the overtaking vehicle during the overtaking manoeuvre. The overtaking vehicles are assumed in AASHTO practice to take 9.3 to 10.4 seconds for this purpose. The dimension d_3 is the distance between the overtaking vehicle at the end of its manoeuvre and the opposing vehicle and is known as the clearance length. It is assumed to vary from 34 to 90 m as per AASHTO practice (Ref 4). The distance d_4 represents the distance travelled by an opposing vehicle at the design speed while the overtaking manoeuvre is taking place. Strictly speaking it should be $d_1 + d_2$, but in practice it is taken as $\frac{2}{3} d_2$, on the reasoning that the overtaking vehicle can always return to its lane if the driver sees an opposing vehicle.

As per latest Indian practice (Ref 12), the calculation have been simplified by assuming that it takes about 9 to 14 seconds for the overtaking manoeuvre. About 2/3rds of this value is accounted for the distance travelled by the opposing vehicle. The final values as per Indian practice are given in Table 7.14.

Table 7.14 : Overtaking Sight Distance (Ref 12)

Speed Kmph	Safe Overtaking Sight Distance (metres)
40	165
50	236
60	300
65	340
80	470
100	640

The overtaking distance is measured between two points both 1.2 m above the carriageway, representing the position of the eye of the driver and the object.

As per Indian practice (Ref 12), for sections of roads where the overtaking sight distance cannot be provided, as a second preference an intermediate value between the safe stopping sight distance and the safe overtaking sight distance is recommended. An intermediate sight distance is considered to afford reasonable opportunities for overtaking. Recommended values are given in Table 7.15.

Table 7.15 : Intermediate Sight Distance for Two-Lane Highways

Speed Kmph	Intermediate Sight Distance (metres)
20	40
25	50
30	60
40	90
50	120
60	160
65	180
80	240
100	360

7.12.4. Sight Distance for Multi-lane Roads

Divided highways with 4 or more lanes need only be designed for a safe stopping sight distance. Undivided highways with 4 lanes have enough opportunities for overtaking within one-half of the carriageway. Such roads, therefore, need only be designed for safe stopping sight distance.

7.12.5. Set-back Distance at Obstructions of Horizontal Curves

On horizontal curves with obstructions on the inside an important consideration is the lateral clearance so as to obtain the sight distance. Fig. 7.11 indicates the elements of design.

The following formulae give the value of the set-back distances:

$$M = \frac{S^2}{8R} \tag{7.53}$$

(for cases where the required sight distance is wholly within the curved road)

$$M = \frac{E(2L - S)}{8R} \tag{7.54}$$

(for cases where the required sight distance is greater than the available length of curve)

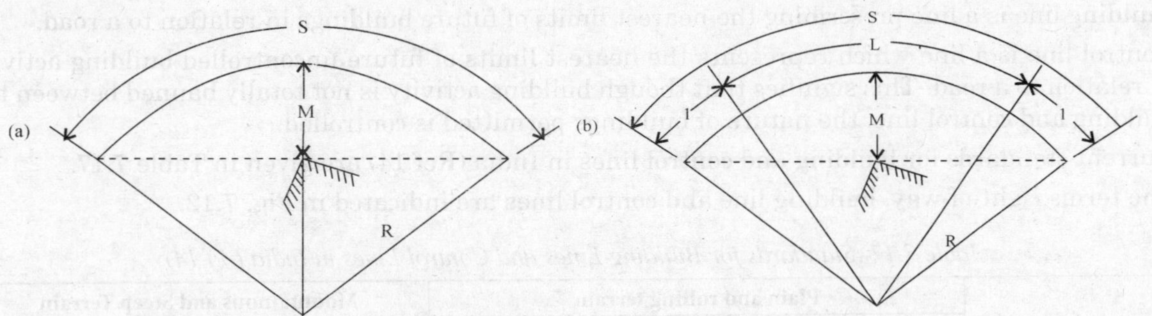

Fig. 7.11: Elements of Set-Back Distance at Obstructions of Horizontal Curves

7.13. Cross-sectional Elements

7.13.1. General

The cross-sectional elements in the highway design pertain to those features which deal with its width. They cover aspects such as right-of-way width, roadway width, pavement width, central reservations (median), shoulder, camber, side slopes, horizontal and vertical clearance etc.

7.13.2. Right-of-way

The road land width or the right-of-way width is the width of land secured and preserved to the public for road purposes. The right-of-way should be adequate to accommodate all the elements that make up the cross-section of the highway and may reasonably provide for future development. Indian practice (Ref 14) is summarised in Table 7.16. In order to prevent ribbon development along highways, it is sometimes necessary to establish building lines and control lines with the following definitions:

Table 7.16 :Right-of-Way Width of Different Classes of Roads in India (Ref 1, 10)

Class of Road	Right-of-way width in metres							
	Plain and rolling terrain				Mountainous and steep terrain			
	Rural Areas		Urban Areas		Rural Areas		Urban Areas	
	Normal	Range	Normal	Range	Normal	Exceptional	Normal	Exceptional
1. National and State Highways	45	30-60	30	30-60	24	18	20	18
2. Major District Roads	25	25-30	20	15-25	18	15	15	12
3. Other District Roads	15	15-25	15	15-20	15	12	12	9
4. Village Roads	12	12-18	10	15-20	9	9	9	9

Building line is a line prescribing the nearest limits of future buildings in relation to a road.

Control line is a line which represents the nearest limits of future uncontrolled building activity in relation to a road. This signifies that though building activity is not totally banned between the building and control line, the nature of buildings permitted is controlled.

Current standards for building and control lines in India (Ref 14) are given in Table 7.17.

The terms right-of-way, building line and control lines are indicated in Fig. 7.12.

Table 7.17 :Standards for Building Lines and Control Lines in India (ref 14)

Class of Road	Plain and rolling terrain			Mountainous and Steep Terrain			
	Rural Area		Urban and industrial area	Distance between Building Line and Boundary (metres)			
				Rural Areas		Urban Areas	
	Width between building lines (overall width) (metres)	Width between control lines (overall width) (metres)	Distance between building line and road boundary (metres)	Normal	Exceptional	Normal	Exceptional
1. *National and State Highways*	80	150	3-6	5	3	5	3
2. *Major District Roads*	50	100	3	5	3	5	3
3. *Other District Roads*	25/30*	35	–	5	3	5	3
4. *Village Roads*	25	30	–	5	3	5	3

* If the land width is equal to the width between building lines indicated in this column, the building lines shall be set 2.5 m from the road boundary lines.

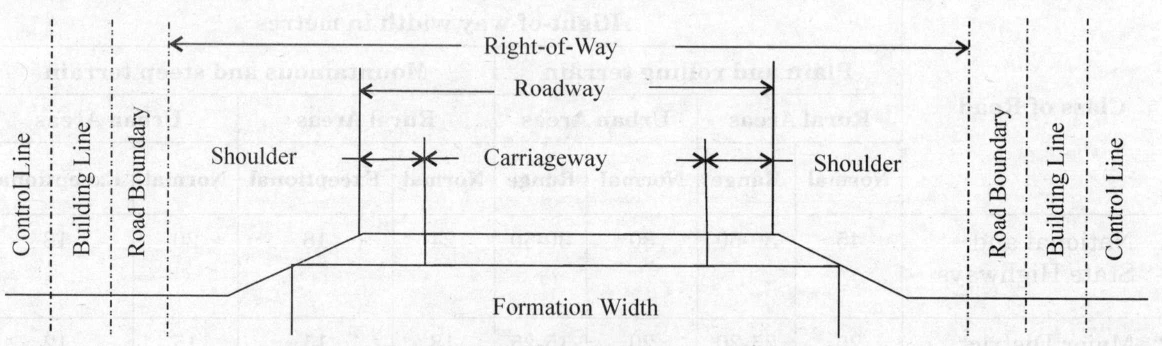

Fig. 7.12: Right of Way, Building Line and Control Line

7.13.3. *Space standards for urban roads*

In urban situations the space available is generally restricted and costly. Table 7.18 below represents current Indian practice (Ref 2, 15).

Table 7.18 : Space Standards for Urban Roads (Ref 2, 15)

S. No.	Category of Road	Minimum Space in Metres
1.	Expressway (6 lane divided)	60
2.	Arterial Street (4 lane divided with cycle tracks)	50-60
3.	Sub-arterial street (4 lane divided)	30-40
4.	Collector street (4 lane)	20-30
5.	Local street such as main residential street, shopping street in residential area, minor residential street and cul-de-sac	10-20

7.13.4. Roadway width

Current Indian practice as regards roadway width is summarised in Table 7.19.

Table 7.19 :Standards for Roadway Width—Indian Practice

Highway Classification	Width of Roadway (in metres)	
	Plan and Rolling Terrain	Mountainous and Steep Terrain
1. National and State Highways		
(a) Single lane		6.25
(b) Two lane	12.0	8.8
2. Major District Roads	9.0	
3. Other District Roads		
(i) two lanes	9.0	4.75
(ii)) single lane	7.5	4.75
4. Village Roads	7.5	4.0

For roads with more than two lanes, the roadway width has to be worked out to suit the number of lanes, medians (if any) and shoulders.

7.13.5. Carriageway Width

The width of a traffic lane governs the safety and convenience of traffic and has a profound influence on the capacity of a road. Generally, a width of 3–4 metres is adopted for each traffic lane. In India, single lane pavements are generally 3.75 metres wide, whereas two lane pavements (without raised curbs) are 7.0 metres wide (Ref 1, 16). When raised curbs are provided, two lane pavements are made 7.5 meters in width. An intermediate width of 5.5 metres is commonly used in India for less important roads with two lanes. For multi-lane pavements Indian practice is to provide 3.5 meter per lane. Single lane pavements in India are often provided with 0.9 metre wide pavement on either side to facilitate crossing and overtaking manoeuvres.

For village roads, a carriageway width of 3.0 meters is generally deemed sufficient.

7.13.6. Central reservation (Median)

A central reservation or a median is the longitudinal space separating dual carriageways. The functions of the central reservation are:

(i) To separate the opposing streams of traffic

(ii) To minimise head light glare

(iii) To include space for safe operation of crossing and turning vehicles at intersections at grade, e.g., right turning pockets.

(iv) To provide a stopping area in case of emergencies.

(v) To provide space for future widening of the carriageways.

Indian practice is to provide a minimum width of 5.0 metres for rural areas, which is relaxed to 3.0 metres in restricted conditions (Ref 1). The current Indian practice is to provide raised median with curbs, but experience has shown that raised median results in poor safety. For urban situations, a width of 5.0 metres is considered desirable, but a minimum width of 1.2 metres is accepted.

American practice is to have a minimum value of 1.2 metres and a greater desirable dimension of 18.3 metres or more for rural highways (Ref 4). For urban freeways, a minimum of 7.3 metres is recommended, which may be further lowered to 5 metres in extremely restricted locations (Ref 5). On severely restricted arterial streets, where a narrow separator 0.6–1.2 metres is alone feasible, it may be desirable to have few, if any, openings in the median except at intersections.

In India, it is recommended (Ref 1) that the minimum width of medians on rural highways should be 5 metres, but this could be reduced to 3 metres where land is restricted. On long bridges and viaducts, the width of median may be reduced to 1.5 metres, but in any case it should not be less than 1.2 m. As far as possible, the median should be of uniform width in a particular section of the highway. However, where changes are unavoidable, a transition of 1 in 15 to 1 in 20 must be provided.

7.13.7 Shoulders

The shoulder is that portion of the roadway contiguous with the travelled way and is intended for accommodation of stopped vehicles, emergency use and lateral support of the pavement courses.

The width of the shoulder should be adequate for giving working space around a stopped vehicle. American practice (Ref 4) recommends a metre width for high type facilities and a width 1.2 to 2.4 metres for low type facilities. U.K. practice for rural roads recommends widths ranging from 1.2 to 3.65 metres depending upon the type of road and the nature of the kerb side treatment. Current Indian practice for 2-lane roads suggests a shoulder width of 2.5 metres for roads in rural areas (Ref 1).

7.13.8 Curbs (Kerbs)

A curb is a vertical or sloping member along the edge of a pavement or shoulder forming part of a gutter, strengthening or protecting the edge, and clearly defining the edge to vehicle users. Its functions are:

(i) To facilitate and control drainage

(ii) To strengthen and protect the pavement edge

(iii) To delineate the pavement edge

(iv) To present a more finished appearance

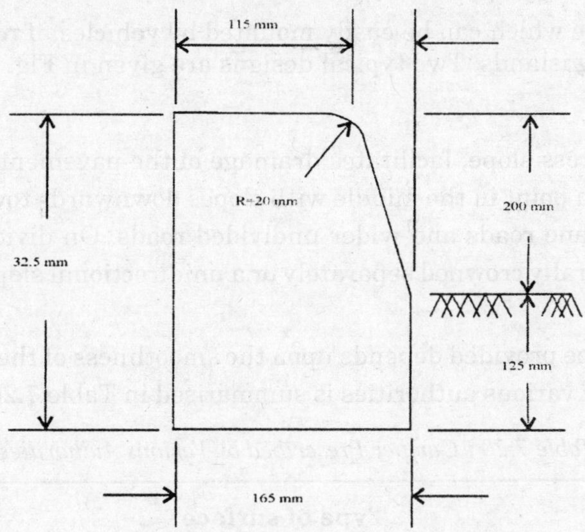

Fig. 7.13: Nearly Vertical Barrier Curb

Curbs are generally needed on urban roads where raised footpaths are a common feature. Curbs are classified as barrier and mountable. Barrier curbs are designed to discourage vehicles from leaving the pavement. The face may be vertical or sloping and the height may vary from 15 to 25 cm. A typical design for a nearly vertical barrier curb is given in Fig. 7.13. In Fig. 7.13 a smaller height (say 15 cm) may be adopted for pedestrian refuge islands. A larger height (say 22.5 cm) is adopted for bridges (Ref 17).

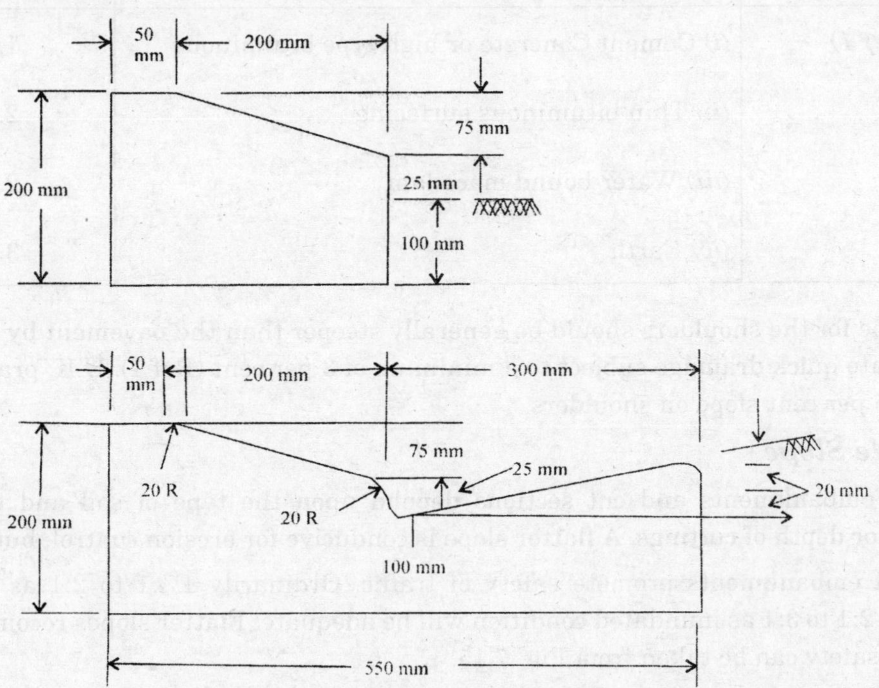

Fig. 7.14: Typical Designs of Mountable Curbs

Mountable curbs are those which can be easily mounted by vehicles, if required. They are used at medians and channelizing islands. Two typical designs are given in Fig. 7.14.

7.13.9. Camber

Camber, also known as cross-slope, facilitates drainage of the pavement laterally. The pavement can have a crown or a high point in the middle with slopes downwards toward both the edges. This type is favoured on two-lane roads and wider undivided roads. On divided roads, the individual carriageways may be centrally crowned separately or a unidirectional slope may be provided across the carriageway width.

The amount of camber to be provided depends upon the smoothness of the surface and the intensity of rainfall. The practice by various authorities is summarised in Table 7.20.

Table 7.20 : Camber Prescribed by Various Authorities

Authority	Type of surface	Camber (%)
1. AASHTO (Ref 4)	*(i)* High	1–2
	(ii) Intermediate	1.5–2
	(iii) Low	2–4
2. U.K. (Ref 18)		2.5
3. I.R.C. (Ref 1)	*(i)* Cement Concrete or high type bituminous	1.7–2.0
	(ii) Thin bituminous surfacing	2.0–2.5
	(iii) Water-bound macadam	2.5–3.0
	(iv) Earth	3.0–4.0

The cross-slope for the shoulders should be generally steeper than the pavement by about 0.5 per cent to facilitate quick drainage subject to a minimum of 3 per cent (Ref 1). U.K. practice (Ref 18) is to provide 5 per cent slope on shoulders.

7.13.10. Side Slope

The slope of embankments and cut sections depend upon the type of soil and the height of embankment or depth of cuttings. A flatter slope is conducive for erosion control, but is costly.

Flat slopes of embankments promote safety of traffic. Ordinarily 1½:1 to 2:1 as uninundated condition and 2:1 to 3:1 as inundated condition will be adequate. Flatter slopes recommended (Ref 19) for traffic safety can be taken from Fig. 7.15.

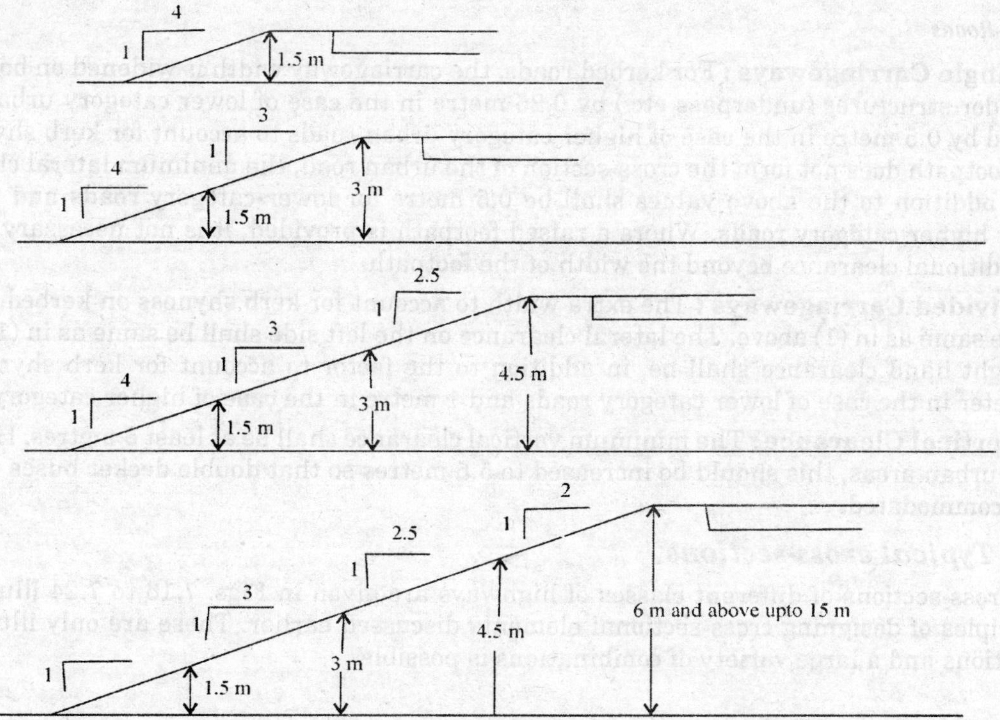

Fig. 7.15: Side Slopes for Road Embankments (Ref 19)

7.13.11. Lateral and Vertical Clearances

Lateral clearance is the distance between the extreme edge of the carriageway to the face of a structure. Vertical clearance is the height above the highest point of the travelled way to the lowest point of the overhead structure.

Current Indian standards (Ref 20) for lateral clearance are summarised below:

Type of Roads		Minimum Lateral Clearance
A, **Rural Roads**		
1. Single Carriageway		
(i) National and State Highways	:	Normal 2.5 metres, exceptional 2.0 meters
(ii) Major District Roads and Other District Roads	:	Normal 2.0 metres, exceptional 1.5 meters
(iii) Village Roads	:	Normal 1.5 metres, exceptional 1.0 meters

If footpaths are provided, the lateral clearance should be the width of the footpaths plus one metre.

2. Divided Carriageway		
(i) Left hand clearance as in 1 above (ii) Right hand side clearance	:	2.0 metres desirable 1.5 metres minimum

B. Urban Roads

1. **Single Carriageways :** For kerbed roads, the carriageway width is widened on both sides under structures (underpass etc.) by 0.25 metre in the case of lower category urban roads and by 0.5 metre in the case of higher category urban roads to account for kerb shyness. If a footpath does not form the cross-section of the urban road, the minimum lateral clearance in addition to the above values shall be 0.5 metre for lower category roads and 1 metre for higher category roads. Where a raised footpath is provided, it is not necessary to have additional clearance beyond the width of the footpath.
2. **Divided Carriageways :** The extra width to account for kerb shyness on kerbed roads is the same as in (1) above. The lateral clearance on the left side shall be same as in (1) above. Right hand clearance shall be, in addition to the factor to account for kerb shyness, 0.5 meter in the case of lower category roads and 1 metre in the case of higher category roads.
3. **Vertical Clearance :** The minimum vertical clearance shall be at least 5 metres. However, in urban areas, this should be increased to 5.5 metres so that double decker buses could be accommodated.

7.13.12. Typical cross-sections

Typical cross-sections of different classes of highways are given in Figs. 7.16 to 7.24 illustrating the principles of designing cross-sectional elements discussed earlier. These are only illustrative cross-sections and a large variety of combinations is possible.

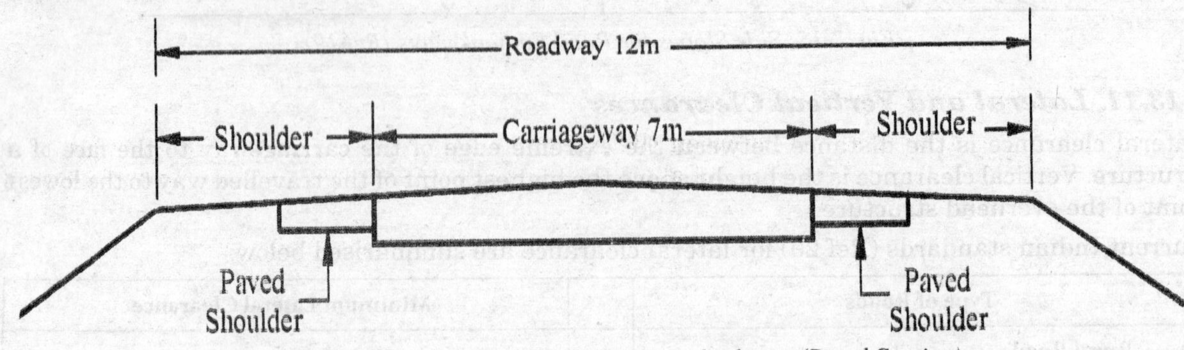

Fig. 7.16: Typical 2-lane National or State highway (Rural Section)

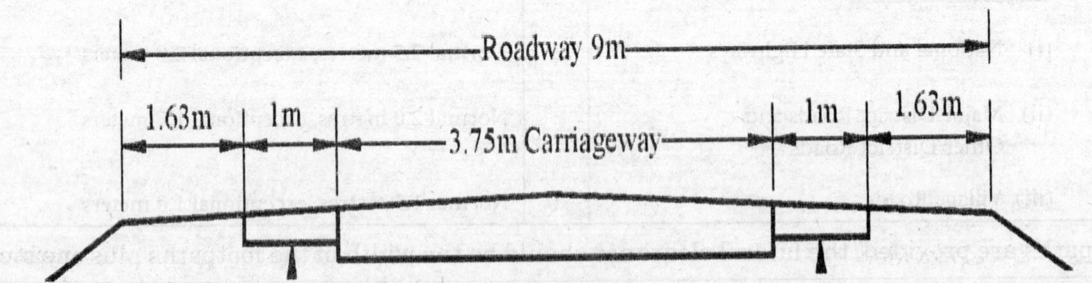

Fig. 7.17: Typical Single Lane Road with Paved Shoulders (Major District Road or other District Road)

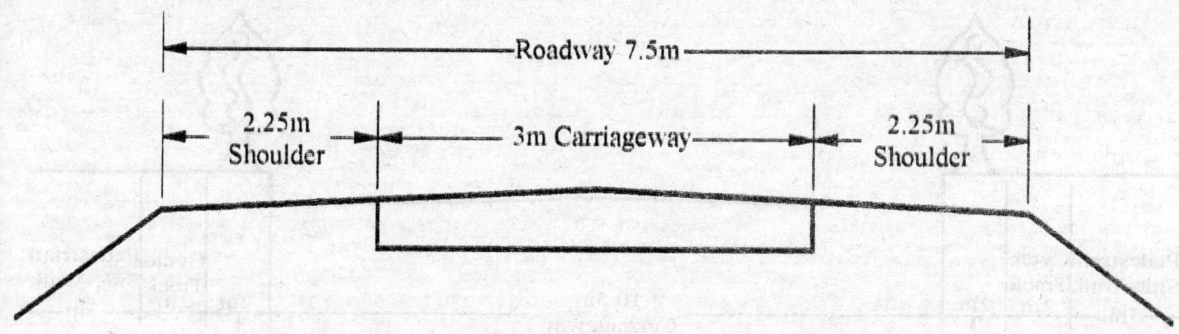

Fig. 7.18: Typical Village Road

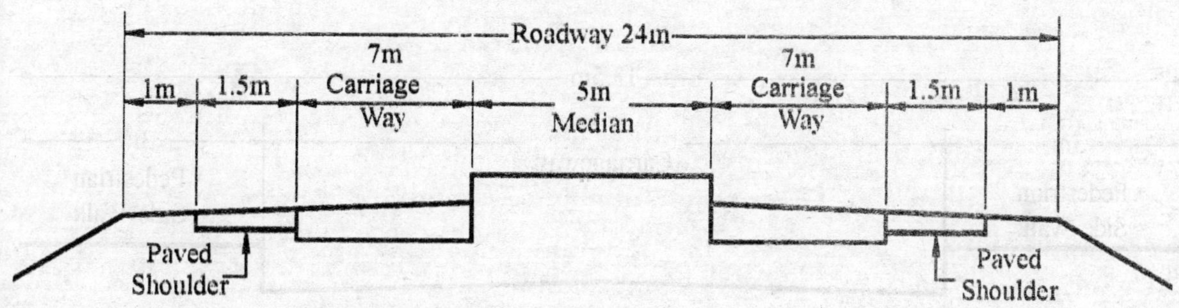

Fig. 7.19: Typical Dual Carriageway (2-Lane Each)

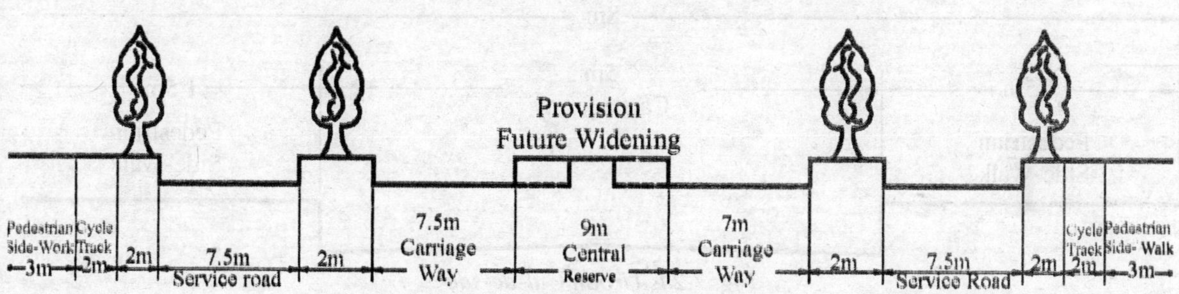

Fig. 7.20: Urban Arterial Street

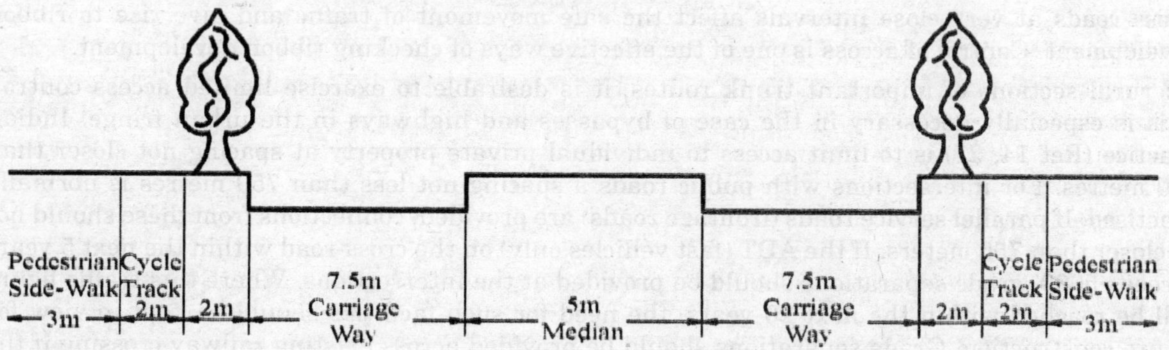

Fig. 7.21: Urban Sub-Arterial Street through open Area

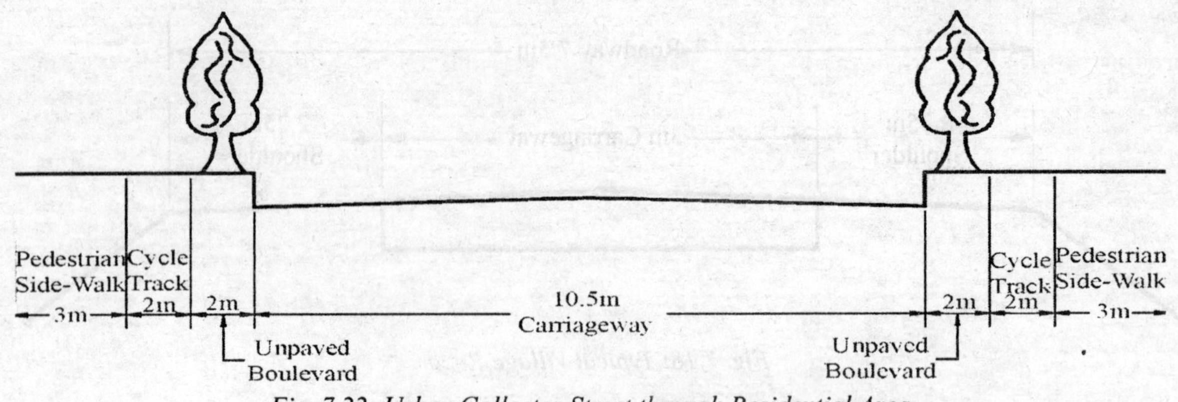

Fig. 7.22: Urban Collector Street through Residential Area

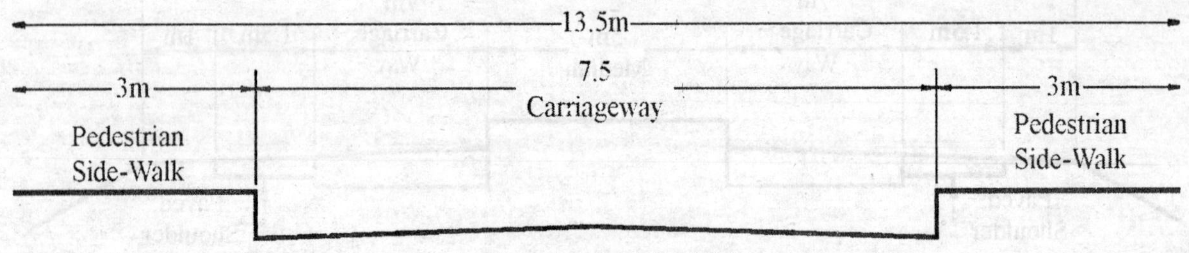

Fig. 7.23: Urban Residential Street

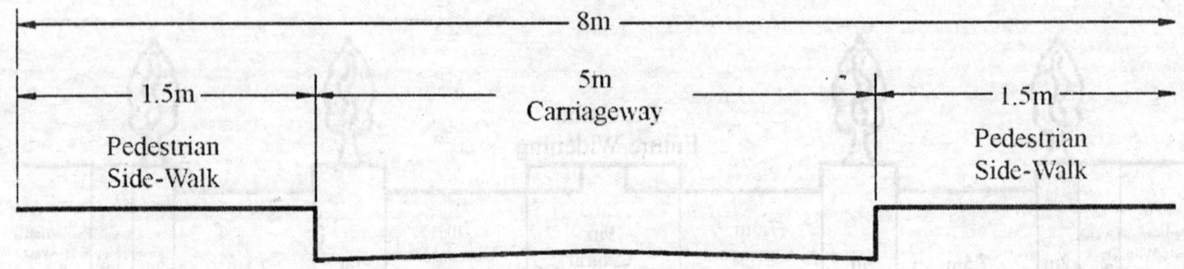

Fig. 7.24: Urban Cul-de-sac

7.14. Control of Access from Cross Roads to Main Road

Cross roads at very close intervals affect the safe movement of traffic and give rise to ribbon development. Control of access is one of the effective ways of checking ribbon development.

On rural sections of important trunk routes, it is desirable to exercise limited access control. This is especially necessary in the case of bypasses and highways in the urban fringe. Indian practice (Ref 14, 21) is to limit access to individual private property at spacing not closer than 300 metres. For intersections with public roads a spacing not less than 750 metres is normally practised. If parallel service roads (frontage roads) are provided, connections from these should not be closer than 750 meters. If the ADT (fast vehicles only) on the cross-road within the next 5 years exceeds 5000, grade separations should be provided at the intersections. Where this traffic figure will be reached within the next 20 years, the need for such facilities should be kept in view for future construction. Grade separations should be provided across existing railway crossing if the product of ADT and the number of trains per day exceeds 50,000 within the next 5 years. For new constructions such as bypasses, grade separations should be provided when this figure is greater than 25,000.

On urban sections, the current Indian practice is to provide the following minimum spacing between access points:

Type of Street	Suggested Minimum Spacing
Expressways	1000 metres
Arterial streets	500 metres
Sub-arterial streets	300 metres
Collector streets	150 metres
Local streets	Free access

7.15. Intersections

7.15.1. General

An intersection (also called road junction) is defined as the general area where two or more highways join or cross. An intersection leg is that part of any one of the roadways radiating from an intersection which is outside of the area of the intersection proper.

The importance of design of the intersection stems from the fact that efficiency of operation, safety, speed, cost of operation and capacity are directly governed by the design. Since an intersection involves conflicts between traffic in different directions, its scientific design can control accidents and delay and can lead to orderly movement of traffic. Intersections represent potentially dangerous locations from point of view of traffic safety. It is believed that well over half the fatal and serious road accidents in built-up areas occur at junctions.

The following principles should be considered in a good design:

(i) The number of intersections should be kept at minimum. If necessary, some minor roads may be connected with each other before joining a major road.

(ii) The geometric layout should be so selected that hazardous movement by drivers are eliminated. This can be achieved by various techniques such as channelizing and staggering.

(iii) The design should permit the driver to discern quickly, either from the layout or from traffic signs, the path he should follow and the actions of merging and diverging. This can be achieved by a good layout, traffic islands, signs and carriageway markings. Good visibility improves safety.

(iv) The layout should follow the natural vehicle paths. Smoothness, in contrast to abrupt and sharp corners, should guide minor streams into stopping or slowing down positions.

(v) The number of conflict points should be minimised by separating some of the many cutting, merging or diverging movements.

(vi) Vehicles that are forced to wait in order to cross a traffic stream should be provided with adequate space at the junctions.

7.15.2. At-grade and grade separated junctions

An intersection where all roadways join or cross at the same level is known as an at-grade intersection. An intersection layout which permits crossing manoeuvres at different levels is known as grade-separated intersection.

The choice between an at-grade and grade-separated junctions at a particular site depends upon various factors such as traffic, economy, safety, aesthetics, delay etc. Grade-separated junctions generally are more expensive initially and are justified in certain situations. These are:

(i) On high type facilities such as expressways

(ii) Certain at-grade intersections which have reached the maximum capacity and where it is not possible to improve the capacity further by retaining the at-grade crossing.

(iii) At certain locations which have a proven record of bad accident history when functioning as at-grade junctions.

(iv) At junctions where the traffic volume is heavy and the delays and loss caused justify economically the provision of grade-separation.

(v) At certain specific topographical situations where it is logical to provide a grade-separated structure rather than an at-grade intersection, which may involve considerable earthwork or acquisition of land.

7.15.3. Basic forms of at-grade intersections

Fig. 7.25 gives the basic forms of at-grade intersections.

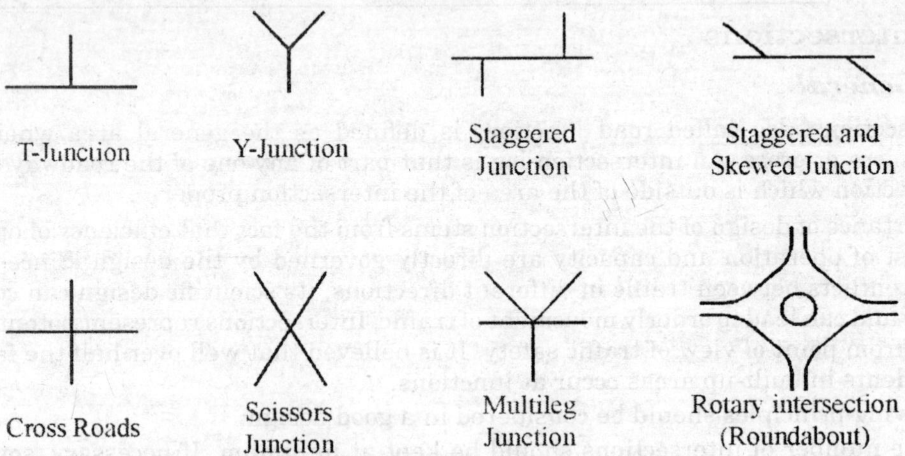

Fig. 7.25: Basic forms of At-grade Intersections

7.15.4 Alignment and vertical profile

In view of the conflict and hazard involved at intersections, the alignment and vertical profile of the intersecting roads should permit the drivers to discern the danger and make the manoeuvres necessary for safety. This requires that the alignment be as straight and gradients as low as practicable.

It is desirable that the intersection roads meet at or nearly at right angles. For this purpose, it may be necessary to modify the alignment of the roads as shown in Fig. 7.26.

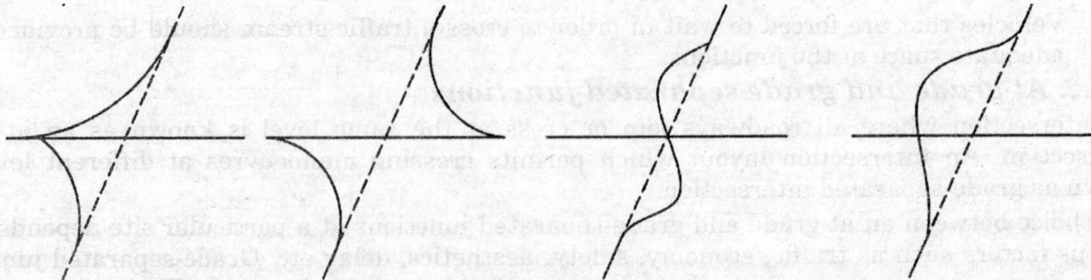

Fig. 7.26: Modification in Alignment of Intersecting Roads to Give a Nearly Right-Angle Crossing

7.15.5. Visibility at Intersections

7.15.5.1. **The safety of traffic can be ensured only if the visibility is full and unimpeded along both roads. Any obstructions should be clear of the minimum visibility triangle for a height of 1.2 metres above the roadway.**

7.15.5.2. Rural Road Intersections

At intersection where the intersecting roads are of more or less equal importance and there is no established priority (generally known as uncontrolled intersections) visibility should be provided on the principle that a driver approaching the intersection on either highway must be able to perceive a hazard and halt the vehicle, if that becomes essential, before reaching the intersection. The stopping distances, d_1 and d_2 appropriate for the design speeds of V_1 and V_2 respectively in the two intersecting highways, may be taken from Table 7.12 which gives the current Indian practice (Ref 12). Fig. 7.27 explains the requirement of the visibility triangle.

On intersections involving a major road and a minor road, it is customary to control traffic on the minor road by STOP or GIVE WAY sign or road markings. Such intersections are known as priority intersections (Ref 22). On priority intersections, the visibility provided should be such that drivers approaching from the minor road are able to see the vehicles on the major road in good time and judge whether the required gap is available in the main road traffic stream for a safe crossing. For this purpose current Indian practice is to specify a visibility distance of 15 metres along the minor road (Fig. 7.28).

V_1, V_2 – Design Speed of Road

d_1, d_2 – Safe Stopping Sight Distance

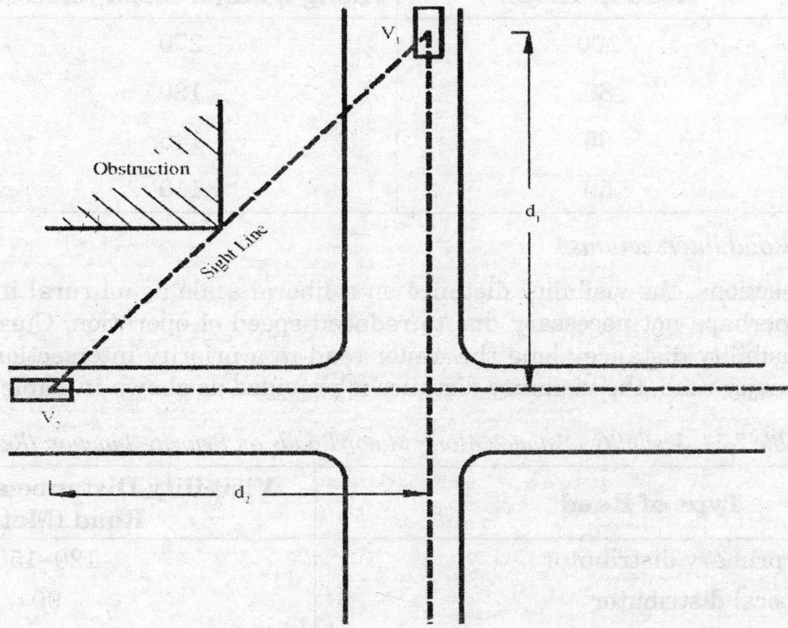

Fig. 7.27: Visibility Triangle at Uncontrolled Intersections

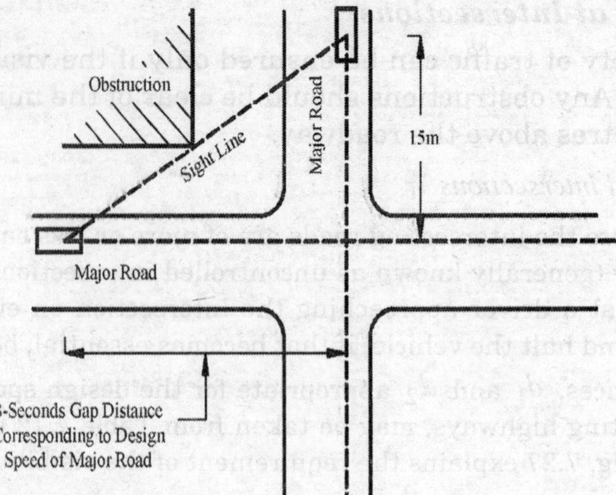

Fig. 7.28: Visibility Triangle at Priority Intersections

On the major road of priority junctions, the visibility needed along the major road depends upon the time required by the driver on the minor road to perceive the traffic conditions on the intersection, evaluate the gaps in the vehicle stream, take a decision about the actual crossing and finally accelerate the vehicle to complete the manoeuvre. A time of 8 seconds may be considered appropriate for this purpose. On this basis, Table 7.21 gives the visibility distance along the major road (Ref 23).

Table 7.21 :Minimum Visibility Distance Along Major Roads At Priority Inter-Sections On Rural Roads (Ref 23)

Design Speed of Major Road in Kmph	Minimum Visibility Distance Along a Major Road (metres)
100	270
80	180
65	145
50	110

7.15.5.3. Urban Road Intersections

On urban intersections, the visibility distance on a liberal scale as on rural intersections is not possible and is perhaps not necessary due to reduced speed of operation. Current U.K. practice is to limit the visibility distance along the minor road in a priority intersection to about 4.5–9.0 metres. On the major road, the visibility distance is provided as shown in Table 7.22.

Table 7.22 :Visibility Distance Along Major Roads on Priority Junction (Ref 22)

Type of Road	Visibility Distance along a Major Road (Metres)
1. All purpose primary distributor	120–150
2. District or Local distributor	90
3. Access Road	60

7.15.6. Radii of Curves at Intersections

7.15.6.1. The design of intersection curves has profound influence on speed. Curve radii are therefore, selected to control speed and ensure safety of operation. Generally, the approach to the design of curves falls under one of the following categories:

(i) Design of curves for a minimum speed and sharpest turn

(ii) Design of curves to permit different operating speeds

7.15.6.2. Sharpest Turn Curves for Minimum Speed

When it is necessary to provide for turning vehicles within minimum space, the minimum turning paths of design vehicles apply.

The turning radii for AASHTO design vehicles have been given earlier in Table 7.3. Considering these, the minimum edge of pavement designs for sharpest turns for an angle of turns of 30^0 and 45^0 for AASHTO vehicles are given in Table 7.23 (Ref 4).

Table 7.23 : Minimum Edge of Pavement Designs for Turns at Intersections (Ref 4)

AASHTO Design Vehicle	Angle of Turn (degrees)	Radius of Simple Curve (m)	3-central compound curve symmetric		3-centred compound curve symmetric	
			Radii, (m)	Offset, (m)	Radii, (m)	Offset, (m)
P	30	18	–	–	–	–
SU		30	–	–	–	–
WB-40		45	–	–	–	–
WB-40		60	–	–	–	–
P	45	15	–	–	–	–
SU		23	–	–	–	–
WB-40		36	–	–	–	–
WB-40		32	60-30-60	0.9	–	–

Compound curves are generally three-centred, as in Fig. 7.29.

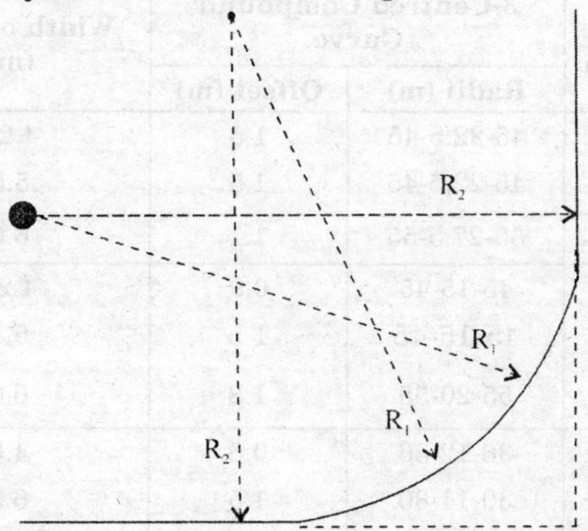

Fig. 7.29: Compound Curve for Corner Design

The sharpest turns are usually provided for unchannelized intersections and for curbs of urban streets at intersection. In residential streets, a curb radius of 6 metres is regarded as the minimum, though radii of 4.5–7.5 metres are satisfactory. For junctions using commercial vehicles, a radius of 9-15 metres will be necessary.

7.15.6.3. Minimum Design for Turning Roadways

Where the inner edges of pavement for left turns at intersections are designed to accommodate semi-trailer combinations, or where the design permits passenger vehicles to turn at speeds 25 K.P.H. or more, the pavement area at the intersections may become excessively large for proper control of traffic. To avoid this a corner island is provided to form a separate turning roadway. A turning roadway is a connecting roadway for traffic turning between two intersection legs, Fig. 7.30.

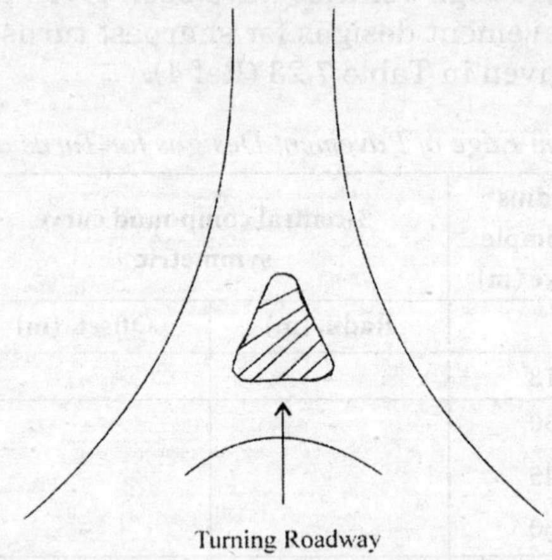

Fig. 7.30: Turning Roadway

AASHTO standards for turning roadways are summarised in Table 7.24 (Ref 3).

Table 7.24 : Minimum Design for Turning Roadways (Ref 4)

Angle of Turn (Degrees)	Design Designation	3-Centred Compound Curve		Width of Lane (m)	Approximate Island Size (Sq. m)
		Radii (m)	Offset (m)		
75	A	45-22.5-45	1.5	4.25	5.5
	B	45-22.5-45	1.5	5.5	4.5
	C	55-27.5-55	1.1	6.0	4.5
90	A	45-15-45	0.9	4.25	4.5
	B	45-15-45	1.5	5.5	7.5
	C	55-20-55	1.8	6.0	11.5
105	A	36-12-36	0.6	4.5	6.5
	B	30-11-30	1.5	6.5	4.5
	C	55-14-45	2.5	9.0	5.5
120	A	30-9-30	0.75	5.0	11.0

Angle of Turn (Degrees)	Design Designation	3-Centred Compound Curve		Width of Lane (m)	Approximate Island Size (Sq. m)
		Radii (m)	Offset (m)		
	B	30-9-30	1.5	7.5	8.5
	C	55-12-55	2.6	10.5	20.0
135	A	30-9-30	0.75	5.0	43.0
	B	30-9-30	1.5	8.0	34.0
	C	50-11-50	2.75	11.0	60.0
150	A	30-9-30	0.75	5.0	130.0
	B	30-9-30	1.8	9.0	110.0
	C	50-11-50	2.2	11.5	160.0

Note: A – Primarily passenger vehicles; permits occasional design single unit trucks to turn with restricted clearances

B – Provides adequately for SU; permits occasional WB-50 to turn with slight encroachment on adjacent traffic lanes

C – Provides fully for WB-50

7.15.6.4. Curves Designed for Different Speeds

Vehicles turning at intersections designed for the sharpest turn by design vehicles, as discussed in para 7.15.6.2 above, have to operate at low speed, perhaps less than 15 Kmph It is often desirable and feasible to design for vehicles operating at higher speeds, as in the case of rural intersections. The speeds for which intersection curves should be designed depend largely upon the vehicle speeds on the approach legs, type of intersection, and volumes of through and turning traffic. Generally, a desirable turning speed for design is the average running speed, which may be roughly taken as 0.7 of the design speed. Design at such speeds offers little hindrance to smooth flow of traffic and may be justified on some intersections for certain movements involving little or no conflict with pedestrians or other vehicular traffic.

Table 7.25 gives the AASHTO practice (Ref 4) for the design of such curves.

Table 7.25 : Minimum Radii for Intersection Curves for Various Turning Speeds (Ref 4)

Design turning speed V (Kmph)	25	30	40	50	60	65
Side friction factor, f	0.32	0.27	0.23	0.20	0.18	0.16
Assumed minimum superelevation, e	0.00	0.02	0.04	0.06	0.08	0.09
Total e+f	0.32	0.29	0.27	0.26	0.26	0.25
Calculated min. R (m) $R = \frac{V^2}{127(e+f)}$	14	28	47	70	96	129
Suggested minimum R (m)	15	27	45	70	95	130

7.15.7. Widths of Carriageways in Junctions

The carriageway width needs to be increased in curves as already explained in 7.9.8. The following values of carriageway widths are as per IRC practice (Ref 23).

Table 7.26 : Widths of Carriageways at Intersections (Ref 23)

Inner Radius (m)	Design Speed (Kmph)	Single Lane Width (m)	Two Lane Width for One or Two Way Traffic (m)
10.5	18	5.5	11.5
15	23	5.5	10.5
20	27	5.0	10.0
30	32	4.5	9.0
40	37	4.5	9.0
50	41	4.5	8.0
75	50	4.5	8.0
100	57	4.5	8.0
125	62	4.5	8.0
150	64	4.5	8.0
Straight		4.5	7.0

7.15.8. Speed Change Lanes

Drivers entering into or leaving an intersection have to necessarily change their speed. Drivers entering an intersection will reduce the speed to a safe value at which the intersection can be negotiated. While leaving the intersection, they will increase the speed until the desired open highway speed is reached. When such deceleration or acceleration is accomplished on the highway travelled way, traffic is disrupted and hazard may be caused. To avoid this, speed change lanes are normally provided on superior facilities such as expressways and are also frequently provided on other main highways. IRC practice is to provide acceleration lanes where the future traffic on the acceleration lane is expected to be more than 1000 PCUs per day (Ref 23). These lanes also increase the capacity of the junction.

Speed change lanes are either acceleration lanes or deceleration lanes. An acceleration lane is an auxiliary traffic lane to enable a vehicle entering a road to increase its speed so that it can merge more safely with through traffic. A deceleration lane is an auxiliary traffic lane to enable a vehicle leaving the through traffic stream to reduce speed without interfering with other traffic.

Acceleration and deceleration lanes are usually provided with taper, vide Fig. 7.31.

Table 7.27 gives the lengths of deceleration and acceleration lanes as per IRC practice (Ref 23).

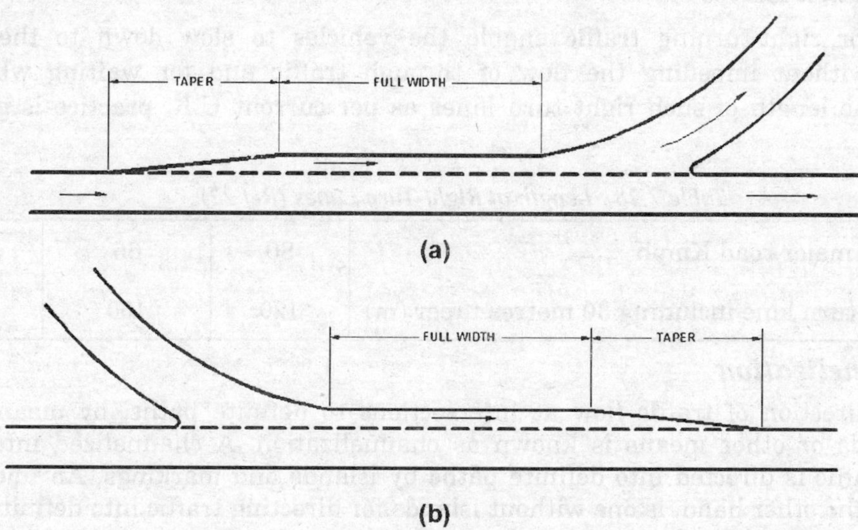

Fig. 7.31: Acceleration and deceleration lanes

Table 7.27 : Design Length of Deceleration Lanes (Grade 2 Per Cent or Less) (Ref 23)

Design speed of turning roadways curve, Kmph		Stop condition	25	30	40	50	60	65	75	80
Minimum curve radius (m)		–	15	27	45	70	95	130	170	210
Design speed of highway (Kmph)	Length of taper (m)	Total length of deceleration lane, including taper (m)								
65	60	100	90	85	75	60	–	–	–	–
80	70	130	120	115	110	100	90	–	–	–
100	80	150	150	145	140	130	120	100	90	–
110	90	165	165	160	150	145	140	115	100	–
Design speed of highway (Kmph)	Length of taper (m)	Total length of acceleration lane, including taper (m)								
65	60	–	100	75	70	–	–	–	–	–
80	70	–	210	190	180	150	120	–	–	–
100	80	–	340	320	300	270	240	180	120	–
110	90	–	465	450	420	400	370	300	250	175

The width of a speed change lane should not be less than a normal traffic lane, viz., 3.75 metres.

7.15.9. *Right Turn Lanes*

Storage lanes for right-turning traffic enable the vehicles to slow down to the necessary turning speed without impeding the flow of through traffic and for waiting when signals are provided. The length of such right turn lanes as per current U.K. practice is given below (Ref 22).

Table 7.28 : Length of Right-Turn Lanes (Ref 22)

Design speed of major road Kmph	80	65	50
Length of right-turn lane including 30 metres taper (m)	120	100	80

7.15.10. *Channelization*

7.15.10.1. The direction of traffic flow at intersections to definite paths, by means of traffic markings, islands or other means is known as channelization. A channelized intersection is one in which traffic is directed into definite paths by islands and markings. An unchannelised intersection, on the other hand, is one without islands for directing traffic into definite paths. An unchannelised intersection is the most simple type but is the most dangerous and inefficient. All important junctions should, therefore, provide for channelizations.

7.15.10.2 Channelization Serves the following purposes:

(i) **Separation of Conflicts:** To diminish the number of possible vehicle conflicts, to reduce the possible area of conflicts in the carriageway and to present drivers with only one decision at a time.

(ii) **Control of Angle of Conflict:** Small angles of crossing cause severe accidents if they occur. Severity is reduced if the angle of conflict is controlled.

(iii) **Control Speed:** To reduce the speed of traffic entering the intersection and increase the speed of traffic leaving the intersection, bending or funnelling by suitable channelization techniques is resorted to, vide Figs. 7.32 and 7.33.

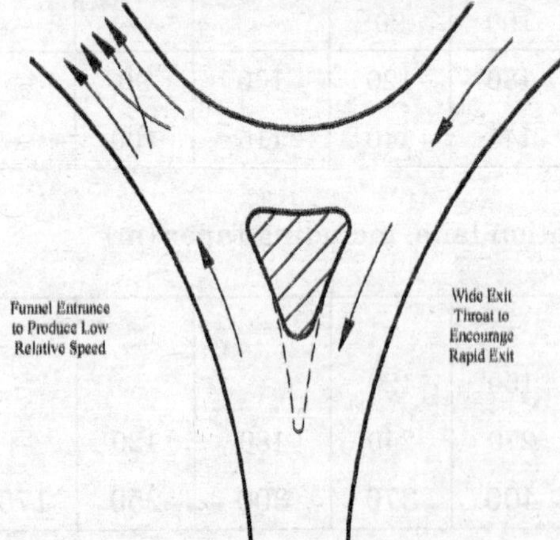

Fig. 7.32: Channelizing island

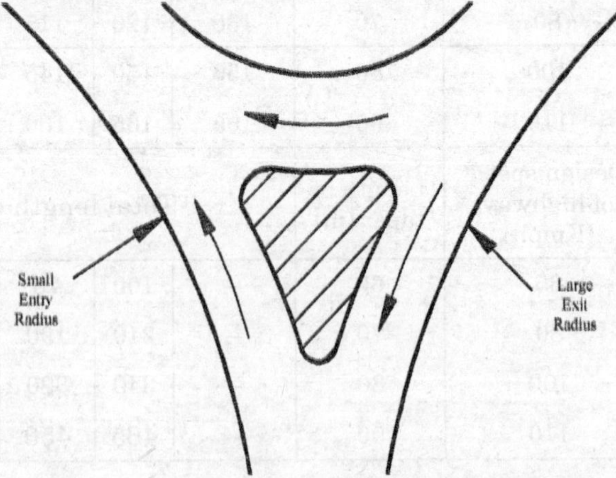

Fig. 7.33: Control of Radius of Entry and Exit for Control of Speed

(iv) **Protection of traffic for vehicles leaving or crossing the main traffic stream:** Separate storage pockets for right turning traffic at an intersection and the adjacent island, vide Fig. 7.34 can increase safety.

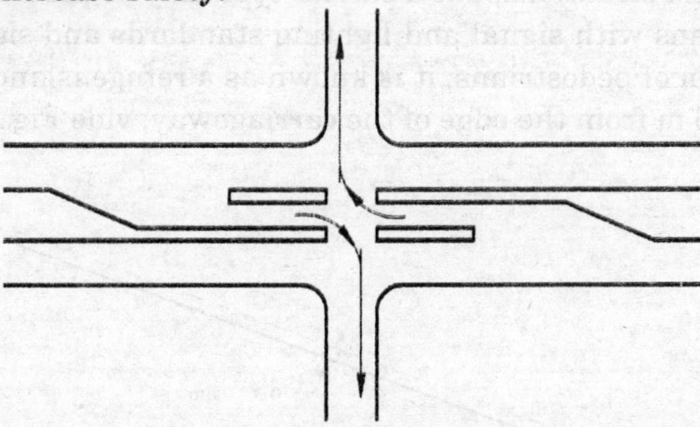

Fig. 7.34: Channelising to Protect Right Turning Traffic

(v) **Protection of Pedestrians:** To provide a refuge for pedestrians between traffic flows. A channelizing island such as in Fig. 7.34 serves as a refuge and makes the crossing much safer.

(vi) **Elimination of Excessive Intersectional Areas:** Intersections with large corner radii and those at oblique angles have large paved areas, which permit and encourage hazardous uncontrolled vehicle movements. If these unused paved areas are converted into channelizing islands, orderly movement results and hazards are reduced vide Fig. 7.35.

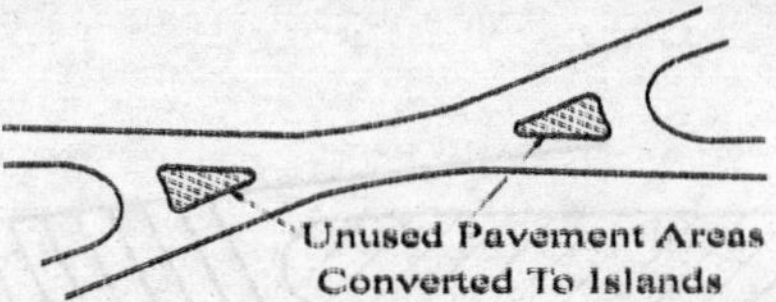

Fig. 7.35: Unused Pavement Areas Converted to Channelizing Islands

(vii) **Location of Traffic Control Devices:** To provide space for traffic control devices such as direction indicators, reflectors, signs, etc.

7.15.10.3. Features of Channelising Islands

They should be sufficiently large to command attention. An area of about 5 sq, m preferably 7 sq. m, should be the minimum. Elongated or divisional islands, introduced on undivided highways to alert drivers and regulate traffic through the intersections, should be at least 1.2 m wide and 3.5 to 6 m long (Fig. 7.36).

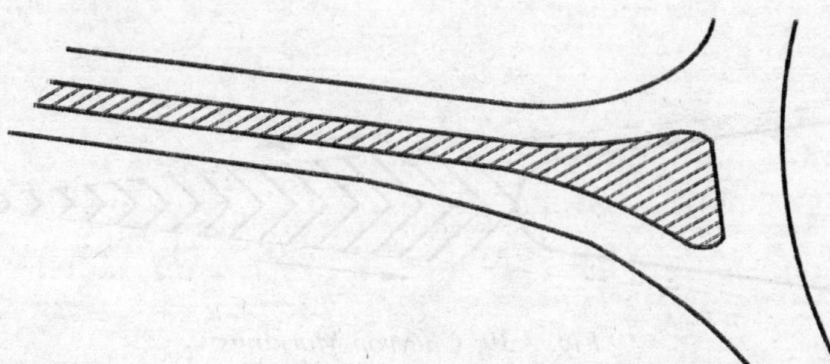

Fig. 7.36: Divisional Island Combined with a Triangular Channelizing Island at Entrance to a Rotary Junction

In rural intersections, island curbs should always be of a mountable type, except where there is a need for a barrier, as at structures or pedestrian crossings. For urban intersections mountable curbs are preferable for most situations, but a barrier type may be provided, where speeds are low and protection of pedestrians with signal and lighting standards and signs is needed. Where the island is used for protection of pedestrians, it is known as a refuge island. The edges of the island are usually offset 0.3 to 0.6 m from the edge of the carriageway, vide Fig. 7.37 to reduce the risk of overriding the islands.

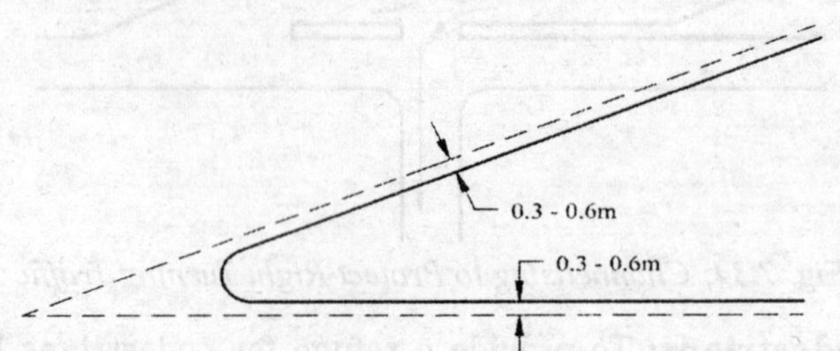

Fig. 7.37: Edge Offset at Channelizing Island

The approach noses of islands should be rounded off to a minimum radius of 0.6–0.9 m and the merging end nose should be rounded off to a radius of at least 0.3 m. The approach to the nose should be marked for additional guidance of traffic by means of diagonal markings and chevrons (Figs. 7.38 and 7.39).

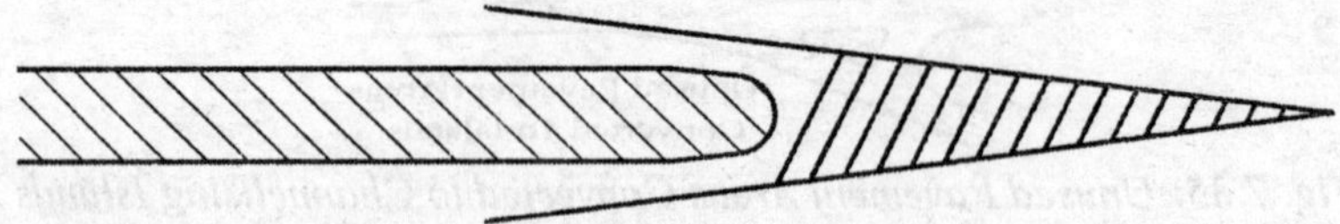

Fig. 7.38: Diagonal markings in advance of divisional island

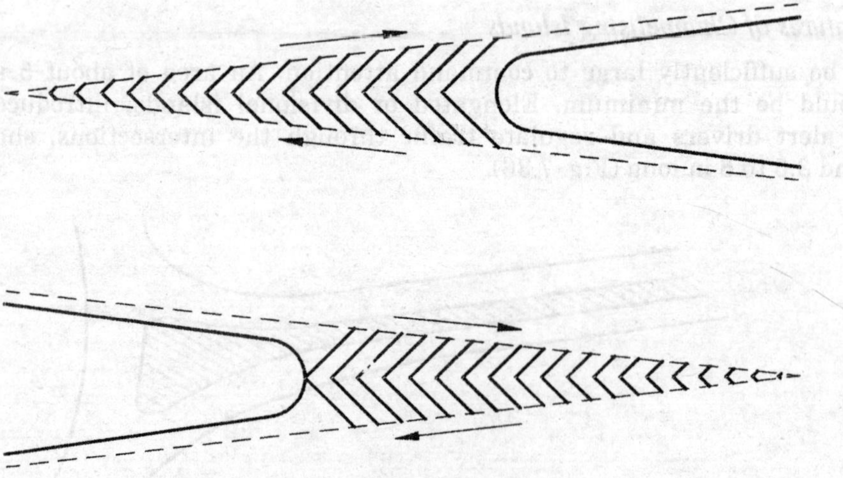

Fig. 7.39: Chevron Markings

7.15.11. Median Openings

7.15.11.1. Minimum Length of Median Openings

The length of the opening depends upon the width of the median (Fig. 7.40). Table 7.29, adapted from AASHTO, gives the dimensions of the opening for SU design vehicles (Ref 3).

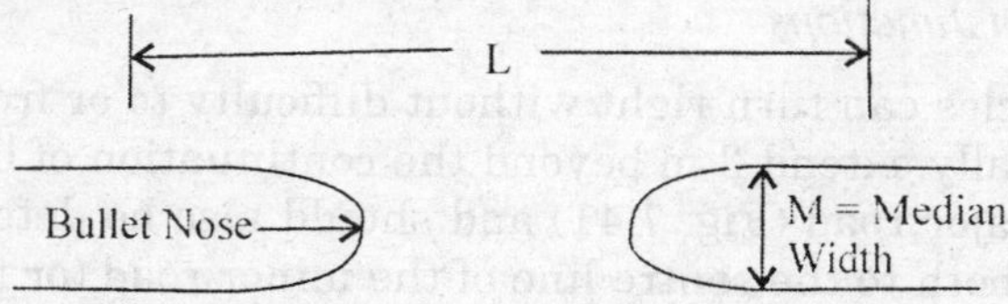

Fig. 7.40: Median Opening with Bullet Nosed Ends

Table 7.29 :Minimum Length of Median Opening

Median width M (in metres)	L= Minimum length of median opening (in m) (bullet nose ends)
1.2	30
1.8	23
2.5	21
3.0	19
3.6	18
4.3	16
5.0	14
6.0	13
7.2 and above	12 minimum

Median opening for making U-turns are needed in the following locations:

(i) Near to all terminal junctions

(ii) At regular intervals, say 1.5 Km, between junctions to facilitate maintenance operation, policing, repair of stalled vehicles etc.

(iii) Beyond important intersections to enable drivers unfamiliar with the locality, to correct any mistake in manoeuvring.

(iv) For accommodating minor turning movements.

A minimum width of 9 m is needed, except for passenger vehicles alone where only 6 m is sufficient. The accommodations for U-turns provided by minimum design as per AASHTO (Ref 4) are summarised in Table 7.30.

Table 7.30 : Median Opening for U-Turns Design (Ref 4)

Median Width (m)	Type of Manoeuvre Accommodated on 4-Lane Divided Highway	AASHTO Design Vehicle Provided for
18	Permits nearly all vehicles to turn from inner to inner lane	All vehicles
12	Permits all passenger vehicles to U-turn from inner to inner lane; permits some trucks to turn from outer to outer lane, with large trucks partially encroaching on shoulder.	P and SU

10	Permits passenger vehicles to U-turn from inner to inner lane, permits all trucks to U-turn with encroachment on shoulders	P and SU
6	Permits passenger vehicles to U-turn from inner to outer lane, U-turn impossible for large trucks without backing.	P only

7.15.11.2. Gap in Median at Junctions

To ensure that large vehicles can turn right without difficulty to or from a major road, the gap in the medians should normally extend 3 m beyond the continuation of both kerb lines of the minor road to the edge of the major road (Fig. 7.41) and should also be determined by 12–15 m radius control circles tangential both to the centre line of the minor road (or to the sides of any refuge or island) and the side of the central reserve away from the minor road.

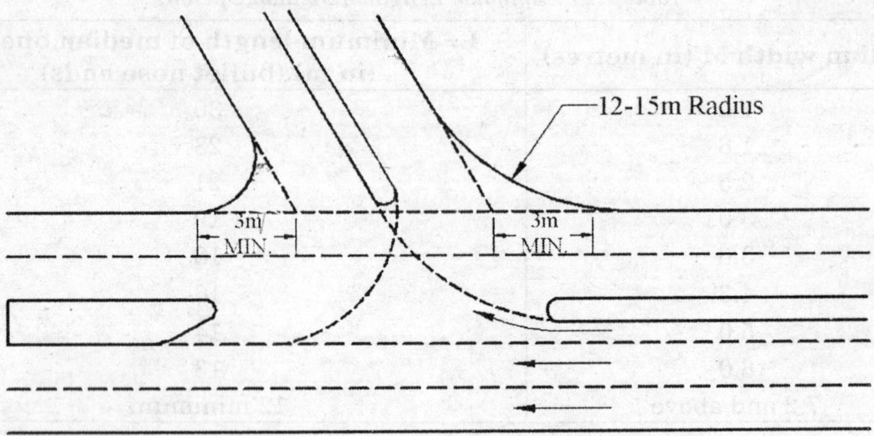

Fig. 7.41: Gap in median at junction

7.15.12 T-Intersection

Fig. 7.42 shows an unchannelised T-junction for very light traffic and minor and local roads. Obviously it is hazardous to use this in locations where speeds are high and turning movements are large.

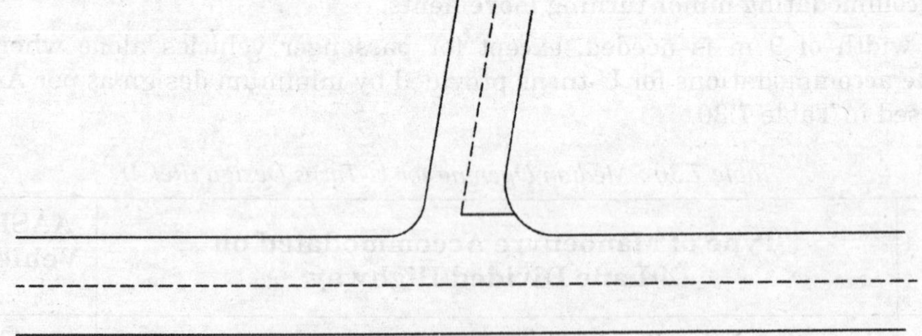

Fig. 7.42: Unchannelised T-junction

Fig. 7.43 is an improvement since the main road is flared to provide an additional lane to facilitate left turns.

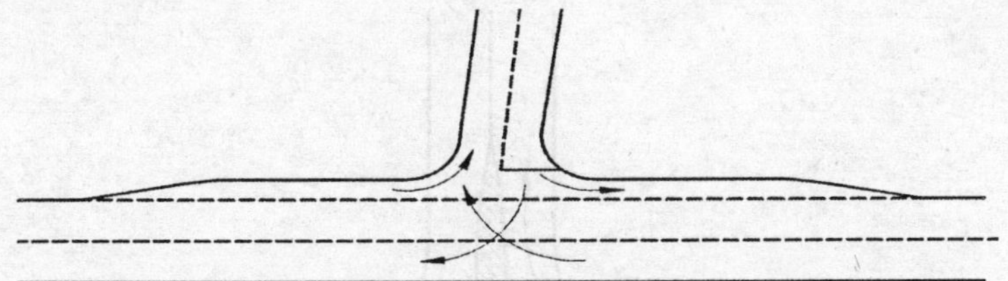

Fig. 7.43: Flared Unchannelised T-intersection to Facilitate Left Turning Movements

If there is heavy right turning movement, the flaring can be done opposite to the intercepted road, as in Fig. 7.44.

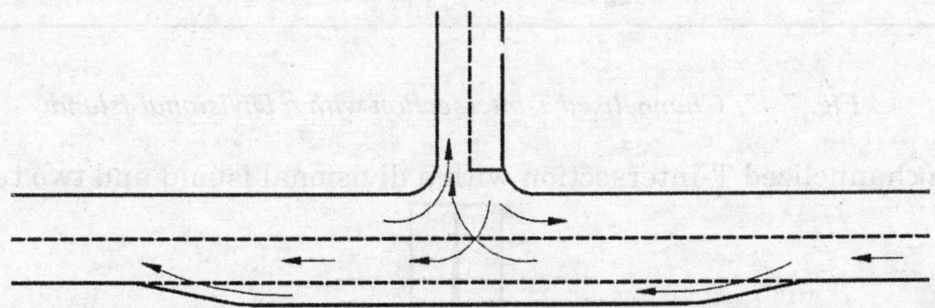

Fig. 7.44: Flared Unchannelised Intersection to Facilitate Right Turning Movements

Fig. 7.45 gives the layout of a channelized T-junction with single turning roadway, providing for large left turning movement from the main road to the intercepted road.

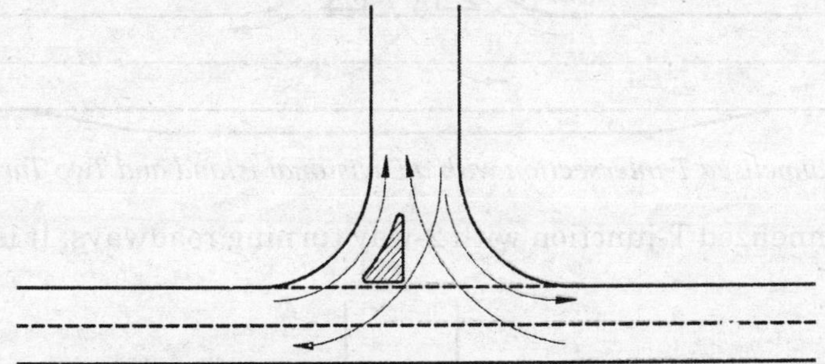

Fig. 7.45: Channelised T-intersection with Single Turning Roadways

Fig. 7.46 gives the layout of a channelized T-intersection with a pair of turning roadways.

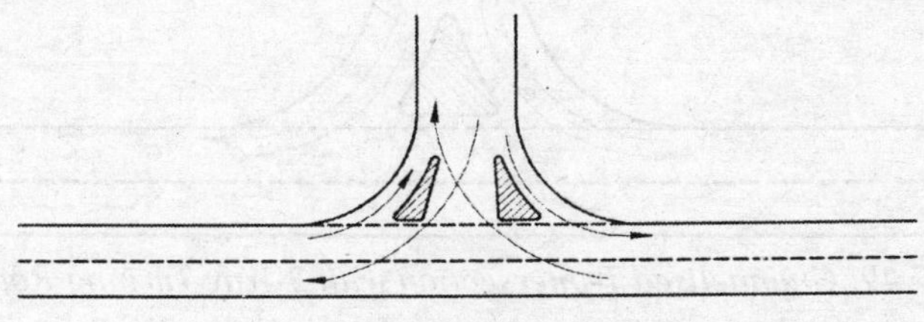

Fig. 7.46: Channelised T-intersection with a Pair of Turning Roadways

Fig. 7.47 shows a channelized T-intersection with a divisional island.

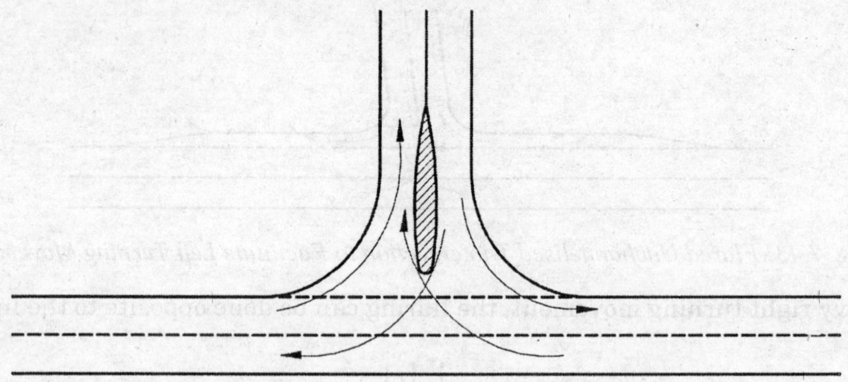

Fig. 7.47: Channelised T-intersection with a Divisional Island

Fig. 7.48 shows a channelized T-intersection with a divisional island and two turning roadways.

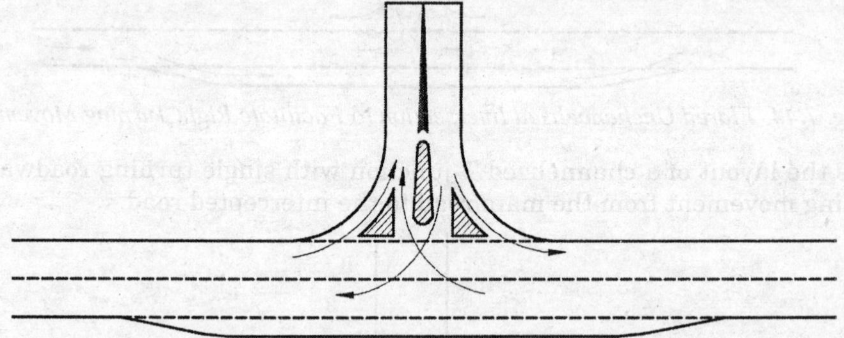

Fig. 7.48: Channelised T-intersection with a Divisional Island and Two Turning Roadways

Fig. 7.49 shows a channelized T-junction with 2-way turning roadways. It is usually provided with signal control.

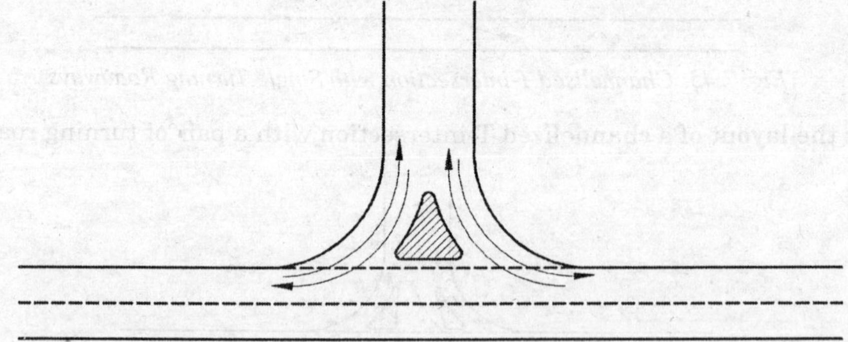

Fig. 7.49: Channelised T-intersection with 2-Way Turning Roadways

Fig. 7.50 shows a channelized T-intersection on a divided highway facility providing for separate right turning lane.

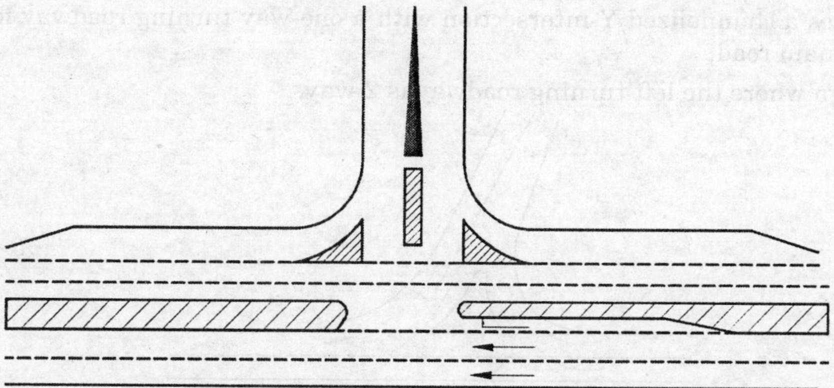

Fig. 7.50: Channelized T-Intersection on a Divided Highway

Fig. 7.51 shows a bulb-type channelized T-intersection suitable for high speeds. The median width on the through highway should be at least 12 m. The inner radius of the bulb should be at least 15 m, preferably 25 m.

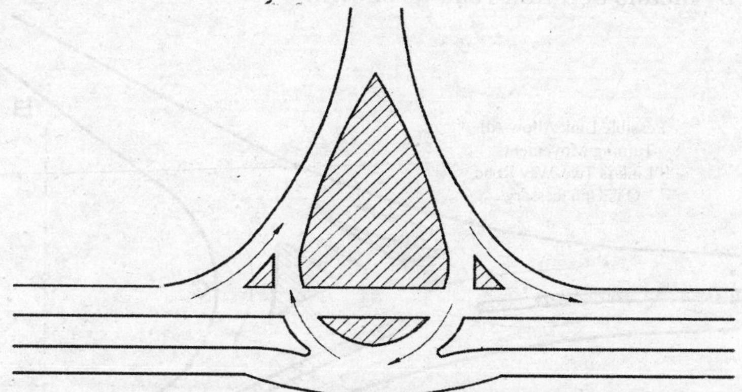

Fig. 7.51: Bulb-Type Channelized T-intersection

7.15.13 Y-Intersection

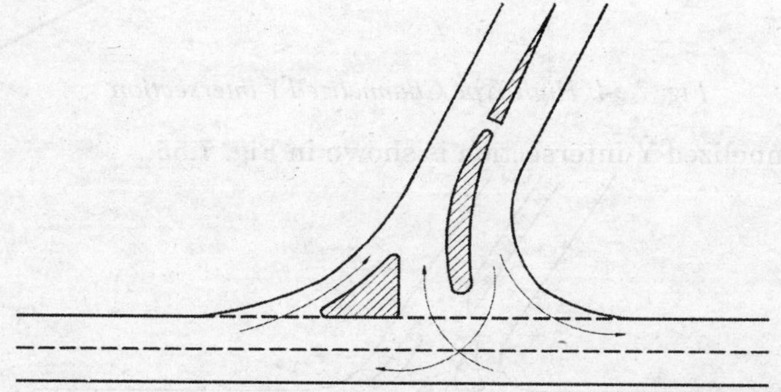

Fig. 7.52: Channelised Y-intersection with One-way Left Turning Roadway from the main Road

For most Y-intersections, the design can suitably follow the same features as for T-intersections, except where the intersection angle is very acute.

Fig. 7.52 indicates a channelized Y-intersection with a one-way turning roadway for left turning traffic from the main road.

Fig. 7.53 is design where the left turning roadway is 2-way.

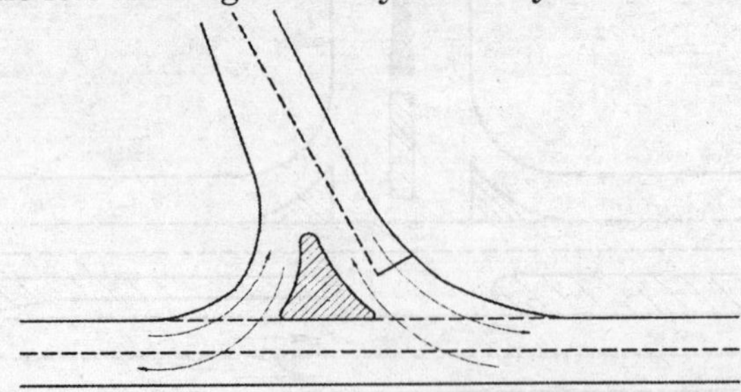

Fig. 7.53: Channelised Y-intersection with 2-way Left Turning Roadway from the main Road

Fig. 7.54 is a high type channelized Y-intersection, where the branches of Y carry equal traffic volume. This layout does not permit turning movements from A to B, but provision can be made for this where necessary by means of a link road as shown.

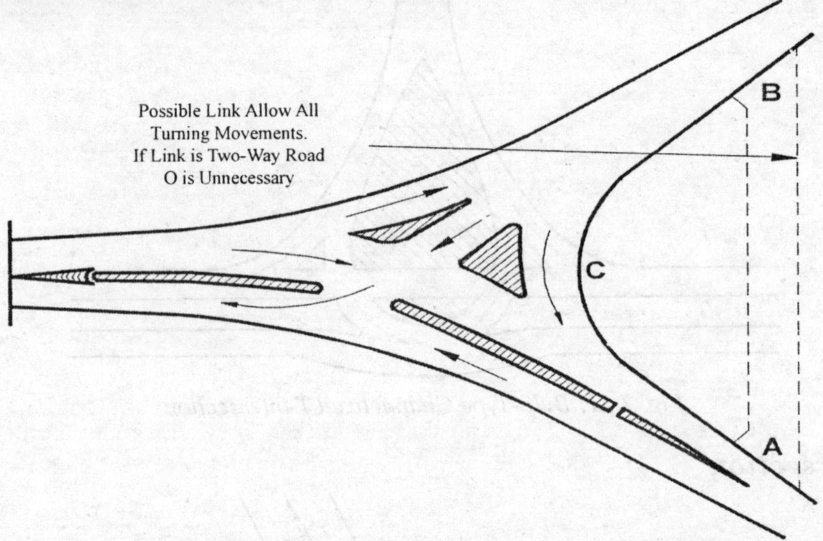

Fig. 7.54: High Type Channelized Y-intersection

Another high type channelized Y-intersection is shown in Fig. 7.55.

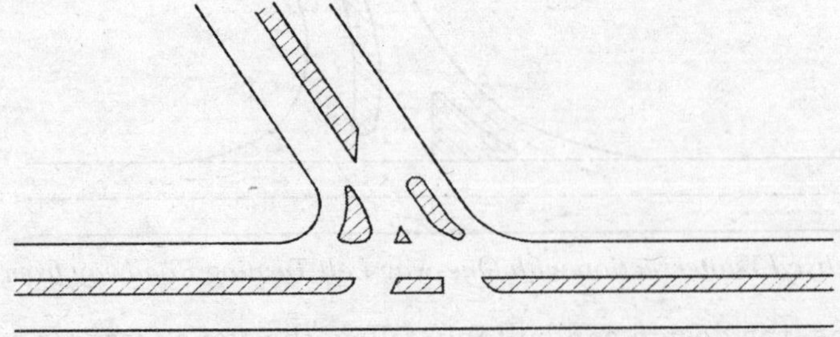

Fig. 7.55: High Type Channelized Intersection On the Y-type

7.15.16. Rotary Intersection (Roundabout)

7.15.16.1. General

A rotary intersection is a specialised form of at grade intersection laid out for movement of traffic in one direction round a central island. The vehicles from the converging areas are forced to move around the central island in a clockwise direction (in countries like India where the left-side rule of traffic prevails) in an orderly manner and weave out of the rotary movement into their desired directions.

7.15.16.2. Definitions of Terms

A sketch of the rotary along with some of its elements is given in Fig. 7.62.

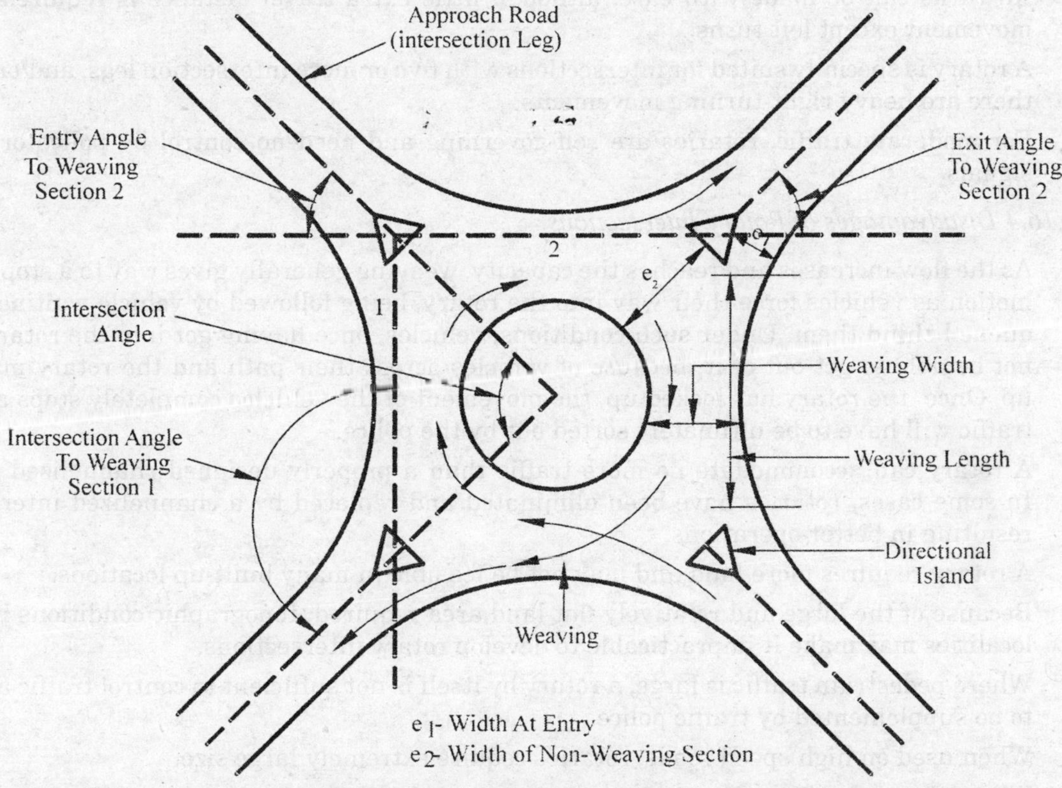

Fig. 7.62: Rotary Elements (Ref 24)

The definitions of some of the common terms in rotary design are given below:

Diverging	:	The dividing of a single stream of traffic into separate streams
Merging	:	The converging of separate streams of traffic into a single stream
Weaving	:	The combined movement of merging and diverging of traffic streams moving in the same general direction
Weaving Length	:	The length of a section of a rotary in which weaving occurs

7.15.16.3. Advantages of Rotary Intersections

The advantages of rotary intersections are:

(i) An orderly and regimented traffic flow is provided by rotary one-way movement. Individual traffic movements are subordinated in favour of traffic as a whole.

(ii) Normally, all traffic proceeds simultaneously and continuously at fairly uniform, though low speed. Frequent stopping and starting are avoided.

(iii) Weaving movements replace the usual angular crossing of typical at-grade intersections. Direct conflict is eliminated, all traffic streams merging of leaving at small angles. Accidents that occur from such movement are of minor nature, usually involving property damage only.

(iv) All turns can be made with ease, although little extra travel distance is required for all movement except left turns.

(v) A rotary is specially suited for intersections with five or more intersection legs, and/or where there are heavy right-turning movements.

(vi) For moderate traffic, rotaries are self-governing and need no control by police or traffic signals.

7.15.16.4 Disadvantages of Rotary Intersections

(i) As the flow increases and reaches the capacity, weaving generally gives way to a stop and go motion as vehicles force their way into the rotary, being followed by vehicle waiting in the queue behind them. Under such conditions, vehicles, once having got into the rotary, may not be able to get out of it, because of vehicles across their path and the rotary may lock up. Once the rotary has locked up, the movement of the vehicles completely stops and the traffic will have to be ultimately sorted out by the police.

(ii) A rotary can accommodate no more traffic than a properly designed channelised layout. In some cases, rotaries have been eliminated and replaced by a channelized intersection resulting in better operations.

(iii) A rotary requires more land and may not be feasible in many built-up locations.

(iv) Because of the large and relatively flat land area required, topographic conditions in some localities may make it impracticable to develop rotary intersections.

(v) Where pedestrian traffic is large, a rotary by itself is not sufficient to control traffic and has to be supplemented by traffic police.

(vi) When used on high speed roads, rotaries require extremely large size.

(vii) Where the angle of intersection between two roads is too acute, it becomes difficult to provide adequate weaving length.

(viii) When provided at close intervals they make travel troublesome.

(ix) Traffic turning right has to travel a little extra distance.

(x) A rotary requires many warning and directional signs for safety. The central island and entrances and exits must be well lighted at night. These tend to make it costly.

(xi) The rotary is not readily adaptable to stage development. Attempts at stage development generally result in some over-design when viewed from immediate traffic needs.

7.15.16.5. Guidelines for Selecting a Rotary Type of Intersection

Considering some of the above advantages and disadvantages, the following general guidelines may be kept in view when selecting a rotary type of intersection:

(i) Rotaries are most apt where volumes entering from different intersection legs are approximately equal.

(ii) A total volume of 3000 vehicles per hour entering from all the intersection legs appears to be the maximum practical capacity of high tyre rotaries.

(iii) A rotary design is most appropriate when the proportion of turning traffic is very high. It is considered that at a four-leg intersection a rotary may need less land than traffic signals if right-turning traffic exceeds about 30 per cent of all approaching traffic.

(iv) A rotary is a good choice when there are more than four approaches to the junction.

(v) When the approach widths are restricted and it is impossible to provide separate lanes for through and turning traffic, a rotary may be desirable.

(vi) Rotaries are not generally warranted for intersections carrying very light traffic. Normally, the lowest traffic volume for which a rotary design should be considered is about 500 vehicles per hour (Ref 24).

(vii) Rotaries should be a good choice for moderately busy intersections in urban and suburban areas, where otherwise the alternative may be to go in for a complicated channelized layout or traffic signals.

7.15.16.6. Shape of Rotary Island

The shape and layout of the rotary island depends upon the layout of the intersecting roads and the traffic flow pattern. The following brief remarks about the different types in common use are intended to be general guidelines.

Type	Remarks
1. Circular	Suited where roads of equal importance carrying approximately equal volumes intersect at nearly equal angles.
2. Squarish with rounded edges	Suited for predominantly straight ahead flows.
3. Elliptical, elongated, oval or rectangular	Intended to favour through traffic. They may be provided to suit the geometry of the intersecting legs or to provide long weaving lengths. Elongated central island will be approximate for a scissors junction.
4. Irregular	Shape is dictated by the existence of a large number of approaches.

Figure. 7.63 indicates the different shapes.

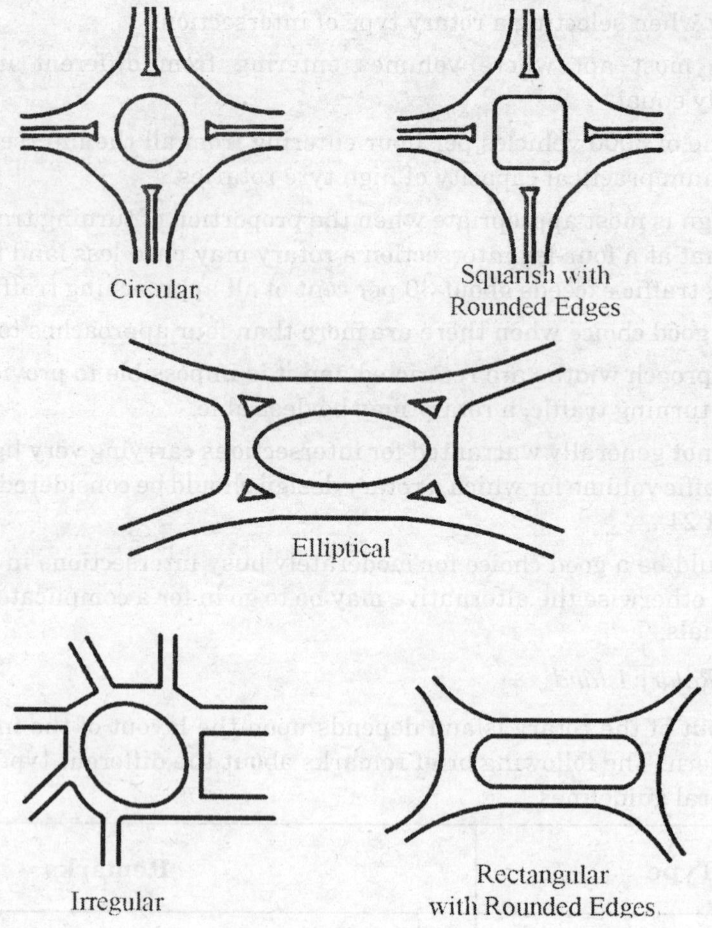

Fig. 7.63: Shape of Rotary Island

7.15.16.7. Rotary Design Elements

7.15.16.7.1. Design Speed

The design speed of a rotary governs the various elements such as radii and weaving lengths. To keep the dimensions within practical limits the design speed has to be necessarily low, as compared to the design speed of the intersecting highways. Current Indian practice is to design rotaries in rural areas for a speed of 40 K.P.H. and those in urban areas to a speed of 30 K.P.H.

7.15.16.7.2 Radius at Entry

The radius at entry is determined by the design speed, super-elevation and coefficient of friction. The radius should result in the drivers lowering their speeds to a safe value, which is the design speed of the rotary. A range of 20-35 m is found to be suitable for rural designs and a range of 15-20 m is suitable for urban design.

7.15.16.7.3 Radiu s at Exit

The exit radius should be higher than the radius of the rotary island so that it favours a higher speed by drivers. This will enable the rotary to be cleared rapidly. The general practice is to keep the radius of exit curves 1½ to 2 times the radius of the entry curves. If, however, there is a large pedestrian traffic across the exit road, radii similar to those at entrances may be provided to keep the exit speeds reasonably low.

7.15.16.7.4 Radius of the Central Island

The radius of the central island is governed by the rotary design speed and theoretically it should be equal to the radius at entry. In practice, however, (Ref 24) the radius of the central island may be kept slightly larger than that of the curve at entry, this being an attempt to give a slight preference to the traffic already on the rotary and to slow the approaching traffic. A value of 1.33 times the radius of entry curve is probably adequate for this purpose.

7.15.16.7.5. Weaving Lengths

The weaving length determines the ease with which the traffic can merge and diverge. The weaving length is decided on the basis of factors such as the width of the weaving section, the average width of entry, total traffic and the proportion of weaving traffic in it. As a general rule, it is desirable to prevent direct traffic cuts and this can be achieved by making the ratio of the weaving length to the weaving width large enough. A ratio of 4:1 is regarded as a minimum (Ref 18). Current Indian practice (Ref 24) as regards weaving length is summarised in Table 7.31.

Table 7.31 : Minimum Length of Weaving Section

Design Speed (Kmph)	Minimum Weaving Length (m)
40	45
30	30

It is also necessary to put a limit to the maximum weaving length to discourage speeding. Twice the values given above can serve as the upper limits.

7.15.16.7.6 Width of Carriageway at Entry and Exit

The carriageway width of the intersection legs is governed by the design year traffic entering and leaving the intersection. A minimum width of carriageway of 5 m, with provision for extra widening due to curvature, may be provided for the entrance and exit. Table 7.32 summarises current Indian practice (Ref 26).

Table 7.32 :Width of Carriageway at Entrance and Exit (Ref 24)

Carriageway width of Approach Road	Radius at Entry (m)	Width of Carriageway at Entry and Exit (m)
7 m (2 lanes)		6.5
10.5 m (3 lanes)		7.0
14 m (4 lanes)		8.0
21 m (6 lanes)	25–35	13.0
7 m (2 lanes)		7.0
10.5 m (3 lanes)		7.5
14 m (4 lanes)		10.0
21 m (6 lanes)	15–25	15.0

7.15.16.7.7 Width of Rotary Carriageway

Fig. 7.64 indicates the design elements.

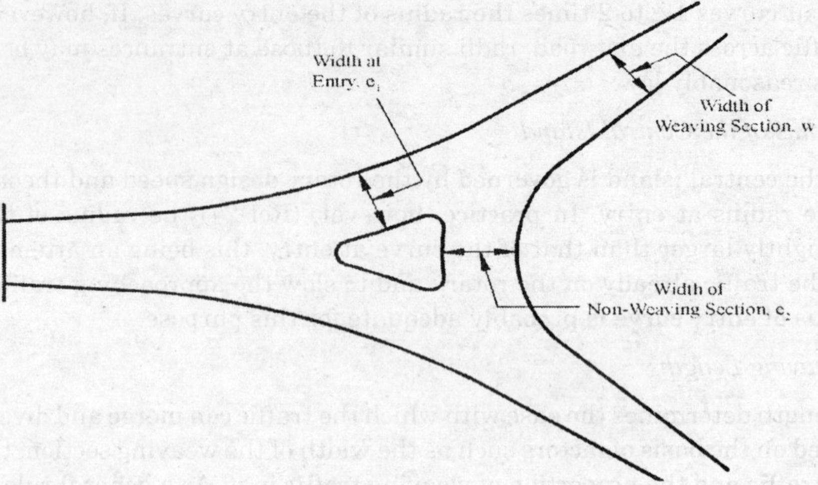

Fig. 7.64: Width of rotary carriageway

The width of the non-weaving section should be equal to be widest single entry into the rotary, and should generally be less than the width of the weaving section. The width of the weaving section (w) should be one traffic lane (3.5 m) wider than the mean entry width thereto.

Thus is: $w = \dfrac{e_1 + e_2}{2} + 3.5$ with reference to Fig. 7.64.

7.15.16.7.8 Entry and Exit Angles

Entry angles should be larger than exit angles and it is desirable that the entry angles should be about 60^0 if possible; the exit angles should be small, even tangential. A design with an entry angle of 60^0 and exit angle of 30^0 is shown in Fig. 7.65.

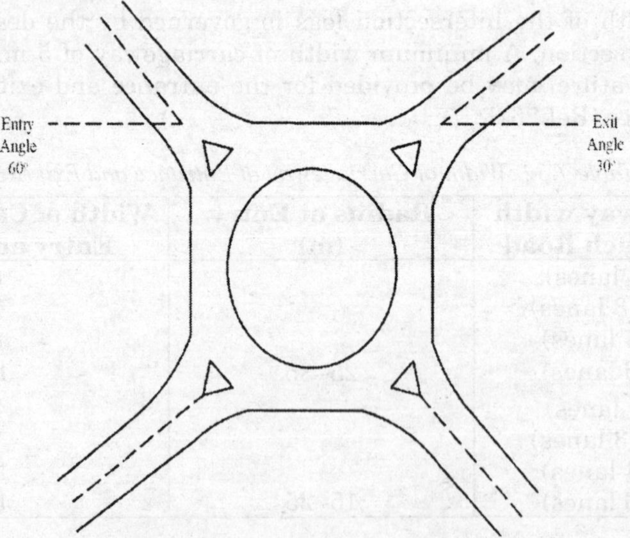

Fig. 7.65: An ideal Layout with 60^0 Entry Angle and 30^0 Exit Angle

7.15.16.7.9. External Curb Line

The external curb line of the weaving sections should not be normally re-entrant, but consist of a straight or large radius curve of the same sense as the entry and exit curves. Such an arrangement eliminates waste of area, which is unlikely to be used by traffic, vide Fig. 7.66.

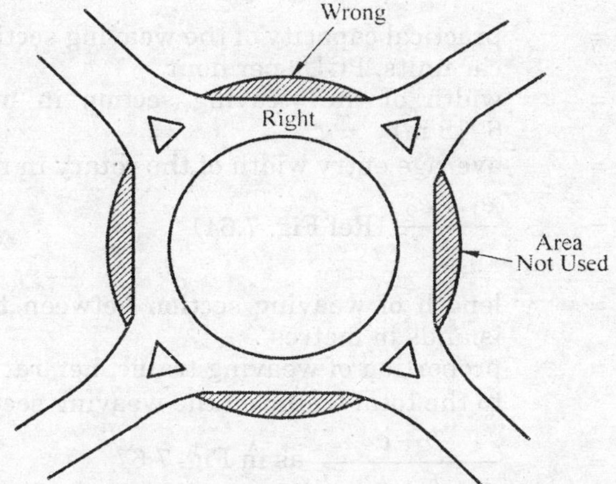

Fig. 7.66: External Curb Line of Weaving Sections

7.15.16.7.10 Super-elevation and Camber

The rotary curvature being opposite to that of entry and exit, the super-elevation in the two portions will be opposite to each other. This causes difficulty to vehicles, especially double-decker buses and top heavy trucks, in changing from one cross-slope to another. To cause the least discomfort, the algebraic difference in the cross-slopes should be limited to about 0.07. The individual superelevations should be limited to the least amount consistent with speed. The crown-line, which is the line of meeting of the opposite cross-slopes should, as far as possible, be located such that the vehicles cross it while travelling along the common tangent to the reverse curve.

Channelizing islands should be situated on the peak with the road surfaces sloping away to all the sides. The cross-slope at entrance should be carried around on the outer edge of the rotary to the adjacent exit making slight adjustments as may be needed from the point of view of curvature and speed.

7.15.16.7.11. Other Miscellaneous Features

Stopping sight distance should be provided for the design speed adopted. Values may be taken from Table 7.12. The rotary should preferably be located on the ground having a slope flatter than 1 in 50. For the channelizing and central islands, mountable type curbs may be provided, whereas for the outer edges of the rotary a barrier type may be provided to discourage pedestrians from crossing over. Landscaping and drainage requirements should be considered.

7.15.16.8. Capacity

The capacity of a rotary is directly determined by the capacity of each weaving section. The capacity of a weaving section is determined by the geometric layout, including entrances and exits, and the percentage of weaving traffic. The Transport and Road Research Laboratory (U.K.) which has pioneered research on this aspect, recommends the following formula (Ref 18), which is a modification of the well-known Wardrop formula (Ref 25):

$$Q_p = \frac{280w\left(1+\frac{e}{w}\right)\left(1+\frac{p}{3}\right)}{1+\frac{w}{l}} \quad (7.55)$$

where Q_p = practical capacity of the weaving section of the rotary in passenger car units, PCUs per hour

w = width of the weaving section in metres (within the range of 6–18 m)

e = average entry width of the rotary in metres

= $\frac{e_1 + e_2}{2}$ (Ref Fig. 7.64)

l = length of weaving section between the ends of the channelizing isiands in metres

p = proportion of weaving traffic, i.e., ratio of sum of crossing streams to the total traffic on the weaving section.

= $\frac{b+c}{a+b+c+d}$ as in Fig. 7.67

In the above formula, the equivalency factors recommended by TRRL (Ref 18, 22) and IRC (Ref 24) are given in Table 7.33.

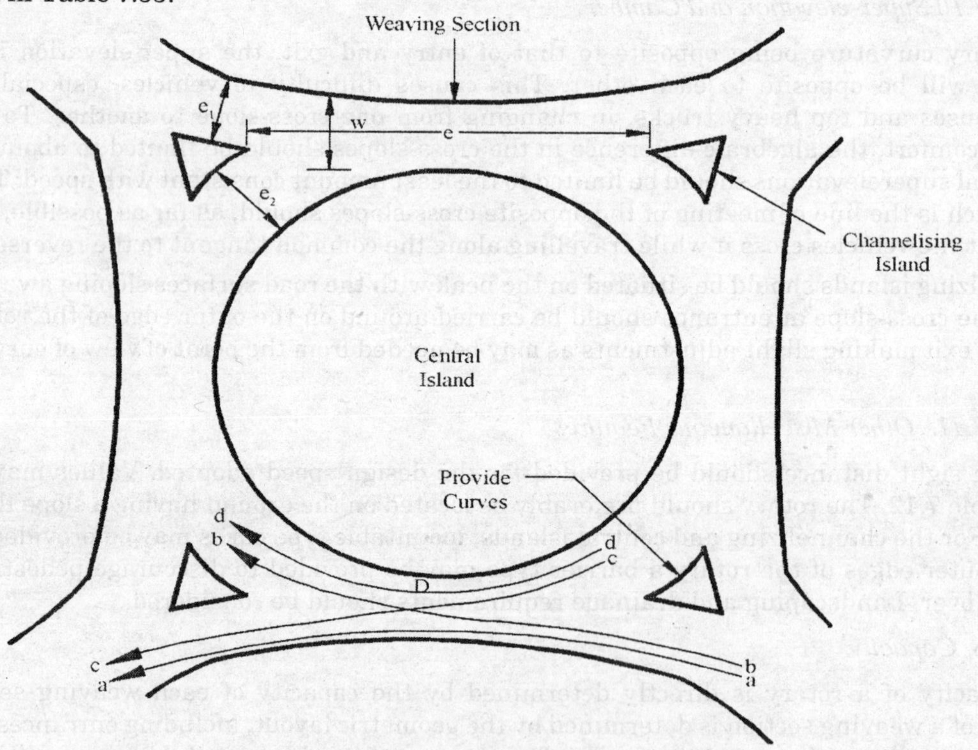

Fig. 7.67: Relevant dimensions of weaving section and proportion of weaving traffic for use in capacity formula for rotaries

Table 7.33 : Equivalency Factors for Use In Rotary Design (Ref 18, 22, 24)

	Vehicle Type	PCU
1.	Cars and light commercial vehicles, including 3-wheelers	1.0
2.	Buses and medium heavy commercial vehicles	2.8
3.	Motorcycles and scooters	0.75
4.	Pedal cycles	0.5
5.	Animal drawn vehicles	4 to 6

The above formula is valid under the following conditions:

(i) There are no standing vehicles on the approaches to the rotary

(ii) The site of the level and approach gradients do not exceed 1 in 25

(iii) $\frac{e}{w}$ should be between 0.4 and 1.0

(iv) $\frac{w}{l}$ should be between 0.12 and 0.40

(v) p should be between 0.4 and 1.0.

(vi) l should be between 18 and 90 m

In the above formula, certain adjustments need to be made to take into account the geometric layout. These are:

(i) Where the entry angle is between 0^0 and 15^0, deduct 5 per cent from the capacity of the weaving section

(ii) Where the entry angle is between 15^0 and 30^0, deduct 2½ per cent from the capacity of the section

(iii) Where the exit angle is between 60^0 and 75^0, deduct 5 per cent from the capacity of the weaving section

(iv) Where the exit angle is greater than 75^0, deduct 5 per cent from the capacity of the weaving section

(v) Where the internal angle is greater than 95^0, deduct 5 per cent from the capacity of the weaving section

(vi) Where the pedestrian flow across an exit is more than 300 per hour, an arbitrary reduction on one-sixth should be made in the practical capacity of the preceding weaving section.

7.15.16.9. Increasing the Efficiency of Rotaries

In U.K., the introduction of the "Give way to traffic from the right" rule has increased the efficiency and safety of rotaries (Ref 22). Locking has been virtually eliminated, delays has been reduced and the accident rate has been reduced by more than 40 per cent. Signals on the approaches can also increase the capacity and improve operation.

7.15.16.10. An Example of the Design of a Rotary Inter-section

Problem: Traffic flow in an urban section at the intersection of two highways in the design year are given below:

Approach	Left turning			Straight Ahead			Right Turning		
	Cars	Commercial	Scooters	Cars	Commercial	Scooters	Cars	Commercial	Scooters
N	200	50	100	250	100	150	150	50	80
E	180	60	80	220	50	120	200	40	120
S	250	80	100	150	50	90	160	70	90
W	220	50	120	180	60	100	250	60	100

The highways at present intersect at right angles and have a carriageway width of 15 m. Design a rotary intersection making suitable assumptions.

Solution. The rotary is located in an urban section and hence a design speed of 30 K.P.H. is appropriate. Since the intersection legs carry almost equal traffic, a round shaped central island will be adopted. The entrance and exit angles will be 45^0 each. A radius of 20 m at entry, 40 m at exit and 80 m for the central island will be adopted. These values conform to current Indian standards discussed earlier. The traffic in terms of PCUs from each leg is depicted in Fig. 7.68.

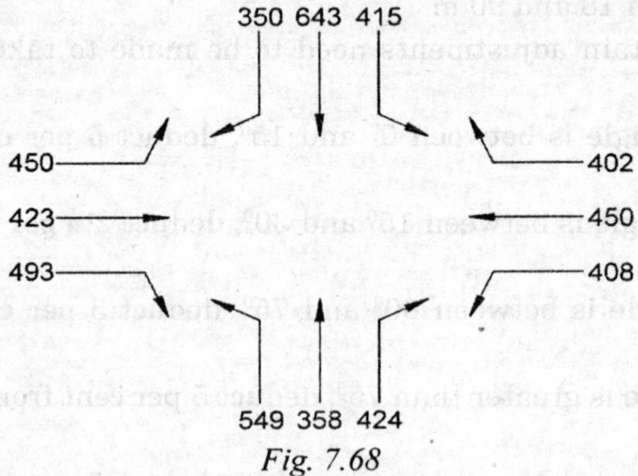

Fig. 7.68

The traffic is assigned to the network in Fig. 7.68

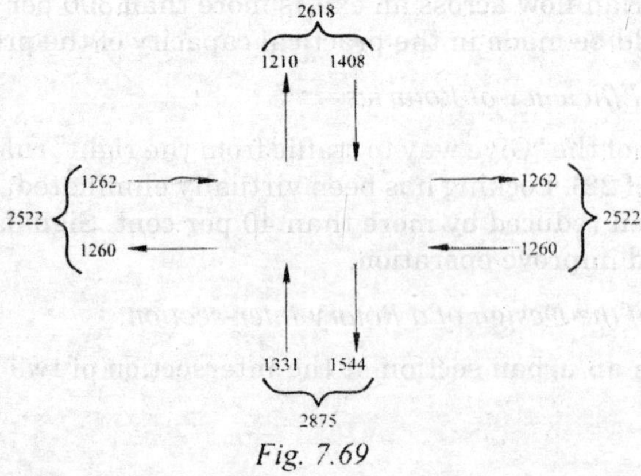

Fig. 7.69

The traffic is assigned to the network in Fig. 7.69

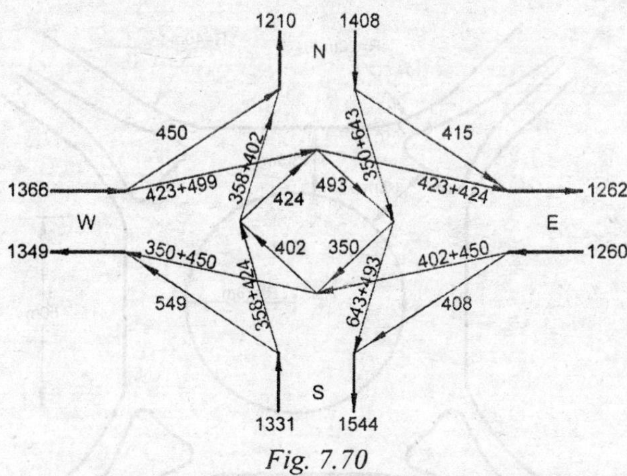

Fig. 7.70

The maximum 2-way flow in the intersections leg (South) is 2875 PCUs/hour, and the maximum in one direction is 1544 PCUs/ hour. A 4-lane road with moderate interference from cross-traffic has a capacity of 1400–1800 PCU/hour in one direction and is good enough.

The width of carriageway at entry and exit may be taken from Table 7.32. For a 4-lane road, a width of 10.0 m is in order. The width of the non-weaving section is also kept 10.0 m. The width of the weaving section is:

$$\frac{e_1 + e_2}{2} + 3.5 = 10 + 3.5 = 13.5 \text{ m}$$

The minimum length of the weaving section should be 30 m from Table 7.31, but this fails the requirement that the ratio l/w should be at least 4. So adopt 55 m, which gives a l/w ratio slightly greater than 4.

$$p = \frac{1136 + 852}{350 + 1136 + 852 + 408}$$

$$= \frac{1988}{2746} = 0.72$$

From Eq. (7.55)

$$Q_p = \frac{280 + 13.5\left(1 + \frac{10}{13.5}\right)\left(1 - \frac{0.72}{3}\right)}{1 + \frac{13.5}{55}}$$

$$= \frac{288 \times 13.5 \times 1.74 \times 0.76}{1.245}$$

$$= 4000 \text{ PCU/ hour}$$

This is very much higher than the traffic flow of 2746 PCU/ hour, and the design is acceptable. A sketch of the design is given in Fig. 7.71.

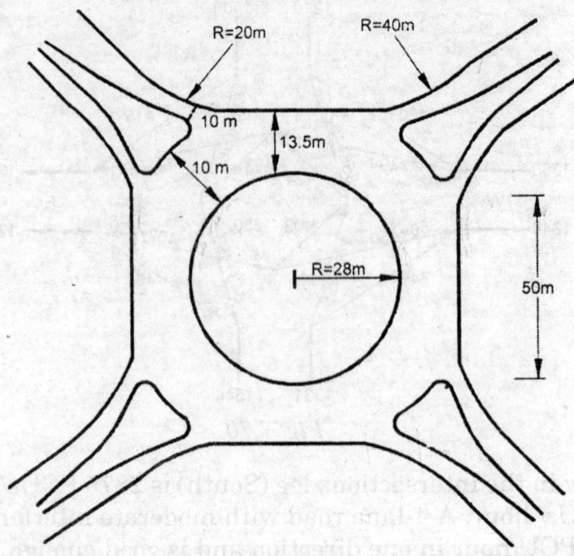

Fig. 7.71:

7.15.17. Mini Roundabouts

7.15.17.1. A serious drawback of the conventional (large diameter island) roundabout is that very often the traffic locks itself. A locked roundabout causes large traffic delays and unlocking can only be achieved by means of police help. The conventional round-about is thus not truly self-regulatory at all times.

7.15.17.2. In the United Kingdom, which is one of the countries where roundabouts still retain their popularity, a major change in the approach to roundabout design was the off-shoot of the introduction of the priority from-the-right rule in 1966. This rule was intended to get over the drawback of locked roundabouts, but other advantages soon began to offer themselves. The absence of weaving under the new rule made possible a drastic reduction in the size of the central island and an availability of greater carriageway space with the existing layout. The width of approach at the entry to the junction could be increased considerably thus rendering multi-vehicle flow from Give-Way line. The ***mini roundabout*** has made its appearance and was soon to become very popular in many cities in the U.K. Experiments were initiated at the T.R.R.L. (Ref 28) and these showed that the use of small diameter islands could bring in higher capacities than large diameter roundabouts. This has formed the basis of current U.K. practice (Ref 27).

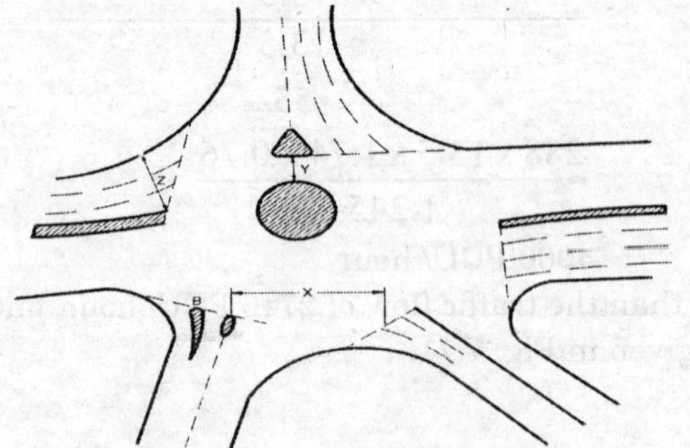

Fig. 7.72: Typical mini-roundabout as per U.K. practice

The basic principles of design of a mini-roundabout illustrated in Fig. 7.72 are:

(i) Provision of small diameter island of a diameter of about one-third that of a hypothetical circle inscribed within the outer carriageway boundaries, but normally not less than 8 m. For new layouts, space for a larger diameter (15 to 25 m) may be reserved if such a provision becomes necessary in future.

(ii) An increase in the number of lanes at the Give-Way line.

(iii) A minimum stopping distance of 26 m between the Give-Way line and the point of conflict with a vehicle from the left. This is the dimension X in Fig. 7.72.

(iv) A width between traffic islands and the roundabout (shown as dimension Y in Fig. 7.72) which is not less than the total lane width at the entry preceding it, i.e., dimension shown as Z in Fig. 7.72.

(v) An entry taper (about 1 in 6) that is about twice sharp as the exit taper (about 1 in 12).

(vi) A deflection island (shown as B in Fig. 7.72) is intended to ensure that straight-through movements do not occur.

(vii) To maintain junction capacity, a single lane approach should be tapered to give at least three lanes at the junction and a 2-lane approach tapered to give at least four lanes at the junction. At the exits, the merge should normally be from four lanes to two lanes and two lanes to one lane.

7.15.17.3. Blackmore (Ref 26) gives the capacity of a mini-round about as:

$$q = K\left(\sum w + a^{1/2}\right) \qquad (7.56)$$

where q = total entry volume in PCU/hour

$\sum w$ = sum of the basic road width (not half widths) used by traffic in both direction to and from the intersection, in metres.

a = area of the junction widening, i.e. the area within the intersection outline (including islands) which lies outside the area of the basic cross-roads, in sq. m.

K = efficiency coefficient which depends upon site conditions. Typical values are:

	K values (PCUs per hour)
3 way junction	80
4 way junction	70
≥ 5 way junction	65

The practical capacity is about 80 per cent of that given by the above equation.

7.15.17.4. Advantages of Mini-roundabouts

(i) ***Increase in capacity:*** Wherever mini-roundabouts have replaced large diameter roundabouts, a general increase in the capacity of the junction has been noted. As compared to signalised junctions too, mini-roundabouts can bring about higher capacities.

(ii) ***Accident rates:*** Typical accident rates before and after conversion of junctions into mini-roundabouts are yet to be fully analysed and evaluated, but a trend in decrease in the number of fatal and serious injury accidents has been noticed.

7.15.17.4.5. Limitation for Use

As indicated earlier, mini-roundabouts are the result of the introduction of the priority-from-the-right rule. They can be introduced only in situations where such rules are in force and are strictly followed. Non-observance of this rule can only lead to chaos and serious accidents.

The mini-roundabout requires careful considerations on the design speed and ways of reducing speeds at junctions to this value. A very high standard of pavement markings and road signing, a great care in provision of guide islands, alignment of the kerbs and location of the central island are necessary for the satisfactory functioning of this type of junctions.

Adequate visibility to the approaching drivers is of prime importance. This alone will enable him to give way to the traffic on his right and to check on the traffic on his left for giving way to him.

7.16 Grade Separated Intersections

7.16.1. Criteria for Provision

The general site conditions favouring a grade-separated intersection have been enumerated earlier in Section 7.15.2. Current Indian practice (Ref 21) requires grade separation across streets and highways to be provided as per the following criteria:

- ***Urban Streets.*** Grade separation to be provided if the estimated traffic volumes within the next 5 years are in excess of the capacity of the intersection; when traffic projections show that volumes within the next 5 years are in excess of the capacity of the intersection; when traffic projections show that volumes within the next 20 years will exceed the capacity of an at-grade intersection, the need for a grade separated facility should be kept in view for future construction.
- ***Rural highways.*** Grade separation should be provided at inter-sections of divided rural highways if the ADT (fast vehicles only) on the cross-road within the next 5 years exceeds 5000. Where this figure will be reached within the next 20 years, the need for such facilities should be kept in view.
- **Grade-separation is of great importance across railway lines.** Current Indian practice (Ref 21) for rural highways requires grade separation to be provided across existing railway lines if the product of ADT (fast vehicles only) and the number of trains per day exceeds 50,000 within the next 5 years. For new constructions such as bypasses, grade separations should be provided when this figure is greater than 25,000.

7.16.2. Types of Grade-separated Intersections

Basically two types are met with:

(i) Grade-separated intersections without interchange

(ii) Grade-separated intersections with interchange

Interchange is a system whereby facility is provided for movement of traffic between two or more roadways at different levels in the grade separated junction. A structure without interchange is an over-bridge or underpass or flyover, whereby the traffic at different levels moves separately without a provision for an interchange between them.

The different forms of a grade-separated junction can be considered under the number of legs the intersection serves. Thus the interchanges can be classified as three-leg, four-leg and multileg, and these, in turn, can be sub-divided into various types as below:

A. Three-leg interchange
   1. T interchange
   2. Y interchange
   3. A partial rotary interchange

B. Four-leg interchange
   1. Diamond interchange
   2. Half clover leaf interchange
   3. Clover leaf interchange
   4. Rotary interchange
   5. Directional interchange

C. Multi-leg interchange
   1. Rotary interchange

These types are discussed below, in detail.

7.16.3. Three-leg Interchange

If one of the intersection legs meets a highway at approximately a right angle but does not cross it, and if the junction is provided with interchanging facility, the junction is called a T interchange. If the intersection angle is acute or obtuse, the junction is called a Y interchange. A widely used form of a T or V interchange is a **trumpet**, shown in Fig. 7.73(a) and (b). (a) is the usual form but (b) can also be adopted under particular site conditions, but this has the drawback that vehicles leaving the major road have to turn through a small radius. These designs can be adopted for junctions of major streets with expressway.

Fig. 7.73: Trumpet Interchange

Designs shown in Fig. 7.74 (a), (b) and (c) are Y-shaped and these permit limited movements.

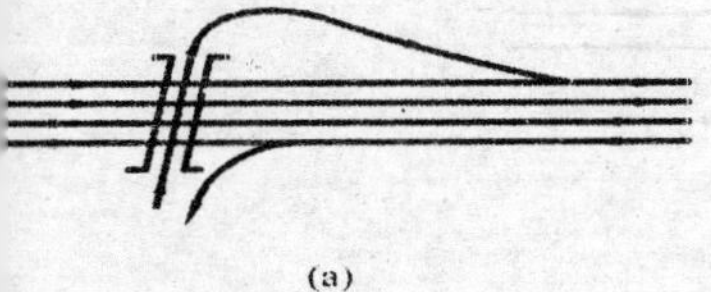

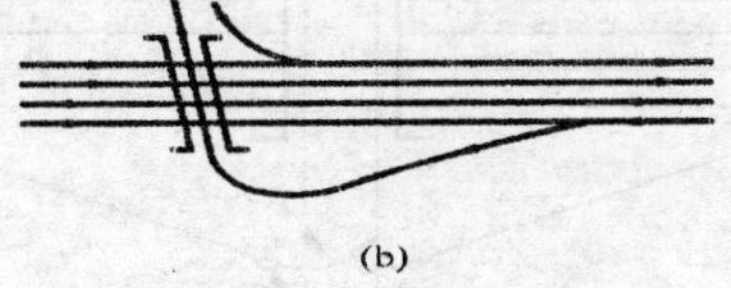

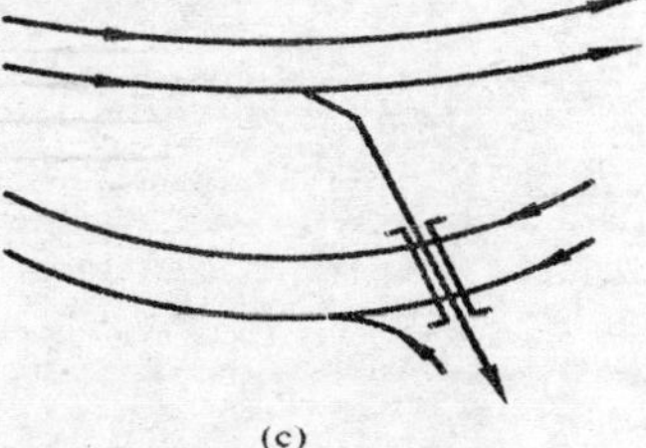

Fig. 7.74: Y-shaped Interchanges

A design which can be converted to full cloverleaf at a later date when the junction is likely to become four-leg, is indicated in Fig. 7.75.

As an alternative to the arrangement suggested in Fig. 7.75, a partial bridged rotary intersection can be adopted, Fig. 7.76.

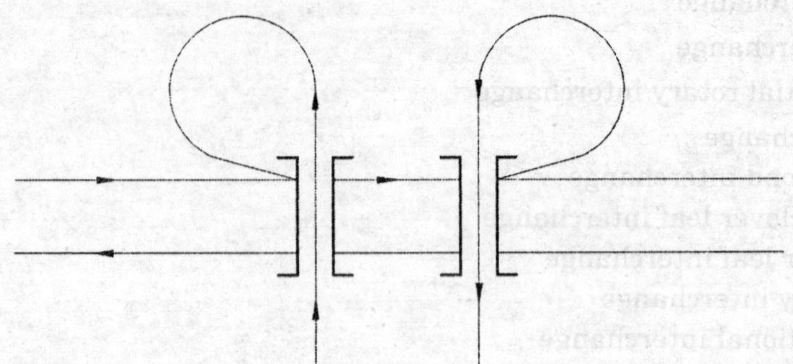

Fig. 7.75: Three-leg Interchange which can be converted to a Cloverleaf

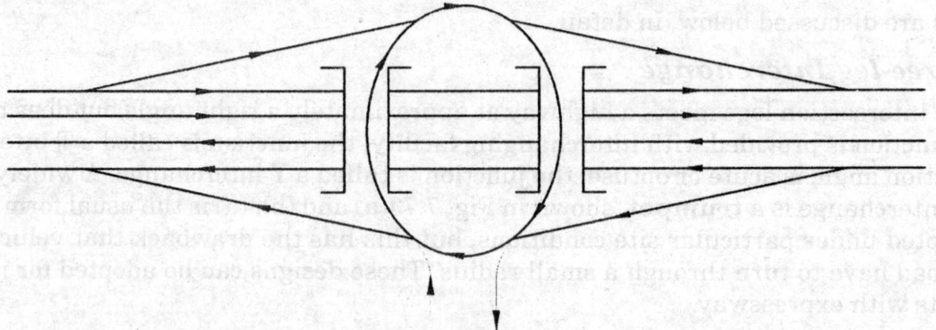

Fig. 7.76: Partial Bridged Rotary

7.16.4. Diamond Interchange

A popular form of interchange in urban locations involving a major-minor crossing is a diamond interchange. It can be designed for a relatively narrow right-of-way of the major road. A simple form of diamond interchange is given in Fig. 7.77. A split-diamond layout Fig. 7.78, can be adopted where two parallel cross-streets are available as connections to the main street.

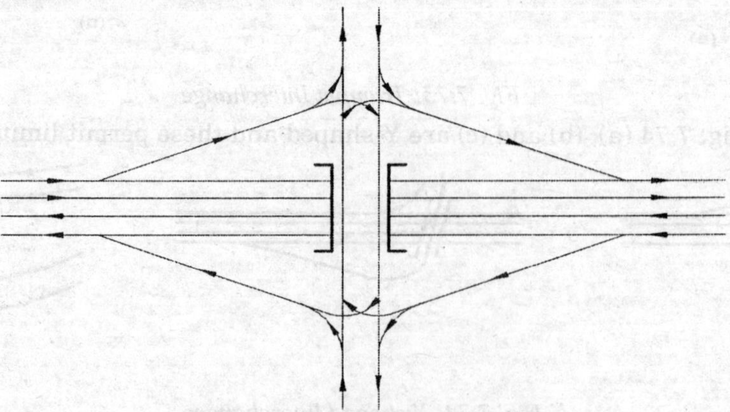

Fig. 7.77: Diamond Interchange

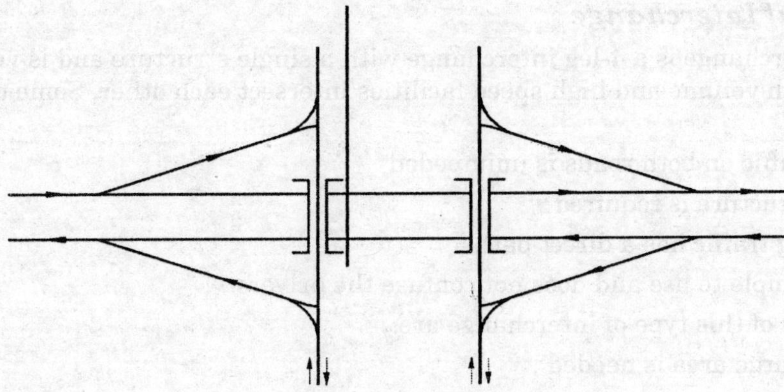

Fig. 7.78: Split Diamond

If the two cross-streets are one-way, the layout in Fig. 7.79 may be adopted.

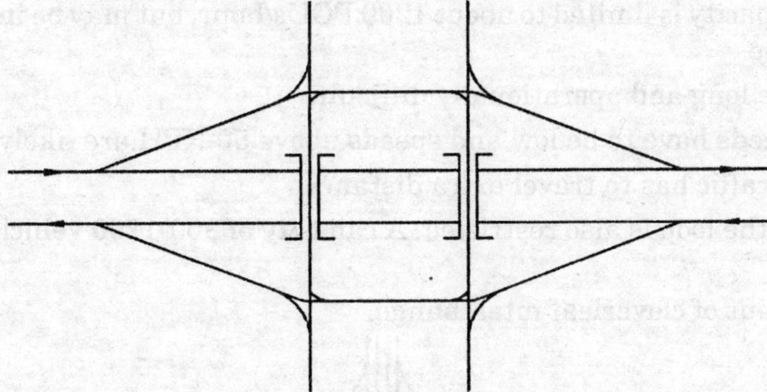

Fig. 7.79: Split Diamond Interchange with One-way Streets

7.16.5. Half Cloverleaf Interchange

When a major road crosses a minor road (not more than 3/lanes), it is possible to provide a partial cloverleaf. Two such designs are given in Fig. 7.80(a) and (b).

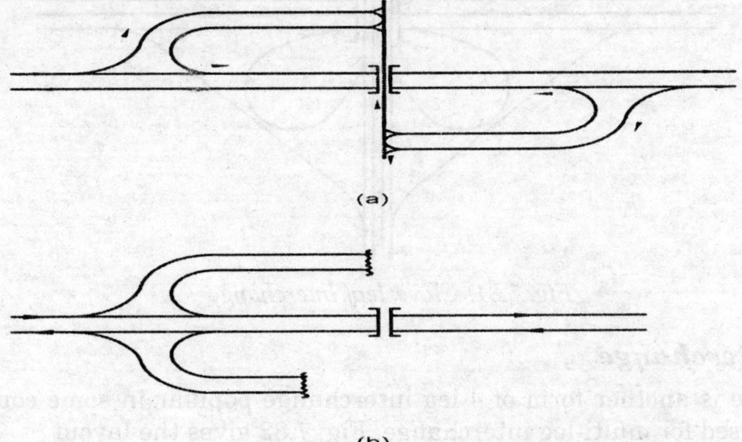

Fig. 7.80: Partial Cloverleaf for Major-Minor Road Crossing

7.16.6 Cloverleaf Interchange

The cloverleaf interchange is a 4-leg interchange with a single structure and is very popular. It is used when two high volume and high speed facilities intersect each other. Some of its advantages are:

(i) Through traffic on both roads is unimpeded

(ii) Only one structure is required

(iii) Left-turning traffic has a direct path

(iv) It is very simple to use and does not confuse the drivers

The disadvantages of this type of interchange are:

(i) Relatively large area is needed

(ii) The carriageway area required is also higher than a rotary interchange

(iii) Weaving manoeuvres are involved, some of them on the roadway of the structure and some underneath the structure.

(iv) The weaving capacity is limited to about 1200 PCUs/hour, but may be increased by providing an ancillary lane.

(v) The U-turns are long and operationally difficult.

(vi) Loop design speeds have to be low and speeds above 50 KPH are likely to increase the cost.

(vii) Right-turning traffic has to travel extra distance.

(viii) The capacity of the loop is also restricted. A capacity of 800-1200 vehicles per hour is almost the limit.

Fig. 7.81 gives the layout of cloverleaf interchange.

Fig. 7.81: Cloverleaf Interchange

7.16.7. Rotary Interchange

A rotary interchange is another form of 4-leg interchange popular in some countries, especially U.K. It can also be used for multi-leg interchange. Fig. 7.82 gives the layout.

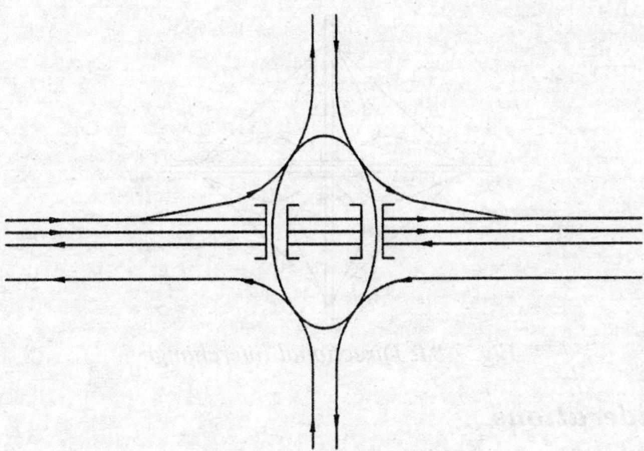

Fig. 7.82: Grade-separated Rotary Interchange

The advantages claimed for this type of interchange are:

(i) It occupies relatively less land area

(ii) The carriageway area is also less than many other types

(iii) U-turns are easy

(iv) Suitable for most sites

The disadvantages are:

(i) The capacity is limited by the capacity of the roundabout itself.

(ii) The straight through traffic on one road is required to weave with turning traffic from the other road.

While the design in Fig. 7.82 above is for a 2-level interchange, a further improvement can be made by having a 3-level rotary interchange (Fig. 7.83).

In this arrangement through traffic on each road is unimpeded and the weaving takes place in the rotary only for the turning traffic. The number of structures required is five.

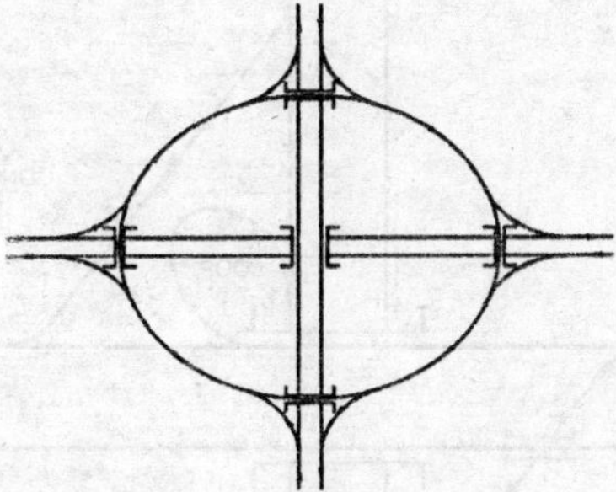

Fig. 7.83: Three-level Rotary

7.16.8. Directional Interchange

An interchange which gives direct or semi-direct connections for the major right-turning movements is called a directional inter-change. These interchanges are rather complex and involve complicated multi-level structures. Fig. 7.84 gives a typical design.

Fig. 7.84: Directional Interchange

7.16.9. Design Considerations

7.16.9.1. Definition of some common terms

In dealing with interchange design, some common terms are met with, which are defined below:

1. ***Interchange Ramp* (Access Ramp or Slip road).** An interconnecting roadway of a traffic interchange or any connection between highway facilities at different levels, on which vehicles may enter or leave a designated roadway. It may be in the form of a loop, outer connection or direct connection.
2. ***Loop.*** A one-way turning roadway that curves about 270^0 to the left to accommodate a right-turning movement.
3. ***Outer Connection.*** A ramp used by traffic destined for left-turn movement from one of the through roadways separated by a grade separation structure to the second through roadway.
4. ***Direct Connection.*** A form of ramp which does not deviate much from the intended direction of travel. It avoids the loop for right-turning movements. An outer connection is a direct connection ramp for left-turning movements.

The above terms are illustrated in Fig. 7.85.

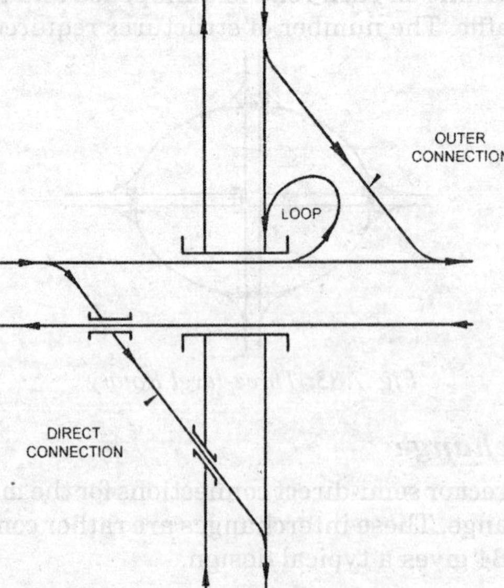

Fig. 7.85: Terms used in Interchange

7.16.9.2. Design speed for ramps

It is rarely, if ever, feasible for designing the ramps for the same speed as the through roads. The design speeds recommended for ramps as per AASHTO (Ref 4, 5) are given in Table 7.34.

Table 7.34 : Guide Values for Ramp Design Speeds (Ref 4, 5)

Highway Design								
Speed K.P.H.	50	65	80	100	105	115	120	135
Ramp Design								
Speed K.P.H.								
Desirable	40	55	70	80	90	100	100	105
Minimum	25	30	40	50	50	50	55	65
Corresponding min radius, m								
Desirable	45	90	170	210	260	320	320	380
Minimum	15	27	45	70	70	70	90	130

For design of loops, the speed should be about 40–45 K.P.H. Higher values are impractical in many locations. For direct connections, a design speed of 60–65 K.P.H. is appropriate.

The standards prescribed for the ramp design elements for interchanges in urban areas in India (Ref 28) are reproduced in Table 7.35.

Table 7.35 : Standards for Ramp Design in India for Urban Locations

Particulars	Design values for major highway				For loop ramps	
	80 K.P.H.		100 K.P.H.			
	Minimum	Desirable	Minimum	Desirable	Minimum	Desirable
1. Ramp Design Speed (K.P.H.)	40	50	50	65	30	40
2. Radius of Curvature (m)	60	90	90	155	30	60
3. Stopping sight distance	45	60	60	90	25	45

7.16.9.3. Radii of Curves on Ramps

Table 7.36 indicates the radii corresponding to the design speed selected.

7.16.9.4. Sight Distance on Ramps

A minimum stopping sight distance appropriate for the design speed selected for ramp design should be provided.

7.16.9.5. Gradients on Ramps

AASHTO (Ref 4, 5) guidelines are:

(i) Ramp gradients should desirably be limited to 6 per cent in areas subject to snow and ice.

(ii) Where heavy truck traffic is likely, the upward gradient should be limited to 4 per cent.

(iii) In exceptional cases, grades may be as steeps as 10 per cent, on minor ramps and low volume ramps.

7.16.9.6. Width of Ramps

Ramps should preferably carry one-way traffic only. The width of pavements should be selected as appropriate for different radii.

7.16.9.7. Capacity

The practical capacity of through lanes can be taken as 1500 PCUs per hour (Ref 22) per lane for interchange design. For ramps, a capacity of 1200 PCUs per hour per lane is appropriate.

7.17 Bus Facilities

7.17.1. Bus Stops

The following guidelines for the location of bus stops may be kept in view:

(i) Since bus traffic near intersections interferes with the traffic flow if not properly designed for, it is desirable that the bus stops should be located at least 75 m away from the intersections (Ref 29).

(ii) It is preferable to locate bus stops on the exit side of the intersection (Ref 29).

(iii) Bus stops on opposite sides of two-way undivided carriageways should be staggered, preferably so that buses stop tail to tail and move off from each other. A suggested stagger is 60-90 m (Ref 22).

(iv) Bus stops should be, as far as possible, so located as to disembark the passengers at safe places such as curbs in islands.

(v) For buses intending to turn right at an intersection, if it becomes necessary to locate the bus stop before the intersection, the stop should be sufficiently away from the intersection so that the bus can be manoeuvred from the curb to the extreme right lane before the intersection.

7.17.2. Bus bays

It is desirable to recess the curb sufficiently for accommodating the bus-bays. Typical designs are given in Figs. 7.86 and 7.87 (Ref 30).

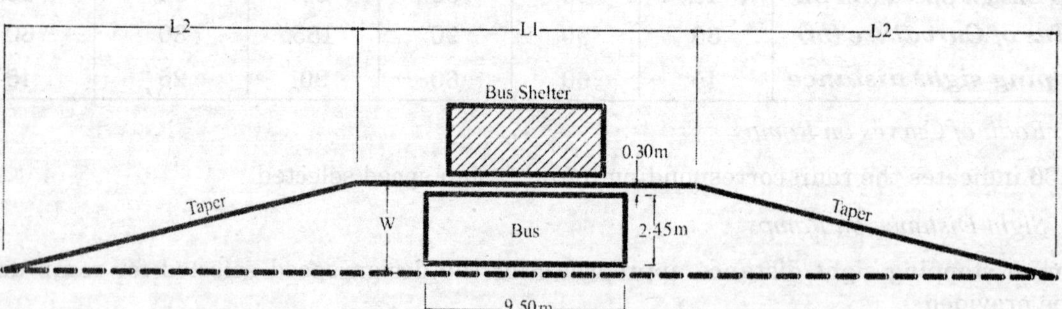

Notes:

1. L = *12 m Minimum for Congested locations for single bus, add 12 m for every additional bus*

   = *15 m desirable for single bus, add 15 m for every additional bus*

2. W = *3 m minimum for congested areas*

= *3.5 m desirable*

3. *L* = *6 W minimum for congested areas*

= *8 W desirable*

Fig. 7.86: Lay Out of Bus-Bay Abutting Footway

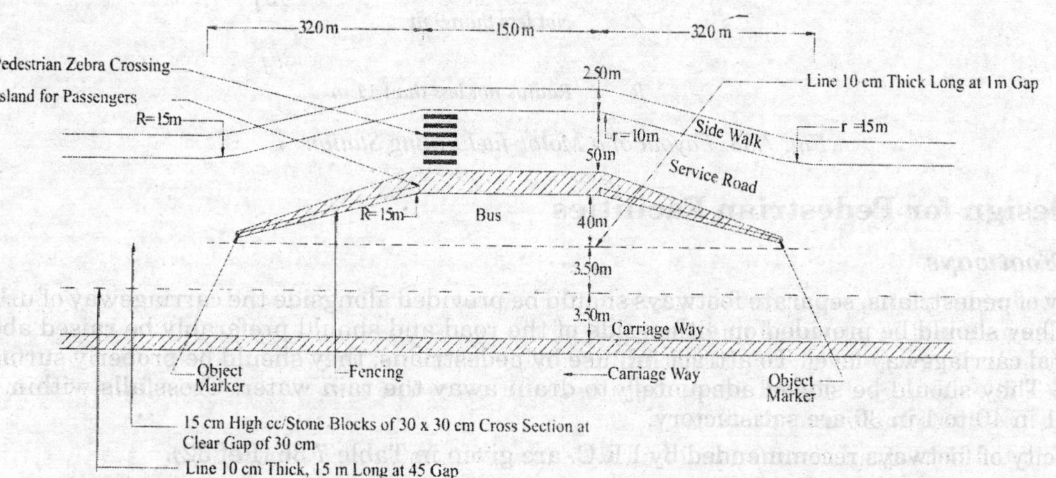

Notes: Width 'W' of island depends on the No. of passengers expected to be present at a particular time for loading and unloading purposes.

No. of Passengers	*Width*
20	*1.0*
45	*1.5*
70	*2.0*
95	*2.5*

Fig. 7.87: Layout of Bus-Bay with a Separate Lane for Slow Moving Traffic

7.18. Driveways

Driveways are roads providing access to abutting property owners and utilities. An important example is the access road to a petrol-filling facility. The geometrics of the driveways can have a profound influence on the traffic flow and safety of travel on the through road. The entrance and the exit should be wide with easy curves. A layout suggested by the I.R.C. (Ref 31), for motor-fuel filling stations is given in Fig. 7.88.

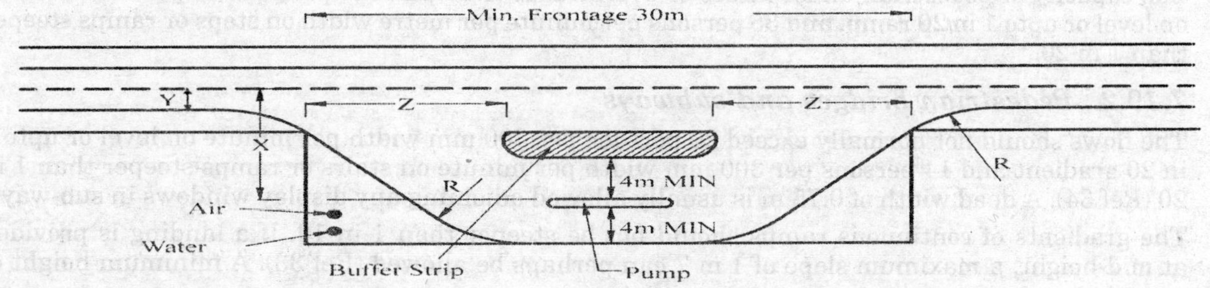

Minimum distance to intersection : 60 m for main roads and 45 m for others

X	*Not less than 13 m where no cycle tracks are required now or in future, not less than 20 m where cycle tracks exist or may be required in future.*
Y	*Half the roadway width but not less than 6 m for national and state highways and 5 m for others.*
Z	*Not less than 9 m*
R	*Radius not less than 11 m*

Fig. 7.88: Layout of a Motor-fuel Filling Station

7.19. Design for Pedestrian Facilities

7.19.1. Footways

For safety of pedestrians, separate footways should be provided alongside the carriageway of urban streets. They should be provided on either side of the road and should preferably be raised above the general carriageway level. To attract full use by pedestrians, they should be properly surfaced or paved. They should be sloped adequately to drain away the rain water. Crossfalls within the range of 1 in 40 to 1 in 30 are satisfactory.

The capacity of footways recommended by I.R.C. are given in Table 7.36 (Ref 32).

Table 7.36 : Capacity of Footways (Ref 32)

Width of Footpath (m)	**Capacity in Number of Persons Per Hour**	
	All in One Direction	**In Both Directions**
1.50	**1,200**	**800**
2.00	**2,400**	**1,600**
2.50	**3,600**	**2,400**
3.00	**4,800**	**3,200**
4.00	**6,000**	**4,000**

The capacity of pedestrian underpasses or overbridges is 50 persons per minute per metre width on level or upto 1 in 20 ramp, and 35 persons per minute per metre width on steps or ramps steeper than 1 in 20.

7.19.2. Pedestrian bridges and subways

The flows should not normally exceed 20 persons per 300 mm width per minute on level or upto 1 in 20 gradient and 14 persons per 300 mm width per minute on stairs or ramps steeper than 1 in 20 (Ref 34). A dead width of 0.75 m is usually allowed adjoining any display windows in sub-ways.

The gradients of continuous ramps should not be steeper than 1 in 10. If a landing is provided at mid-height, a maximum slope of 1 m 7 can perhaps be allowed (Ref 33). A minimum height of

2.30 m and width of 2.45 m should be provided for subways upto about 23 m length, and for longer subways the dimensions need to be increased to 2.60 m height x 2.75 m width.

7.20. Design Criteria for Separate Cycle Tracks

Segregation of cycle tracks promotes safety. In view of the large number of cyclists on the streets of Indian towns and cities, it is desirable that separate cycle tracks be provided wherever feasible. The design criteria for cycle tracks are laid down by the Indian Roads Congress (Ref 34). The design criteria commonly adopted are:

	Design Element	Guideline
1.	Justification for separate cycle tracks	Where the peak hour cycle traffic is 400 or more on routes with a vehicular traffic of 100-200 veh/hr. Where the motor vehicle traffic is more than 200 veh/hr, separate cycle tracks are justified even if the cycle traffic is only 100 per hour.
2.	Capacity (per hour)	Predominantly one-way traffic 2.5 m width 3000
3.	Horizontal curves (Radius)	(i) Minimum 10 m (ii) Where the gradient is steeper than 1 in 40, minimum radius shall be 15 m.
4.	Vertical curves	Minimum Radius Summit curves : 200 m Valley curves : 100 m
5.	Gradients	Gradient — Maximum length 1 in 30 — 90 1 in 35 — 125 1 in 40 — 160 1 in 45 — 200 1 in 50 — 250 1 in 55 — 300 1 in 60 — 360 1 in 65 — 425 1 in 70 — 500 (Gradients of 1 in 20 and 1 in 30 may be allowed for short lengths of 20 m and 50 m respectively.
6.	Sight distance	Clear view 25 m. In gradient of 1 in 40 or steeper, clear view shall not be less than 60 m.
7.	Minimum width	1.8 m
8.	Clearance	(i) Vertical : 2.25 m minimum 2.50 desirable (ii) Horizontal : 0.25 m

Other general considerations for the design of separate cycle tracks are:

(i) Preferably cycle tracks should be provided on both sides of the road.

(ii) They should be separated from the main carriageway by a verge or a berm, which should be preferably 0.5-1.0 m wide.

(iii) Cycle tracks should be constructed and maintained with care and should have good riding qualities. Black-topped and concrete surfaces provide smooth riding qualities, whereas water bound macadam and earthen tracks are not inviting for the cyclists to ride upon.

(iv) The tracks should be clear of obstructions such as hedges, ditches, tree roots, kerbs etc. by at least 0.5 m.

Design speeds for cycles is about 32 km/hr (Ref 35).

7.21. Geometric Design Standards for Expressways

1. Right of Way

Plain and Rolling Terrain	:	90 m
Mountain Terrain		
Same Level	:	60 m
Split Level	:	30 m

2. Design Speeds KPH

Terrain	*Ruling Speed*	*Minimum Speed*
Plain	120	100
Rolling	100	80
Mountainous	80	60

3. Carriageway Width

Per lane	:	3.75 for plain and rolling terrain 3.5 m for mountainous terrain

Carriageway should be divided

4. Shoulder width

Paved shoulder on median side	:	0.75 m
Paved shoulder on outer (left) side	:	3.0 m for plain and rolling terrain 2.5 m for mountainous terrain
Earthen shoulder (left or outer side)	:	1.5 m

5. Median Width

Without barrier	:	12 m
With barrier	:	4.5 m

6. Median Openings

To be provided with detachable barriers at every 2 to 5 Km spacing for traffic management and maintenance works.

7. Cross-fall

With rainfall 100 cm or more	:	2.5%
With rainfall less than 100 cm	:	2.0%

8. Stopping Sight Distance

Design Speed (KPH)	*Stopping Sight Distance (m)*
120	250
100	180
80	120
60	80

9. Radius of horizontal curves

Design Speed (KPH)	*Radius (m)*
120	670 (absolute min.) 1000 (desirable)
100	440 (absolute min.) 700 (desirable)
80	260 (absolute min.) 400 (desirable)
60	140 (absolute min.) 200 (desirable)

10. Gradient

Terrain	*Ruling Grade (%)*
Plain	3
Rolling	4
Mountainous	5

In addition to the above, the following standards for interchange elements may be noted:

Interchange design speed	:	Desirable 80 KPH
	:	Minimum 60 KPH
Interchange ramp gradient	:	Desirable 4%
		Maximum 6%
Interchange carriageway width		Desirable 7 m
		Minimum 5.5 m

A typical cross-section of an Expressway as adopted for the Taj Expressway (Ref 36) is given in Fig. 7.89.

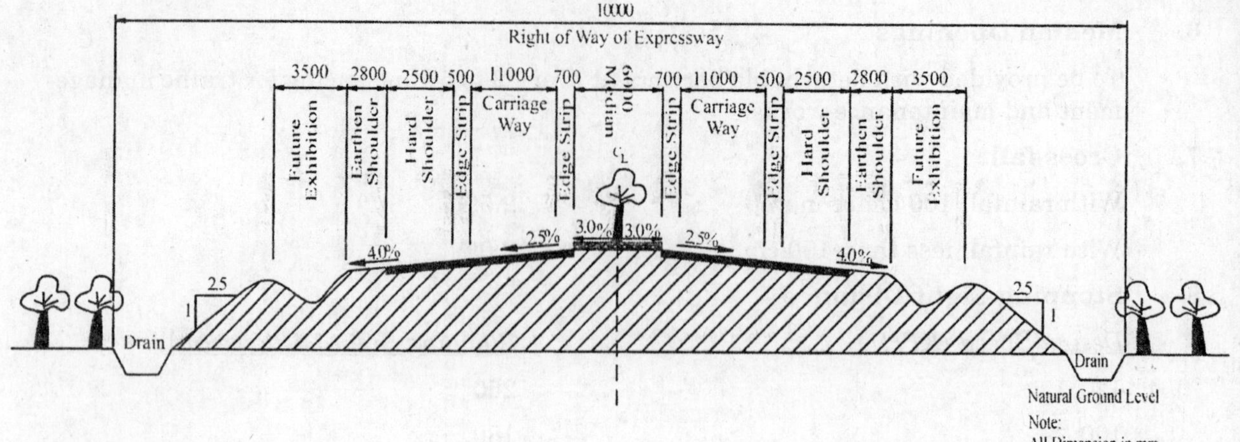

Fig. 7.89: Typical Cross-section of 2 x 3 Lane Expressway (Taj Expressway)

REFERENCES

1. Geometric Design Standards for Rural Highways, IRC:73-1980, Indian Roads Congress, New Delhi, 1980.
2. Geometric Design of Urban Roads, IRC:86-1983, Indian Roads Congress, New Delhi, 1983.
3. Dimensions and Weight of Road Design Vehicles, IRC:3-1984, Indian Roads Congress, New Delhi, 1984.
4. A Policy on Geometric Design of Highways and Streets, AASHTO, 1994.
5. A Policy on Design of Highways and Arterial Streets, American Association of Highway Officials, Washington, 1973.
6. Manual on Uniform Traffic Control Devices, Federal Highway Administration, U.S. Dept. of Transportation, Washington, 1971.
7. Kadiyali, L.R., Rolling resistance and air resistance factors for Indian conditions and vehicles, Highway Research Bulletin No. 18, Indian Roads Congress, New Delhi, 1982.
8. Babkov, V. and Zamakhayev, Highway Engineering, MIR Publishers, Moscow, 1967.
9. Study for Updating Road User Cost Data, Ministry of Surface Transport, Report prepared by L.R. Kadiyali and Associates, New Delhi, 1992.
10. Recommendations about the Alignment Survey and Geometric Design of Hill Roads, IRC:52-1973, Indian Roads Congress, New Delhi, 1973.
11. Guidelines for Design of Horizontal and Transition Curves for Highways and Design Tables, IRC:38-1988, Indian Roads Congress, New Delhi, 1988.
12. Recommended Practice for Sight Distances on Rural Highways, Indian Roads Congress, IRC:66-1976, New Delhi, 1976.
13. Vertical Curves for Highways, IRC Special Publication 23, Indian Roads Congress, New Delhi, 1983.
14. Ribbon Development Along Highways and its Prevention, Special Report, Indian Roads Congress, New Delhi, 1974.
15. Space Standards in Urban Areas, Indian Roads Congress, New Delhi, 1977.

16. Geometrics of Roads, Transport and Communications Monthly Review, May 1966, Indian Roads Congress, New Delhi, 1966.
17. Standard Specifications and Code of Practice for Road Bridges, Section I, IRC:5-1970, Indian Roads Congress, New Delhi, 1970.
18. Layout of Roads in Rural Areas, Department of Environment, U.K., H.M.S.O., London, 1968.
19. Recommended Practice for the Construction of Earth Embankment for Road Works, IRC:36-1970, Indian Roads Congress, New Delhi, 1970.
20. Lateral and Vertical Clearances at Underpasses for Vehicular Traffic, IRC:54-1974, Indian Roads Congress, New Delhi, 1974.
21. Guidelines for Control of Access on Highways, IRC:62-1976, Indian Roads Congress, New Delhi, 1976.
22. Roads in Urban Areas, Department of Environment, (U.K.); H.M.S.O., London, 1966.
23. Guidelines for the Design of At-grade Intersections in Rural and Urban Areas, IRC:SP:41, Indian Roads Congress, New Delhi, 1994.
24. Recommended Practice for Traffic Rotaries, IRC:65-1976, Indian Roads Congress, New Delhi, 1976.
25. Wardrop, J.G., The Traffic Capacity of Weaving Sections of Roundabouts, Proceedings of the First International Conference on Operations Research, English University Press, London, 1957.
26. Blackmore, F.C., Capacity of Single Level Intersections, Ministry of Transport, R.R.L. Report LR 356.
27. Junction Design (Interim), Department of Environment, (U.K.) Tech. Memo No. H 7/71, June, 1971, London.
28. Guidelines for the design of interchanges in urban areas, IRC:92-1985, Indian Roads Congress, New Delhi, 1985.
29. Guidelines for the Regulation and Control of Mixed Traffic in Urban Areas, IRC:70-1977, Indian Roads Congress, New Delhi, 1977.
30. Type Designs for Pick-up Bus Stops on Rural (Non-urban) Highways, IRC:80-1981, Indian Roads Congress, New Delhi, 1981.
31. Recommended Practice for Location and Layout of Roadside Motor Fuel Filling and Motor Fuel Filling-cum-Service Stations, IRC:12-1983, New Delhi, 1983.
32. Guidelines for Pedestrian Facilities, IRC:103-1988, Indian Roads Congress, New Delhi, 1989.
33. A Manual for Planning, Pedestrian Facilities, U.S. Dept. of Transportation, Washington, 1974.
34. Recommended Practice for the Design and Layout of Cycle Tracks, IRC:11-2015, Indian Roads Congress, New Delhi, 2015.
35. Guide for Department of New Bicycle Facilities, American Association of State Highway and Transportation Officials, Washington, 1981.
36. Geometric Design Standards for Expressways, Draft under consideration by the H-1 Committee of the Indian Roads Congress, New Delhi, 2008.
37. Taj Expressway, Detailed Project Report, Jaiprakash Industries Ltd, New Delhi, 2006.

8

Chapter

Surveys, Investigations and Project Preparation

8.1. Introduction

Accurate surveys and investigations are needed for preparing a highway project which incorporates economy, safety, environmental sustainability and user comfort. Surveys and investigations of the following types are needed.

1. Traffic Surveys
2. Topographic Surveys
3. Drainage Studies
4. Soil Surveys
5. Material Survey
6. Pavement Evaluation Studies

8.2. Traffic Surveys

8.2.1. Traffic Counts

Information about traffic forms the basis for the design of the pavement, fixing the number of traffic lanes and economic appraisal of the project. Classified volume counts for a seven-day period should be conducted. Applying a seasonal correction factor, the Average Annual Daily Traffic (AADT) can be computed. Since traffic on Indian roads is of a mixed nature, it becomes necessary to convert the flow into equivalent Passenger Car Units (PCUs) by using certain equivalency factors. The PCU equivalents as per Indian practice are given in Table 8.1.

Table 8.1 : PCU Equivalency Factors (Ref 1)

S.No.	Vehicle Type	PCU
1.	Passenger car, Pick-up van or autorickshaw	1.00
2.	Motor Cycle, Scooter or Cycle	0.50
3.	Agricultural Tractor, Light Commercial Vehicle	1.50
4.	Truck or Bus	3.00

(Contd...)

S.No.	Vehicle Type	PCU
5.	Tractor-trailer, Agricultural Tractor-trailor	4.50
6.	Cycle Rickshaw	2.00
7.	Hand Cart	3.00
8.	Horse-drawn vehicle	4.00
9.	Bullock Cart*	6.00–8.00
* For smaller bullock-carts, a value of 6 will be applicable		

For geometric design, the commonly used unit for traffic is the 30th HV, defined as the hourly traffic that is exceeded by 29 hourly volumes in a year. For Indian conditions, 30th HV may be assumed to be 8-10 per cent of the AADT. For the design of carriageway width, the traffic in both directions is taken for a 2-lane road. For roads with more than 2 lanes, the design is done on the approximate assumption that the traffic in the predominant direction can be taken to be 67 per cent of the total traffic. The field data sheet prescribed for classified volume count in India is given in Table 8.2 (Ref 4).

Table 8.2: Field Data Sheet for Traffic Census (Ref 4)

Date and day of the week .... Road Classification ....

Direction of traffic Kilometrage/mileage

From .... To .... Route number (if any) ......

District .... State ....

Type of Vehicles / Hour of Count	Cars, Jeeps, Vans, Three-wheelers	Buses	Trucks	Motor-cycles and scooters	Animal drawn vehicle**	Cycles	Others (specify)	Remarks, Including weather conditions
(1)	(2)	(3)	(4)	(5)	(6)	(7)	(8)	(9)
From– Hours* To– Hours								
Hourly Total								
From– Hours* To– Hours								
Hourly Total								
From– Hours* To– Hours								
Hourly Total								
From– Hours* To– Hours								
Hourly Total								

Notes:

1. Record traffic-volume in columns 2 to 8 by making tallies in the form of vertical strokes for first four vehicles and drawing an oblique stroke for every 5th as shown in within brackets (1111).
2. Some roads carry appreciable volume of other traffic like cycle rickshaws. Record the volume of such vehicles in column 8 after specifying the vehicle type.

* The hour of counts should be entered before the start of the enumeration. P.M. hours should be recorded after adding 12 to the actual hour, for example, 2 P.M. should be recorded as 14.00 hours.

** If felt necessary by Highway Authority, this column could be subdivided into two for recording the volume of 'pneumatic—tyred' vehicle separately.

Name and signature of enumerator with date

....

Name and signature of supervisor with date

....

8.2.2 Prediction of Traffic for the Design Year

Highway facility is designed for a design period of 15-25 years, and a design period of 20 years is widely used as the basis. Future traffic estimates should consider:

1. Normal traffic growth
   - Diverted Traffic
   - Induced Traffic
2. Generated Traffic
3. Development Traffic

Rate of normal growth of traffic can be estimated by analysing the historical trend and fitting a best fit line (Ref 2). Thus:

$$T_n = T_0 (1+r)^n \tag{8.1}$$

where T_n = Traffic in the nth year

n = number of years

r = rate of growth

Taking logs,

$$Log_e T_n = Log_e T_0 + n(\log 1 + r) \tag{8.2}$$

If the past data is available on traffic for a number of years and the corresponding data on some economic indicators such as GNP, population etc. is also available, an econometric model of the following type can be fitted.

$$Log_e T = A_0 + A_1 \log_e GNP \tag{8.3}$$

where T = Traffic Volume

GNP = Gross National Product

A_0 = Regression Constant

A_1 = Regression Coefficient

The value of A_1 is known as the Elasticity Coefficient, and is the factor by which the GNP growth rate is to be multiplied to arrive at the growth rate of traffic. The following values of Elasticity Coefficient are appropriate for Indian conditions.

Vehicle Type	Elasticity Coefficient
Car	1.8
Two-wheeler	2.1
Bus	1.5
Truck	1.3
Slow-moving vehicle	1.5

Thus if the country's GNP is expected to grow at 8 per cent per annum, the car population will grow at 8 x 1.8 = 14.4 per cent per annum.

The estimation of diverted, induced and development traffic requires a detailed study.

8.2.3. Speed and Delay Surveys

To understand how the road is performing under traffic conditions, it is necessary to conduct a speed and delay survey. The most popular method is to adopt the "Moving Observer" method, under which a car is driven along the road, noting its travel time, number of vehicles met with in the opposite direction and the number of vehicles overtaken (Ref 3). The survey yields both the journey speed and volume.

8.2.4. Axle Load Survey

Trucks in India are generally overloaded and cause severe damage to the pavement. In order to assess the damaging effect of such trucks, it is necessary to stop the trucks and move them on to a portable weigh bridge. Sampling has to be done since all trucks cannot be managed. The Vehicle Damage Factor is calculated by adopting the Fourth Power Rule as under:

$$F = \left(\frac{L}{L_s}\right)^4 \qquad (8.4)$$

where F = Equivalence factor of load L

L = Load L in kN

L_s = Standard axle load (80 kN)

Thus, an axle load of 120 kN causes a damage equivalent to:

$$F = \left(\frac{120}{80}\right)^4 = 5.06 \text{ times the standard axle load of 80 kN}$$

Axle Load surveys can also be done by using a modern technology, Weigh-in-Motion (WIM) where the trucks do not have to be stopped.

8.2.5. Origin-Destination Survey

For assessing the bypassable traffic at towns or for planning new facilities like Expressways, bypasses, new ring roads etc.

8.3. Soil Surveys and Properties

8.3.1. Need for soil surveys

Road pavements are built on embankments which are constructed using soil borrowed from nearby locations. The properties of soil determine their suitability for use in the embankment or for use in soil-stabilisation. In hill roads the soil strata influences the stability of the cuts and their vulnerability of slips and landslides. In high altitude areas, frost action which can cause failure of pavements depends on the soil properties. Soil surveys thus fulfil the following requirements:

(i) determination of the physical properties of soil for use in embankment and location of suitable borrow areas

(ii) design of slopes of embankments and cuts

(iii) design of high embankments to safeguard against slope failure and settlement

(iv) determination of suitability for stabilisation with other additives

(v) determination of strength of the subgrade which controls the design of pavement thickness and composition

(vi) determination of surface and subsurface drainage requirements

(vii) classification of earthwork into ordinary soil, hard soil, soft rock, hard rock etc. to facilitate cost estimation.

8.3.2. Soil sampling for low embankment sections

Soil samples at intervals of 200 m for new roads and 500 m for road widening projects shall be collected by digging trial pits 0.5 to 1.0 m square and of a depth 1.25–1.5 m more than the depth of borrow pits proposed.

8.3.3. Soil sampling for high embankments

Embankments of height greater than 6 m require soil sampling at 50–100 m intervals. The depth should be 15 m or 1½ times the height of the embankment. Undisturbed soil samples are collected by drilling holes.

8.3.4. Tests to be carried out

The following tests (Ref 5) are to be carried out on soil to be obtained from borrow areas:

1. Classifications and Identification of Soils for General Engineering Purposes (IS:1498-1970)
2. Gradation Test (IS:2720-Part IV)
3. Liquid Limit and Plastic Limit (IS:2720-Part V)
4. Density and Optimum Moisture Content (IS:2720-Part VII and VIII, Part VII being Standard Proctor Test, and Part VIII being Modified AASHTO Compaction)
5. Deleterious Constituents (only in salt infested areas, IS:2720-Part XXVII)
6. Shrinkage and Swelling Properties (IS:2720, Part XXXX)

8.3.5 Soil Classification System as followed in India

Indian soil classification system considers two types of soils:

(a) Coarse Grained

- Gravels (G), with more than half coarser than 4.75 mm
- Sand (S), with size 0.075–4.75 mm

(b) Fine grained (with size less than 0.075 mm)

- Silts (M)
- Clays (C)
- Organic matter (O)

The soils are identified by following a system of two alphabets. The first alphabet G, S, M, C and O refer to the classes indicated above. The second alphabet denotes the basic engineering property of that soil. The following second alphabets describe the property:

W: Well graded
P: Poorly graded
L: Low Plasticity
I: Intermediate Plasticity
H: High Plasticity

Thus, GW stands for well-graded gravel, ML stands for silt with low plasticity and CH stands for clay with high plasticity.

The following definitions of some of the properties (known as Atterberg Limits) are needed to understand the classification of soils:

Liquid Limit (LL) is the minimum water content at which the soil can flow under its own weight.

Plastic Limit (PL) is the minimum water content at which the soil can be rolled into a thread 3 mm in dia without breaking.

Shrinkage Limit (SL) is the maximum water content at which further loss of moisture does not cause a decrease in the volume of the soil.

Plasticity Index (PI) is the range in water content at which the soil exhibits plastic behaviour.

Thus, PI = LL–PL

Fig. 8.1 is a Plasticity Chart which helps in classifying the soil.

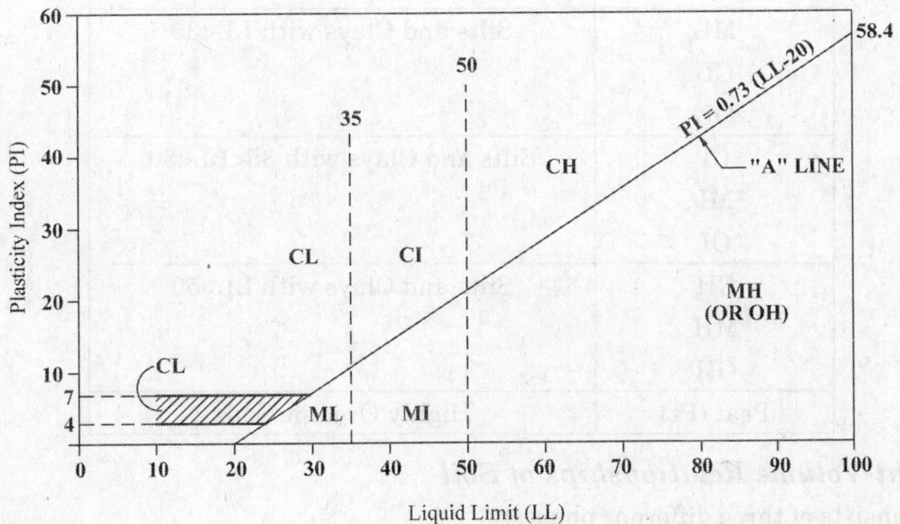

Fig. 8.1: Plasticity Chart

Tables 8.3, 8.4 and 8.5 give the classification of gravels, sands and fine grained soils.

Table 8.3 : Classification of Gravels

Symbol	**Percent Fines Passing 0.075 Sieve**	**Grading**	**Plasticity**
GW	0–5	Cu>4 I<Cc<3	–
GP	0–5	Above conditions not satisfied	–
GM	>12		Below A line (Fig. 8.1) or PI less than 4
GC	>12		Above A line (Fig. 8.1) and PI>7

In the above Table Cu is the Uniformity Coefficient defined as D_{60}/D_{10} where D_{60} is the diameter at which 60% of soil (by weight) is finer and D_{10} is the diameter at which 10% of soil is finer.

Cc = $D_{30}{}^{2}$, / D_{60} x D_{10}, where D_{30} is the diameter at which 30% soil by weight is finer and D_{60} and D_{10} have already been described above (8.5)

Table 8.4 : Classification of Sands

Symbol	Percent Fines passing 0.075 sieve	Grading	Plasticity
SW	0–5	Cu>60 I<Cc<3	–
SP	0–5	Above conditions not satisfied	–
SM	>12	–	Below "A" line in Fig. 8.1 or PI<4
SC	>12	–	Above "A" line and PI>7

Table 8.5: Grading of Fine-Grained Soils

Symbol	**Soil Type**
ML CL OL	Silts and Clays with LL<35
CI MI OI	Silts and Clays with 35<LL<50
CH MH OH	Silts and Clays with LL>50
Peat (Pt)	Highly Organic Soils

8.3.6. Weight-Volume Relationships of Soil

A soil mass consists of three different phases:

(*a*) Solids (mineral particles)

(*b*) Air (gas)

(*c*) Liquid (water)

Fig. 8.2 shows the three phase diagram.

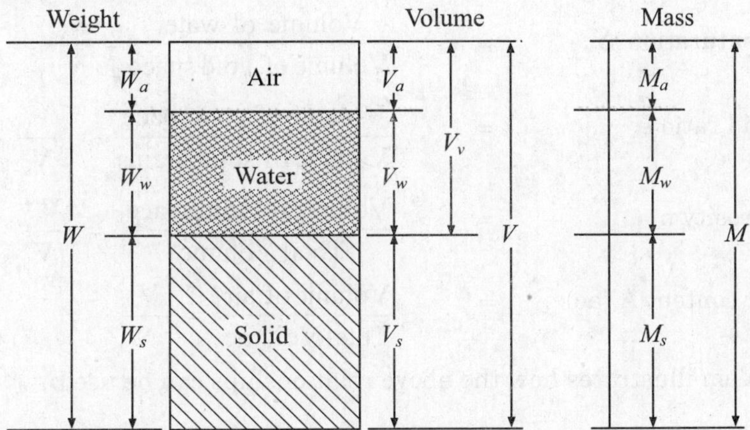

Fig. 8.2: Three Phase Diagram

M	=	Total Mass	M_a	=	Mass of air (=0)
M_w	=	Mass of water	M_s	=	Mass of solids
G_s	=	Specific Gravity of solid grains	W	=	Total Weight
W_a	=	Weight of air (=0)	W_w	=	Weight of water
W_s	=	Weight of solids	V	=	Total volume
V_a	=	Volume of air	V_v	=	Volume of void space
V_w	=	Volume of water	V_s	=	Volume of solids
g	=	Gravitational acceleration = 9.81 m/sec²			

The following definitions need to be noted:

$$\text{Bulk density } \rho = \frac{\text{Total Mass}}{\text{Total Volume}} = \frac{M}{V} \quad (8.6)$$

$$\text{Bulk unit weight } \gamma = \frac{\text{Total weight}}{\text{Total Volume}} = \frac{W}{V} \quad (8.7)$$

$$\text{Thus } \gamma = \rho g \quad (8.8)$$

$$\text{Density of solids } \rho_s = \frac{\text{Mass of solids}}{\text{Volume of solids}} = \frac{M_s}{V_s} \quad (8.9)$$

$$\text{Unit weight of solid soil grains} = \frac{W_s}{V_s} = \frac{M_s . g}{V_s} = \rho_s . g \quad (8.10)$$

$$\text{Water content in soil mass (w)} = \frac{\text{Weight of water}}{\text{Weight of solid grains}}$$

Dry density σ_d = $\frac{\text{Mass of solids}}{\text{Total Volume}} = \frac{M_s}{V}$ (8.11)

Dry unit weight σ_d = $\sigma / (1 + w)$ (8.12)

Degree of saturation S_r = $\frac{\text{Volume of water}}{\text{Volume of void space}} = \frac{V_w}{V_v}$ (8.13)

Void ratio e = $\frac{\text{Volume of void space}}{\text{Volume of solid grains}} = \frac{V_v}{V_s}$ (8.14)

Porosity n = $\frac{\text{Volume of void space}}{\text{Total Volume}} = \frac{V_v}{V}$ (8.15)

Air content A (ac) = $\frac{\text{Volume of air}}{\text{Total Volume}} = \frac{V_a}{V}$ (8.16)

The following problem illustrates how the above relationships can be used.

Problem 8.1

A cylindrical specimen of soil sample has a volume of 85 cm³ and weighs 162 gms in its natural condition. After drying completely in an oven, the specimen weighs 128 gms. The specific gravity of the solid particles is 2.70. Calculate the water content, void ratio and the degree of saturation.

W = 162 gms

Ws = 128 gms

Weight of water in the specimen = 162–128 = 34 gms

Volume of water = 34 cc

Water content = $\frac{34}{128}$ = 0.265 or 26.5 per cent

Volume of solids = $\frac{\text{Weight of solids}}{\text{Specific gravity of solids}} = \frac{128}{2.70}$ = 47.4 cm³

Volume of voids = 85–47.4 = 37.6 cm³

Void ratio = $\frac{\text{Volume of voids}}{\text{Volume of solids}} = \frac{37.6}{47.4}$ = 0.79

Degree of saturation = $\frac{\text{Volume of water}}{\text{Volume of void space}} = \frac{34}{37.6}$ = 0.90 or 90 per cent

8.3.7. Compaction of Soil

Soil obtained from the borrow pits is in a loose state and needs to be compacted. The density to which a soil can be compacted depends on the soil type, moisture content and the compaction effort. Sandy and gravelly soils can be compacted to a high dry density (in the range of 1.7 to 2.3 gms/ cc), whereas clays can only be compacted to a low density (in the range of 1.4 to 1.8 gms/cc) and silty soils can be compacted to a density between that of sand and gravel and clayey soil.

The moisture content at which the soil is compacted determines the density that can be achieved. For a given type of soil, and a compactive effort, there is an optimum moisture content (known as OMC) at which a maximum value of density (MDD) can be achieved. The general shape of the compaction curve is given in 8.3.

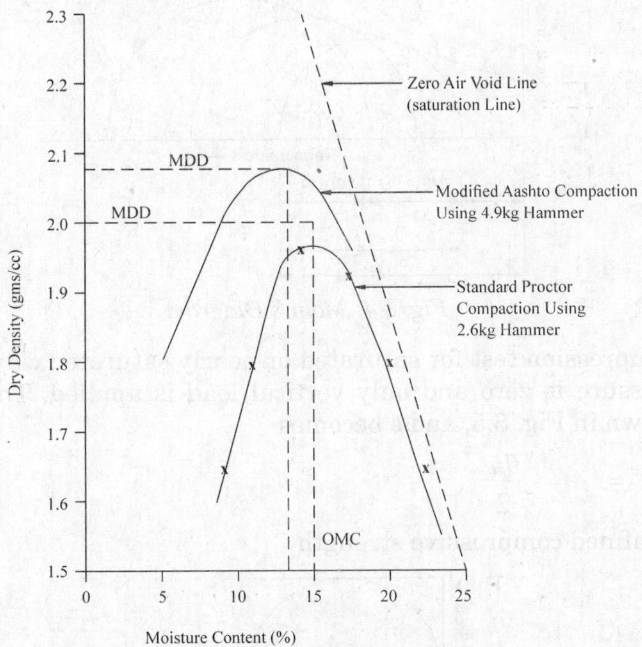

Fig. 8.3: Compaction Curve

Fig. 8.3 shows two curves, one for the Standard Proctor Compaction using 2.6 Kg hammer (IS:2720-Part VII) falling through a height of 310 mm and the other for Modified AASHTO compaction using 4.9 Kg hammer falling through a height of 450 mm (IS:2720-Part VIII) (Ref 5). It is seen that as the compactive effort increases, the MDD increases and the OMC comes down. The Standard Proctor compaction (light compaction) is used for low volume roads (Ref 6), whereas for high volume roads Modified AASHTO compaction (heavy compaction) is used (Ref 7).

8.3.8. Determination of Shear Strength Properties

The shear resistance of soils is determined by the Coulomb's law, which is expressed by the following relationship:

$$s = c + \sigma \tan \phi \quad (8.17)$$

where s = shearing resistance

c = cohesion

σ = normal stress

$\varnothing$ = angle of internal friction

$\tan \varnothing$ = coefficient of internal friction

The shearing resistance is found by:

(i) Direct Shear Test, using a shear box, where a normal load is applied and a horizontal force is applied to the upper box, and finding the horizontal force that causes a shear failure.

(ii) Triaxial Test, in which a cylindrical specimen is subjected to all-round pressure and an axial load is applied till the specimen fails in shear. The analysis is done by plotting the Mohr's diagram, Fig. 8.4.

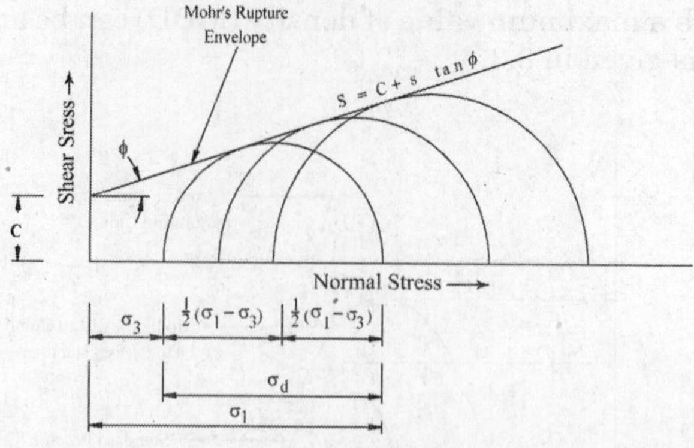

Fig. 8.4: Mohr's Diagram

(iii) Unconfined compression test for saturated or nearly saturated clays is conducted in which the lateral pressure is zero and only vertical load is applied. The Mohr's diagram then becomes as shown in Fig. 8.5, and c becomes

$$C = \frac{q_u}{2} \tag{8.18}$$

where q_u is the unconfined compressive strength

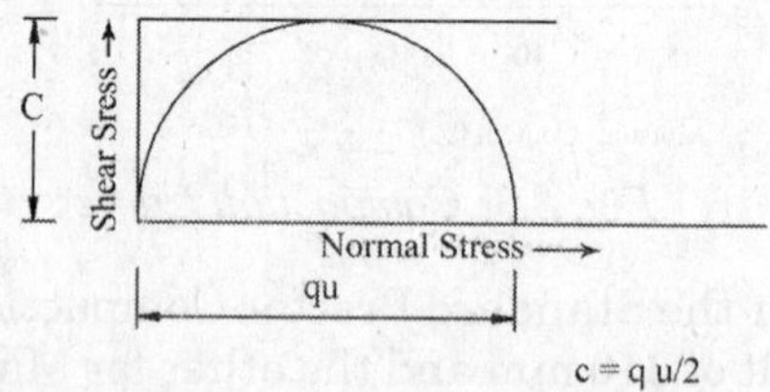

Fig. 8.5: Mohr's circle for φ = 0 in an unconfined compression test

This test is the standard test for measuring the strength of soil-cement and soil-lime stabilised materials.

8.3.9. Plate Bearing Test

For designing rigid pavements, the Modulus of subgrade reaction (K) is needed. This is obtained by a plate load test, using a 75 cm dia plate. Also smaller sizes of plates can be used and proper correction is applied to get results. The loading arrangement is shown in Fig. 8.6.

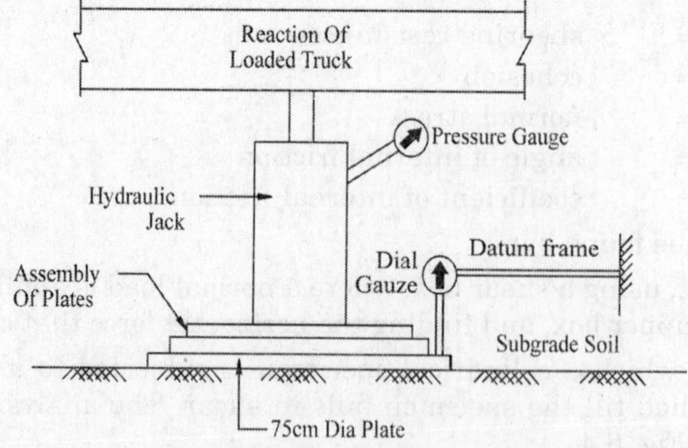

Fig. 8.6: Plate Bearing Test for Coefficient of Subgrade Reaction (K)

The load settlement curve is generally a curve, but a straight line passing through the origin and Formula for K = P/Δ where P = Load Corresponding to 1.25 mm settlement is drawn as shown in Fig. 8.7. The slope of this line gives the Modulus of Subgrade Reaction k. If a 75 cm dia plate is not available, a smaller plate of 30 cm dia is used and the following conversion is adopted (Ref 8).

$$K_{750} = K_{\varnothing}(1.21\varnothing + 0.078)$$

where K_{750} = Modulus of subgrade reaction using 750 mm dia plate

$K_{\varnothing}$ = Modulus of subgrade reaction using plate of dia ∅

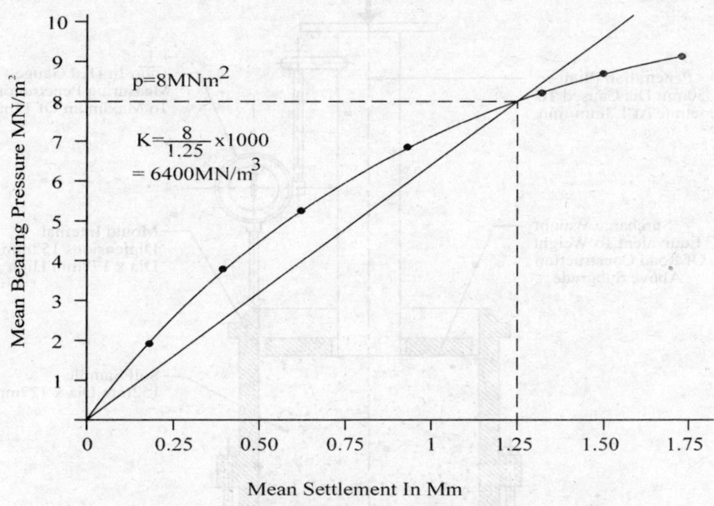

Fig. 8.7: Load settlement curve for plate bearing test

8.3.10. CBR Test

The California Bearing Ratio (CBR) Test is the most popular test for determining the strength of the subgrade on which the flexible pavement rests and is covered by IS:2720-Part XVI) (Ref 5). It consists of a mould of 150 mm dia and 175 mm height in which the soil specimen is prepared to have the desired density d (gm/cc). The weight of the wet soil is calculated as under:

$$\frac{100+m}{100} \text{ x Volume of mould x d}$$

The volume of the standard mould is 2244 c.c., and thus, the weight of the wet soil is:

$$\frac{100+m}{100} \text{ x } 2244 \text{ x d} \qquad (8.19)$$

The density required for roads is given in the specification (Ref 6 and 7). Some specifications require that the specimen be soaked for 4 days and then the test done. The test set up is given in Fig. 8.8. Standard load should be in Kg as we get test load in Kg. CBR 2.5 and 5.0.

CBR Value = Test load/Standard Load

The load is applied through a plunger 50 mm dia and the plunger is made to penetrate at a rate of 1.3 mm/min. The loads at 2.5 mm and 5 mm penetration are recorded. The standard crushed stone penetration curve which gives a penetration of 2.5 mm at 6.895 MN/m^2 load and 5.0 mm at 10.343 MN/m^2 load is shown in Fig. 8.9. If actual loading curve is conclave upwards, a correction is required. The CBR value is expressed as a percentage of the actual load causing penetration of 2.5 mm and 5 mm to the standard crushed stone load values (6.895 MN/m^2 and 10.343 MN/m^2 as mentioned earlier). If the value at 2.5 mm penetration is greater than at 5 mm penetration, the former is adopted. If not, the test is repeated and if the new value at 5 mm is still greater, that value is adopted.

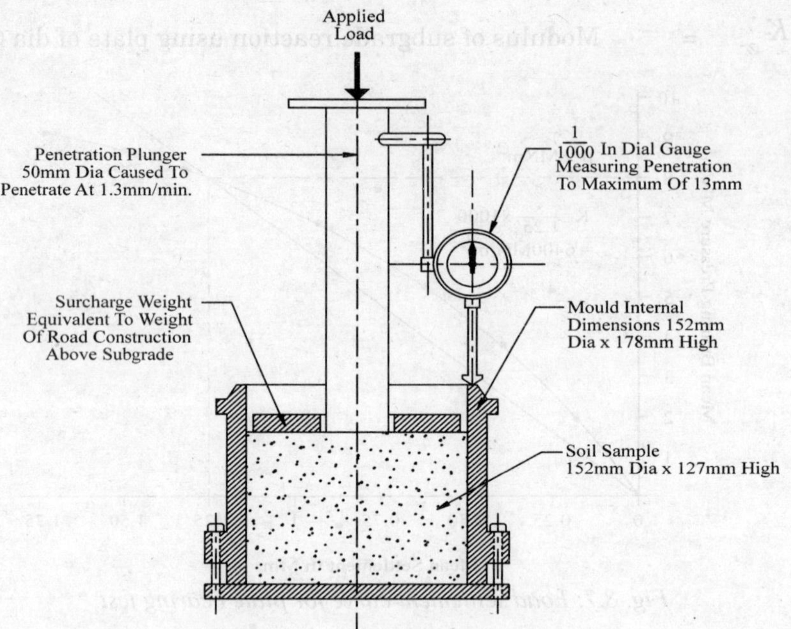

Fig. 8.8: CBR Test Apparatus

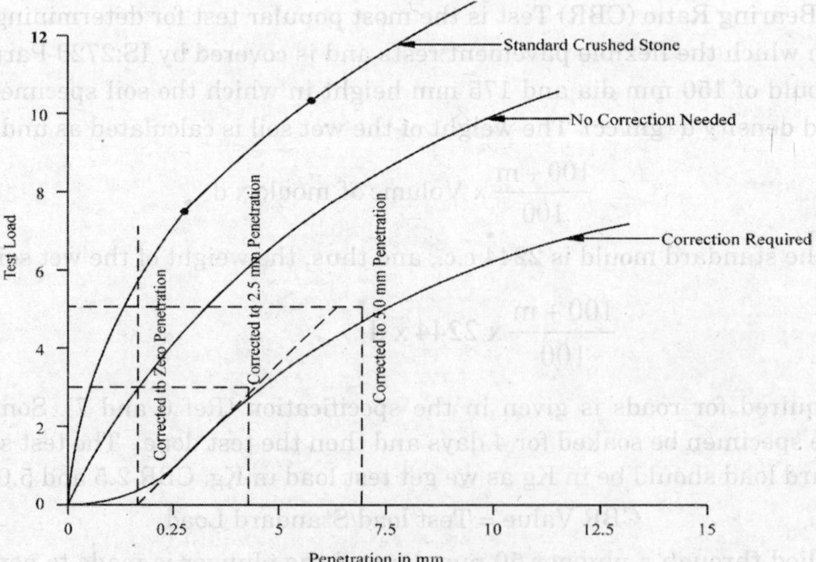

Fig. 8.9: Load-penetration Curve

The typical values of CBR of Indian soils are given in Table 8.6.

Table 8.6 : Typical CBR Values of Soils Met With in India

Soil Type	CBR
Black Cotton Soil and Heavy Clays	1 to 2
Alluvial Silt	2 to 5
Silty Clay	3 to 5
Sandy Clay	4 to 6
Dune Sand	6 to 8
Well-graded Sand	8 to 20
Gravel (Murram)	8 to 20

8.4. Material Survey

8.4.1. Apart from soil, the other material widely used in road construction is stone. Stone aggregates are used in drainage layers, sub-bases, non-bituminous bases, bituminous layers and cement concrete. Road-making aggregates are obtained by crushing rock. The types of rock met with in India are:

Igneous Rocks

- Granite
- Basalt

Metamorphic Rock

- Quartzite
- Limestone
- Sandstone
- Kankar (impure form of limestone)

Where stone materials are costly, as in the Gangetic Plains of North India, broken brick ballast from over burnt bricks are used. Slag from steel plants are also used as aggregates.

Aggregates are obtained by crushing stone obtained by blasting in quarries.

8.4.2. Tests on stone aggregates

The following are the tests normally carried out on samples of stone aggregates:

1. Particle size distribution
2. Deleterious Materials content
3. Water absorption and Bulk Specific Gravity
4. Flakiness and Elongation
5. Aggregate Impact Test
6. Los Angeles Abrasion Test
7. Aggregate Crushing Test
8. Aggregate Polishing Test
9. Soundness Test
10. Stripping Test

8.4.3. Particle Size Distribution

To determine the gradation of stone aggregates, sieving is done. The standard set of sieves is prescribed in IS:460-1985 (Part 3) (Ref 9). The sieving is done with dry aggregates if they do not contain dust or clay. If they do contain dust or clay, a wet sieve analysis is carried out.

8.4.4. Deleterious Materials Content

Deleterious materials in aggregates include clay lumps, organic matter and soft particles which are undesirable for road works. Normally, most specifications require that not more than 5 per cent of the sample should be finer than 0.075 mm. A test usually done for determining the proportion of silt/ clay/ organic matter is the Sand Equivalent Test, IS:2720 (Part XXXVII) (Ref 5).

8.4.5. Water Absorption and Bulk Specific Gravity Tests

Specific gravity of aggregates is a good indicator of its strength. The higher the specific gravity, the denser is the rock and hence stronger is the aggregate. Pores in the aggregates absorb water and when water enters the strength is adversely affected. The test is conducted by immersing the sample (2-3 Kg) in distilled water enclosing the sample in a wire mesh container. The weight of the container with the aggregates immersed in water is noted (W_1). The material is then surface dried and weighed in air, giving saturated weight (W_2). The material is then oven dried at a temperature 100–110⁰C and its dry weight noted (W_3).

$$\text{The percentage of water absorption} = 100 \times \frac{(W_2 - W_3)}{W_3} \quad (8.20)$$

$$\text{Bulk Specific Gravity of aggregates} = \frac{\text{Dry Weight of Aggregates}}{\text{Volume of Aggregates}}$$

$$= \frac{W_3}{\text{Weight of Water Displaced} \div \text{Specific Gravity of Water}}$$

$$= \frac{W_3}{W_2 - W_1}, \text{ since the specific gravity of water} = 1 \quad (8.21)$$

Water absorption above 1.0 per cent is unsatisfactory for use in wearing courses, and values above 2.0 per cent are unsatisfactory for use in base courses.

8.4.6. Flakiness and Elongation

Aggregates which are flaky and elongated are not conducive for interlocking. Flaky and elongated aggregates are difficult to compact and affect the workability and stability of mixes. Flakiness and elongation percentages are determined by testing the individual particles through a standard Flakiness gauge and Elongation gauge. The test is covered by IS:2386 (Part I) (Ref 10). The latest MORTH specifications (Ref 7) prescribe a combined Flakiness and Elongation Index. In this test procedure, the Flakiness Index is first found out and the flaky particles are removed and the balance quantity of aggregates are subjected to Elongation Test. The Elongation Index is then calculated. The sum of the two values (Flakiness Index and Elongation Index) is the Combined Flakiness and Elongation Index. Normally, a combined Flakiness and Elongation Index value of 30 is specified for important works (Ref 7).

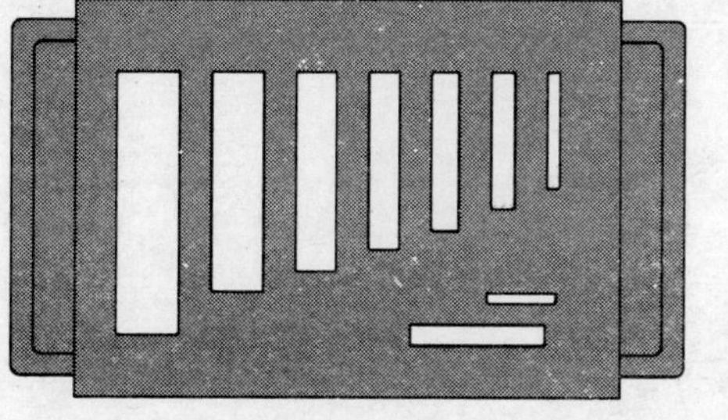

Flakiness Gauze

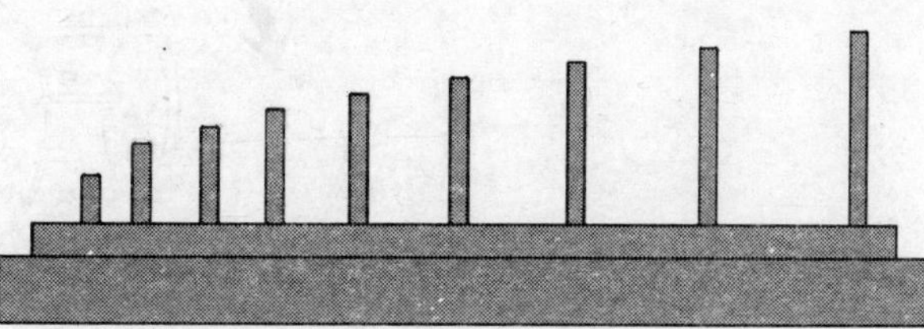

Elongation Gauze

8.4.7. *Aggregate Impact Test*

In road pavements, stone aggregates in the various layers are subjected to impact. Hence, the aggregates should have adequate resistance to impact and should not get crushed. The test has been standardised (Ref 10), and the equipment consists of an arrangement of making a weight of 13.5–14.0 Kg fall through a height of 380 mm over a cylindrical cup of dia 102 mm containing aggregates passing through 12.5 mm sieve and retained on 10 mm sieve. 15 blows through the hammer are given and the crushed aggregates are sieved through 2.35 mm sieve. The weight of material passing through this sieve, expressed as a percentage of the total weight of the sample gives the Aggregate Impact Value (AIV). For base courses, the AIV should not exceed 40, and for surface courses it should not exceed 30. The testing equipment is shown in Fig. 8.10.

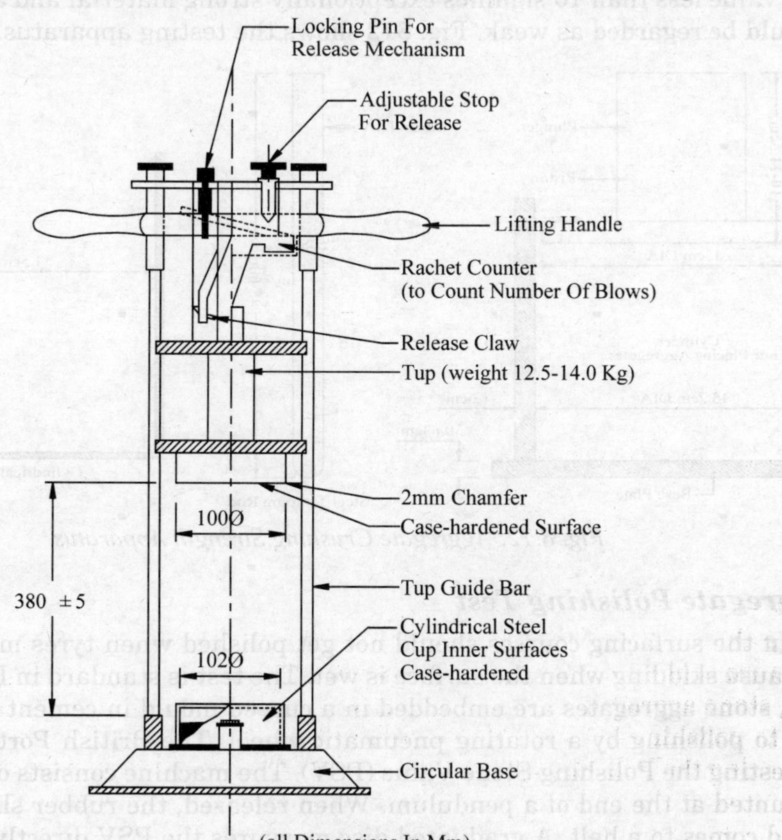

Fig. 8.10: An Aggregate Impact Test Equipment

Fig. 8.11: Los Angeles Abrasion Test Apparatus

8.4.9. Aggregate Crushing Test

In this test, a compressive load of 40 Tonnes is applied for 10 minutes on a coarse aggregate (passing through 12.5 mm and retained on 10 mm sieves) placed in a cylindrical mould 11.5 cm dia and 18 cm high. The crushed material is then sieved through a 2.36 mm sieve, and the weight of fines expressed as a percentage of total weight is known as the Aggregate Crushing Value. The MORTH Specifications (Ref 7) do not specify any value for the various pavement sources. Generally, a value less than 10 signifies exceptionally strong material and aggregates with values above 35 would be regarded as weak. Fig. 812 shows the testing apparatus.

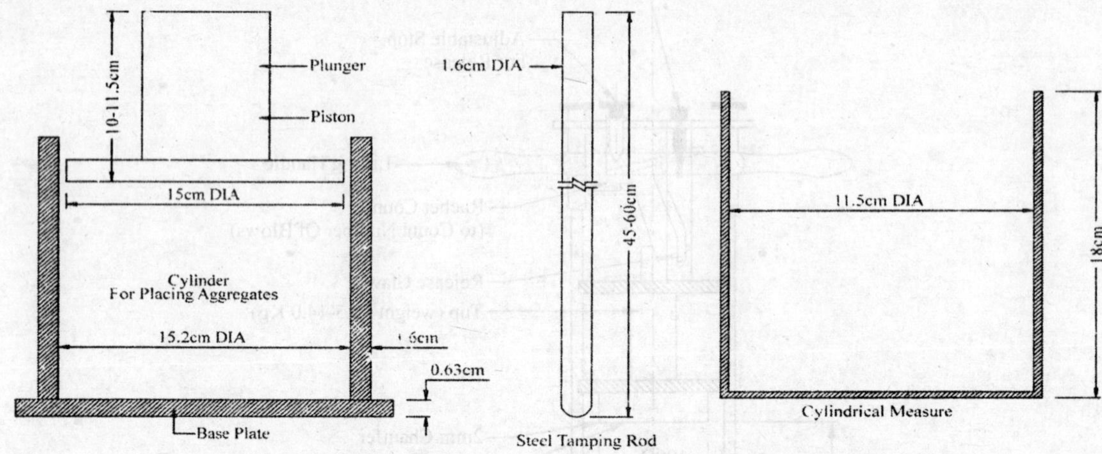

Fig. 8.12: Aggregate Crushing Strength Apparatus

8.4.10. Aggregate Polishing Test

Aggregates in the surfacing courses should not get polished when tyres move on them. Polished aggregates cause skidding when the surface is wet. The test is standard in BS:812-114 (Ref 11). In this method, stone aggregates are embedded in a curved mould in cement mortar and the mould is subjected to polishing by a rotating pneumatic wheel. The British Portable Tester is a quick method for testing the Polishing Stone Value (PSV). The machine consists of a rubber sliding shoe which is mounted at the end of a pendulum. When released, the rubber sliding brushes past the specimen and comes to a halt. A graduated disc measures the PSV directly. A minimum value of 55 is specified for surface courses.

8.4.10. Aggregate Polishing Test

Aggregates in the surfacing courses should not get polished when tyres move on them. Polished aggregates cause skidding when the surface is wet. The test is standard in BS:812-114 (Ref 11). In this method, stone aggregates are embedded in a curved mould in cement mortar and the mould is subjected to polishing by a rotating pneumatic wheel. The British Portable Tester is a quick method for testing the Polishing Stone Value (PSV). The machine consists of a rubber sliding shoe which is mounted at the end of a pendulum. When released, the rubber sliding brushes past the specimen and comes to a halt. A graduated disc measures the PSV directly. A minimum value of 55 is specified for surface courses.

8.4.11. Soundness Test

Porous aggregates which absorb water to an extent of more than 2 per cent do not have good durability. They are then subjected to soundness test. Under this test, aggregates of a specified size are subjected to alternate cycles of wetting and drying, in a saturated solution of either sodium sulphate or magnesium sulphate. After 5 cycles, loss in weight due to fragmentation is determined by sieving all undersize particles. Generally the loss shall not be more than 12 per cent if sodium sulphate solution is used or 18 per cent if magnesium sulphate solution is used (Ref 7).

8.4.12. Stripping Test

Bitumen coats the aggregates and imparts binding property. In the presence of water, the bond between the bitumen film and the stone aggregates gets affected, depending upon the type of aggregate. This phenomenon is known as stripping. The test for determining stripping prescribed by the Indian Standard (Ref 12) is to immerse the freshly coated specimen in water at a controlled temperature for 24 hours. By visual inspection, the percentage of area uncoated is assessed. The maximum retained coating specified by IRC for Dense Bituminous Macadam and Bituminous Concrete is 95 per cent (Ref 7). Fig. 8.13 shows the testing set-up.

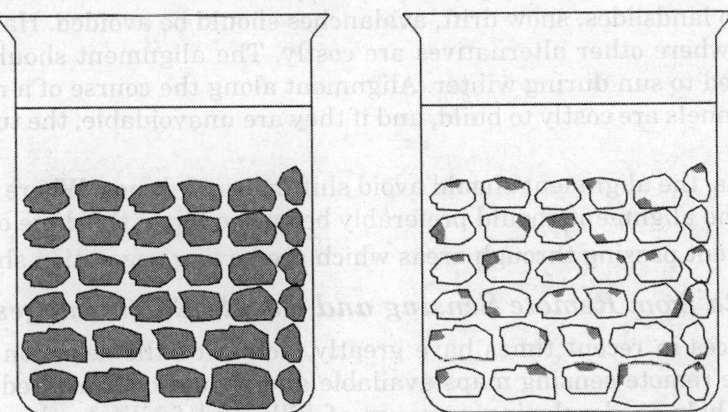

Fig. 8.13: Stripping Test

8.5. Topographic Survey

8.5.1. Guidelines for Selecting the Alignment

The first stage in topographic survey is the selection of the alignment and locating the route. The following guidelines should be followed in selecting the alignment (Ref 13, 14):

1. A direct alignment results in the least length of road, and hence the lowest initial cost, maintenance cost and road user cost.
2. The alignment should serve the maximum number of villages and towns, though it may involve construction of a few bypasses.
3. The alignment should keep the curvature and gradient to a low value consistent with the terrain.
4. The use of agricultural land should be minimal. If a road already exists and adequate right-of-way is available, the improvements suggested should be as far as possible within the available land.
5. The alignment should have minimal impact on the environment including fauna, flora and drainage system.
6. The alignment should be located so that suitable borrow areas are available close by and sources of gravel and stone materials are also within short leads.
7. Frequent crossing and recrossing railway lines should be avoided.
8. Religious places of worship, burial grounds and cemeteries should not be disturbed by the alignment.
9. Places of archaeological and historical importance should not be interfered with.
10. Proximity to educational institutions, hospitals, playgrounds should not be interfered with.
11. Where major rivers have to be crossed, a right-angled crossing results in the most economical bridge.
12. Areas liable to flooding and land-slides and which are low-lying and marshy should be avoided.
13. The alignment should not interfere with utility services like electric high tension lines, main water pipes and optical fibre cables which already exist.
14. In aligning hill roads, obligatory points such as passes, saddles and valleys dictate (Ref 15). Areas liable to landslides, snow drift, avalanches should be avoided. Hair-pin bends should be used only where other alternatives are costly. The alignment should be on the side of the hill exposed to sun during winter. Alignment along the course of a river will have easy gradients. Tunnels are costly to build, and if they are unavoidable, the tunnel length should be minimal.
15. In desert areas, the alignment should avoid shifting sand dunes. Where longitudinal dunes are present, the alignment should preferably be on the top of the dune or in the interdunal space. Alignment passing through areas which are having vegetation should be preferred.

8.5.2. *Use of Data from Remote Sensing and Google Earth Images*

Technological advances in recent times have greatly facilitated the selection of alignment. The Survey of India have remote sensing maps available and should be consulted. The Google earth images are a great help in developing contours of hills and facilitate the selection of proper alignment of hill roads. They are also of great use in selecting alignment in areas other than hilly terrain since they give information on location of settlements, existing tracks, drainage channels, wooded areas and other environmentally sensitive areas.

8.5.3. *Use of Topo Sheets*

In India, the Survey of India publish Topo sheets of a scale of 1:50,000. These should be studied to select possible alignments keeping in view the guidelines given in para 8.5.1.

8.5.4. Great Reconnaissance

The map study described in para 8.5.2 and 8.5.3 above results in various possible alignments and route location. They need to be examined in the field by ground reconnaissance. The engineer should walk along the possible alignments collecting the following information:

1. Topography (plain, rolling or hilly)
2. Length along various alternatives
3. Bridging requirement
4. Railway line crossings
5. Right-of-way available, if there is already a track or road, and constraints on account of built-up area, monuments, educational institutions, places of religious importance etc.
6. Geology of the area, giving soil type, nature of hill slopes, landslides, rocky faces, places susceptible to flooding, snow-drift, avalanches, sand dunes and their disposition.
7. Temperature variation, annual rainfall, wind direction, exposure to sun, water-table
8. Availability of borrow area and stone aggregates
9. Availability of water
10. Value of land
11. Important towns and villages, and need for bypasses
12. Location of existing utilities
13. Availability of labour
14. Availability of suitable contractors
15. Traffic counts from existing records

The reconnaissance survey shall enable the highway authority to select the best route to be followed.

8.5.5. Preliminary Survey

Once the best route to be adopted is known, the next stage is the Preliminary Survey. It consists of running a traverse line along the selected route. In hilly areas, it is necessary to cut a trace 1.0 to 1.2 m wide so that the instruments for the survey can be conveniently placed. The survey should be done by modern instruments such as theodolite or total station. Levels are taken at 50 m intervals along the chosen traverse line. The levels should be connected to GTS datum. Distances can be measured by metallic tape or chain, or with modern distance measuring instruments. An accuracy of 1 in 2000 should be aimed at. Cross-sections are taken at intervals of 100-250 m in plain terrain. Bench Marks should be established at every 250-500 m, connected to the GTSS datum.

8.5.6. Preparation of Drawings

Based on the survey details collected, plans and longitudinal sections are prepared. Contours at 1-3 m intervals can also be shown in the drawings. The scales to be adopted are:

(i) Built-up areas and stretches in hilly terrain

1:1000 for horizontal scale

1:100 for vertical scale

(ii) Plain and Rolling Terrain

1:2500 for horizontal scale

1:250 for vertical scale

8.5.7. *Determination of Final Centreline in the Office*

Firstly, the maps prepared as part of Reconnaissance Survey are studied and a few alternatives are prepared. The best is then selected which satisfies the design standards, environmental impact and economy.

For the selected alignment, the horizontal curves are designed including transitions. The vertical profile is then determined. A sample drawing is given in Fig. 8.14.

The estimates based on the above are then prepared.

8.5.8. *Final Location Survey*

The final centreline is then laid out on the ground by a theodolite or Total Station. The Horizontal Intersection Points (HIPs) are fixed on the ground and suitably referenced. The salient points on curves such as Beginning of Spiral (BS), Beginning of Circular Curve (BC), End of Circular Curve (EC) and End of spiral (ES) are fixed on the ground and suitably referred. The final centre line is staked at 50 m intervals in plain terrain and 20 m in hilly terrain. These will facilitate the construction.

8.6. Drainage Surveys

The drainage surveys include:

(i) determination of High Flood Level (HFL)

(ii) determination of water table

(iii) catchment area of rivers and streams where culverts and bridges are to be provided

(iv) cross-sections of the river/ streams at the proposed crossing and upstream and downstream

(v) rainfall details (maximum hourly rainfall)

8.7. Project Preparation

The Detailed Project Report (DPR) shall contain the following:

1. Brief Description of the Project
2. Selection of Route
3. Environmental factors
4. Road Design
    - Design Standards
    - Alignment
    - Vertical Profile
    - Cross-sections
    - Pavement Design
5. Drainage
    - Surface and Subsurface Drainage
    - Pavement Drainage
    - Cross-drainage Structures
6. Availability of Material, Labour, Equipment and Contractors
7. Rates
    - Schedule of Rates
    - Analysis of rates

8. Estimate
9. Construction Programming
10. Drawings

REFERENCES

1. Guidelines for Capacity of Roads in Rural Aras, IRC:64-1990, Indian Roads Congress, New Delhi, 1990.
2. Guidelines for Traffic Prediction on Rural Highways, IRC:108-1996, Indian Roads Congress, New Delhi, 1996.
3. Kadiyali, L.R., Traffic Engineering, and Transport Planning, Khanna Publishers, New Delhi, 2013.
4. Indian Roads Congress, Traffic Census on Non-urban Roads, IRC:9-1972, New Delhi, 1972.
5. Compendium of Indian Standards on Soil Engineering, Bureau of Indian Standards, New Delhi, 1987.
6. Specifications for Rural Roads, Ministry of Rural Development, Published by the Indian Roads Congress, New Delhi, 2004.
7. Specifications for Road and Bridge Works (Fifth Revision), Ministry of Road Transport and Highways, Published by the Indian Roads Congress, New Delhi, 2013.
8. Guidelines for the Design of Rigid Pavements for Highways, IRC:58-2012, Indian Roads Congress, New Delhi, 2012.
9. Specification for Test Sieves: Methods of Examination of Apertures of Test Sieves, IS:460-1985 (Part 3) Bureau of Indian Standards, New Delhi, 1985.
10. IS:2386-1963 (Various Parts), Methods of Tests for Aggregates for Concrete, Bureau of Indian Standards, New Delhi, 1963.
11. Testing Aggregates, Method for determination of Polished Stone Value, British Standards Institution, London, 1975.
12. IS:6241-1971, Methods of Test for Determination of Stripping Value of Stone Aggregates, Bureau of Indian Standards, New Delhi, 1971.
13. Manual for Survey, Investigation and Preparation of Road Projects, Indian Roads Congress, Special Publication 19, New Delhi, 1977.
14. Manual on Route Location, Design, Construction and Maintenance of Rural Roads, Special Publication 20, Indian Roads Congress, New Delhi, 1979.
15. Hill Road Manual, IRC:SP:48-1998, Indian Roads Congress, New Delhi, 1998.

9

Chapter

Embankment Design and Construction

9.1. Importance of Proper Design and Construction of Embankment

The road pavement rests on an embankment which is suitably kept above the natural ground level. In stretches where cutting of the natural ground is involved, the ground after cutting to the desired level also needs attention because it may not have the desired degree of compaction. The type, thickness and composition of the pavement depend upon the nature of the embankment soil. High embankments need special attention to ensure the stability of slopes and failures due to settlement.

9.2. Top Width of Embankment

The top width of the embankment includes the carriageway width, median if provided and shoulders. This aspect has been discussed in Chapter 7.

9.3. Height of Embankment

The height of the embankment is determined by the longitudinal profile selected. The general rules followed is that the bottom of the pavement is at least 0.6 m and preferably 1.0 m above the high flood level or water table.

9.4. Side Slopes

Normally 2:1 (2 horizontal and 1 vertical) slope is adequate for low embankments in uninundated conditions and 3:1 for inundated conditions. For high embankments, the slopes given are in Chapter 7.

9.5. Stability of High Embankment Slopes

Side slopes of high embankments can fail due to rotational failure. The most popular method to evaluate slope failure is by the Swedish Circle method also referred to as the Fellenius method (Ref 1). Two types of analysis are done:

(i) $\phi = 0$ analysis (Total stress analysis)

(ii) c, ϕ analysis (Effective Stress analysis)

The total stress analysis is carried out when the embankment is built on saturated clay and the internal angle of friction $\phi = 0$. Fig. 9.1 gives the forces under $\phi = 0$ analysis.

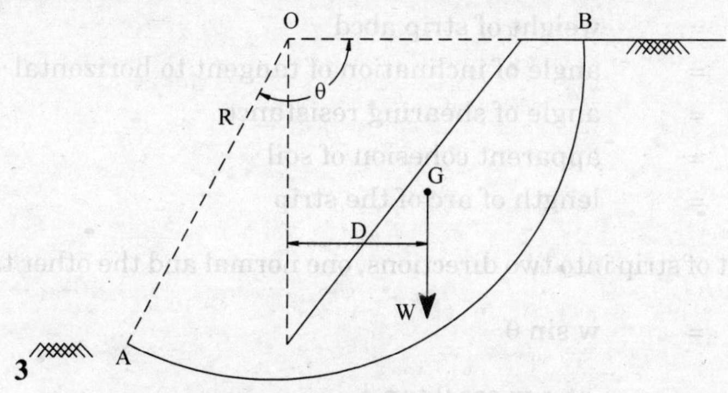

Fig. 9.1: ϕ = 0 Analysis of stability

R	=	radius of arc of circle
W	=	weight of soil in the failure zone acting at centroid G
D	=	Horizontal Distance between centre of circle and centroid G
s	=	shear strength per unit area
θ	=	angle subtended by the arc AB

Then disturbing moment = W.D.

Restoring moment = s x length of arc AB x R

= s x R x θ x R

$$= sR^2\theta \quad (9.1)$$

When failure is about to take place disturbing moment is equal to the restoring moment.

$$\therefore WD = sR^2\theta \quad (9.2)$$

Factor of safety against rotational failure

$$= \frac{\text{Restoring moment}}{\text{Disturbing moment}} = \frac{sR^2\theta}{WD} \quad (9.3)$$

The c-ϕ analysis is done by the method of slices as illustrated in Fig. 9.2.

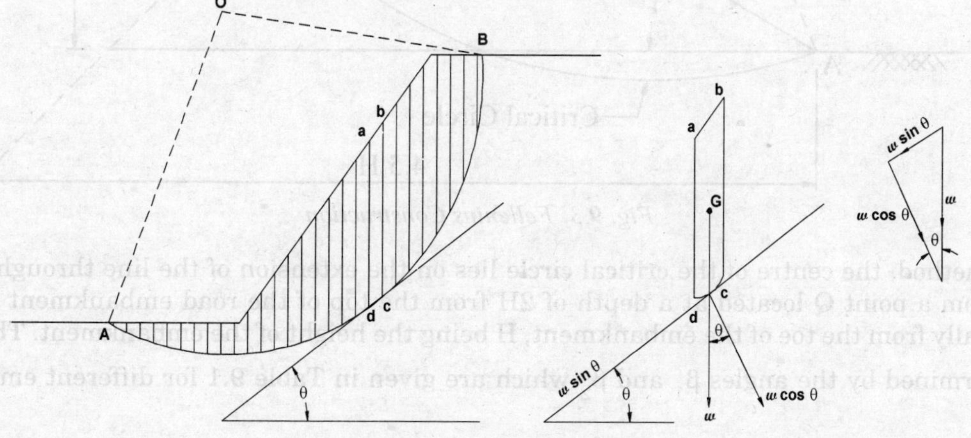

Fig. 9.2: C, ϕ Analysis of Stability

Let w = weight of strip abcd

θ = angle of inclination of tangent to horizontal

φ = angle of shearing resistance

c = apparent cohesion of soil

l = length of arc of the strip

Resolving w the weight of strip into two directions, one normal and the other tangential to the arc cd,

Disturbing force = $w \sin \theta$ (9.4)

Restoring force = $c\ell + w \cos \theta \tan \phi$ (9.5)

Factor of safety = $\dfrac{c\ell + w \cos \theta \tan \phi}{w \sin \theta}$ (9.6)

The critical circle is the one which has the lowest factor of safety. It is determined by the Fellenius method shown in Fig. 9.3.

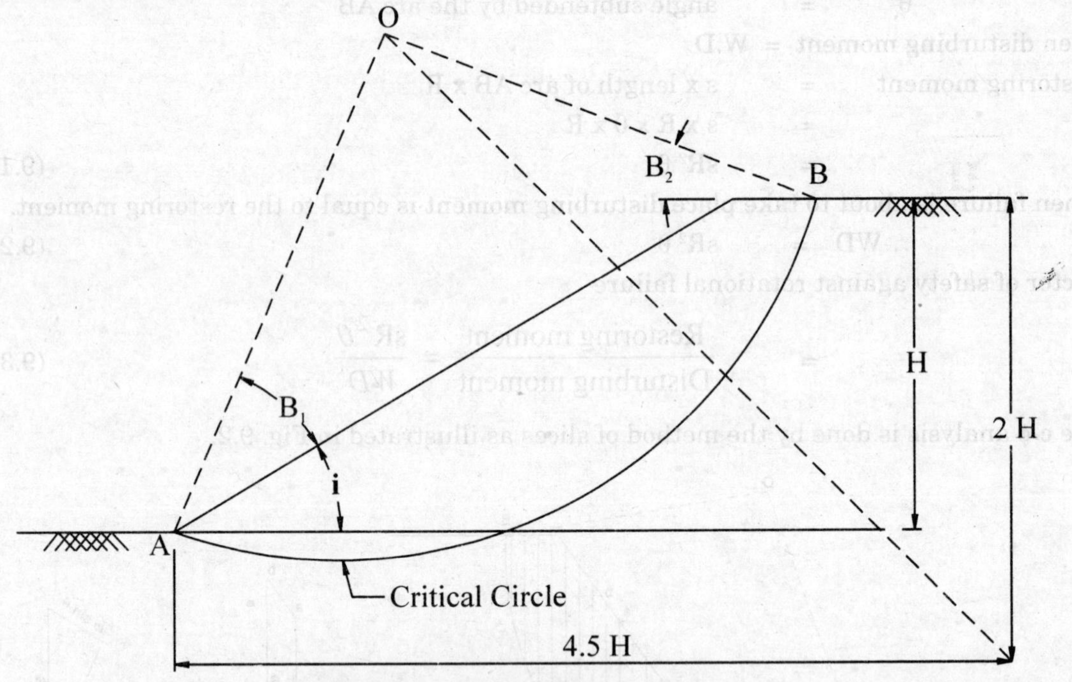

Fig. 9.3: Fellenius Construction

In this method, the centre of the critical circle lies on the extension of the line through O, (when φ = 0) from a point Q located at a depth of 2H from the top of the road embankment and 4.5 H horizontally from the toe of the embankment, H being the height of the embankment. The location 0 is determined by the angles β_1 and β_2 which are given in Table 9.1 for different embankment slopes.

Table 9.1 : Angles β_1 and β_2 for Different Embankment Slopes

Slope	i^0	$\beta 1^0$	$\beta 2^0$
1 : 0.58	60	29	40
1 : 1	45	28	47
1 : 1.5	33.8	26	35
1 : 2	26.6	25	35
1 : 3	18.4	25	35
1 : 5	11.3	25	37

9.6. Settlement of Embankment on Clayey Strata

Clay is a highly compressible soil and hence embankments constructed on clayey strata undergo settlement. Settlement causes undulations in the pavement even though the body of the embankment is well compacted. The amount of settlement depends upon the decrease in the void ratio. The following analysis is used for determining the settlement of clayey strata under the embankment (Ref 2):

Let H be the thickness of the clay strata subjected to consolidation. Let e_0 be the initial void ratio which reduces to e as the pressure on clay increases from po to po + Δp as a result of the load imposed by the embankment.

$$e_0 - e = e = a_v \Delta p$$

where a_v = coefficient of compressibility (9.7)

$$\therefore a_v = \frac{e_0 - e}{\Delta p} \quad (9.8)$$

Decrease in porosity Δ_n is given by:

$$\Delta_n = \frac{\Delta e}{1 + e_0} \quad (9.9)$$

$$= \frac{a_v}{1 + e_0} \Delta p = m_v \Delta p \quad (9.10)$$

where m_v = coefficient of volume compressibility

$$= \frac{a_v}{1 + e_0} \quad (9.11)$$

S = settlement of clay stratum

$$= H.\ \Delta p\ m_v \quad (9.12)$$

Assuming that the field consolidation for clay is a straight line on the semi-log scale (Fig. 9.4),

$$e = e_0 - C_c \log_{10} \frac{po + \Delta p}{po} \quad (9.13)$$

where C_c = Compression Index

Combining 9.13 with 9.8 and 9.11, one gets

$$A_v = \frac{C_c}{\Delta p} \log_{10} \frac{po + \Delta p}{po} \qquad (9.14)$$

and

$$m_v = \frac{C_c}{\Delta p\,(1 + e_0)} \log_{10} \frac{po + \Delta p}{po} \qquad (9.15)$$

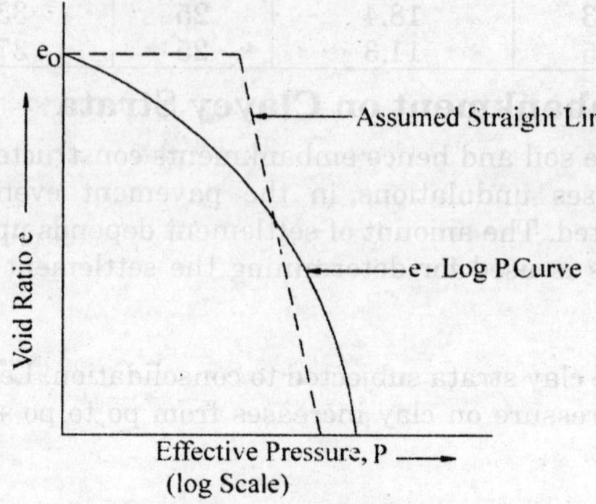

Fig. 9.4: e-log p Curve

Substituting the value of m_v in 9.12,

$$S = H \frac{C_c}{1 + e_0} \log_{10} \frac{po + \Delta p}{po} \qquad (9.16)$$

Time required for consolidation (t) can be found from Ref 1 as under:

$$t = \frac{TH^2}{C_v} \qquad (9.17)$$

where T = Time factor corresponding to the degree of consolidation

C_v = Coefficient of consolidation

H = Length of effective drainage path

= D for one-way drainage

= $\frac{D}{2}$ for two-way drainage

D = Depth of stratum

9.7. Methods of Embankment Construction in Problematic Locations

Construction of embankment in highly slushy or soft soil can be done using one of the following methods (Ref 1):

(i) Use of light weight material like flyash, cinder etc.

(ii) Removal of problematic soil by excavation or bog blasting

(iii) Construction of the embankment in stages so that settlements can take place before the pavement is constructed.

(iv) Placing surcharge fill above the finally desired level of the embankment

(v) Use of sand drains (Fig. 9.5)

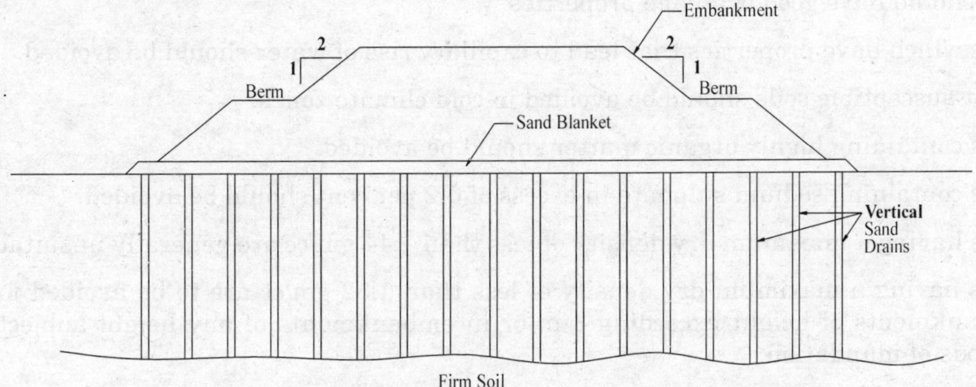

Fig. 9.5: Accelerated Preconsolidation using Sand Drains

9.8. Selection of Soil for Embankment Construction

9.8.1. Materials that can be used

The following materials can be used in forming the embankment (Ref 3, 4):

(i) Soil

(ii) Moorum

(iii) Gravel

(iv) Reclaimed Material from Pavement

(v) Fly ash and Pond Ash

9.8.2. Unsuitable Fill Material

The following materials are unsuitable as fill material:

(i) Materials from swamps, marshes and bogs

(ii) Peat, log, stump and perishable material; any soil classified as OL, OI, OH or Pt in accordance with IS:1498 (Ref 5)

(iii) Materials susceptible to spontaneous combustion

(iv) Materials in a frozen condition

(v) Clay having liquid limit exceeding 50 and plasticity index exceeding 25

(vi) Materials with salts resulting in leaching in the embankment

(vii) Materials with total sulphate content (expressed as SO_2) exceeding 0.5 per cent by mass, if deposited within 500 m, or other distance prescribed in the Contract, of metallic items forming part of permanent works.

9.8.3. Guidelines for selection of materials for embankment construction

(i) A coarse-grained soil is to be preferred to a fine-grained soil.

(ii) Highly plastic soils should be avoided as they are liable to shrink and swell as the moisture content charges.

(iii) The material should have high strength and stability since it economises the cost of pavement.

(iv) Soil should have good drainage properties.

(v) Soils which have properties that lead to capillary rise of water should be avoided.

(vi) Frost-susceptible soils should be avoided in cold climate zones.

(vii) Soils containing highly organic matter should be avoided.

(viii) Soils containing sodium sulphate in excess of 0.2 per cent should be avoided.

(ix) Soils having a maximum dry density of less than 1.44 gm/cc are generally unsuitable.

(x) Soils having a maximum dry density of less than 1.52 gm/cc are to be avoided for use in embankments of height exceeding 3 m or in embankments of any height subject to long periods of inundation.

(xi) The top 500 mm of the embankment (subgrade) shall be constructed with soils having a maximum dry density greater than 1.65 gm/cc.

9.9. Embankment Construction Practice

Before starting the embankment construction, the original ground shall be cleared and grubbed, removing all grass, shrubs and tree stumps. The top soil shall, thereafter, be stripped and stored for use on embankment slopes thus facilitating the growth of turf. The original ground, if soft and loose, and has a density of less than 90 per cent of Proctor's density, shall be loosened, watered and compacted in layers of 250 mm to obtain a density of 100 per cent of Proctor's density. This treatment shall cover a thickness of 500 mm.

Soil shall be obtained from borrow pits observing the following practices (Ref 6):

(i) Borrow-pits shall be rectangular in shape

(ii) They shall be as near to the boundary of the Right-of-Way

(iii) No borrow-pit shall be located with 5 m from the toe of the final embankment after accounting for any future widening.

(iv) The depth of borrow-pits should not exceed 1.5 m.

(v) The depth of borrow-pit and its location should be so regulated that it should not cut an imaginary line starting from the edge of the embankment and having a slope of 1:4 (Fig. 9.6).

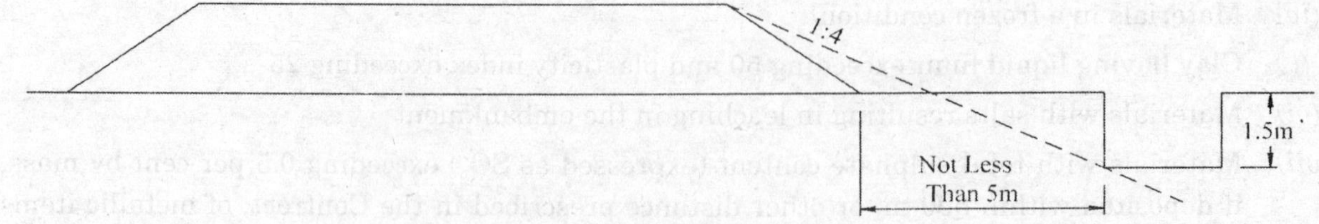

Fig. 9.6: Location of Borrow Pit

(vi) Borrow-pits should not be dug continuously. Ridges not less than 8 m wide should be left at intervals not exceeding 300 m.

(vii) If borrow-pits are dug in temporarily acquired land, their depth should not be greater than 45 cm.

(viii) Borrow-pits shall not be dug within 0.8 Km of habitations since stagnant water in them can breed mosquitoes.

The borrowed soil is deposited in layers and compacted by rollers. The depth of each compacted layer shall not exceed 250 mm. The compaction shall be done at Optimum Moisture Content to achieve the maximum density as under (Ref 3).

Table 9.2: Compaction Requirement of Embankment

Type of Work/ Material	Relative Compaction as per cent of max. lab. dry density as per IS:2720 (Part 8)
1. Sub-grade and Shoulders	Not less than 97%
2. Embankment	Not less than 95%
3. Expansive Clays	
Subgrade and 500 mm below sub-grade	Not allowed
Remaining portion of embankment	90–95%

If the embankment is to be widened, and its slopes are steeper than 1 in 4, continuous horizontal benches at least 300 mm wide shall be cut to ensure adequate bond with the new material. If the slope is flatter than 1 in 4, the slope surface may be scarified before placing the new material.

9.10. Construction of Embankment Using Fly-Ash

Fly-ash is an industrial waste and is available at no cost near thermal plants. Its disposal is an environmental problem. Fly-ash can be used in the construction of road embankment, and Government of India have made it mandatory that it should be used in road construction within 100 Km of a thermal plant. It is a light-weight material and can be used advantageously in high embankments on soft soil. Fly-ash embankments shall be covered on the sides and top by soil to prevent erosion. Typical cross-sections are given in Fig. 9.7 and 9.8 (Ref 7). The side cover is laid along with each layer of fly-ash. The layer thickness can be in the range of 200-400 mm depending upon the compaction equipment used.

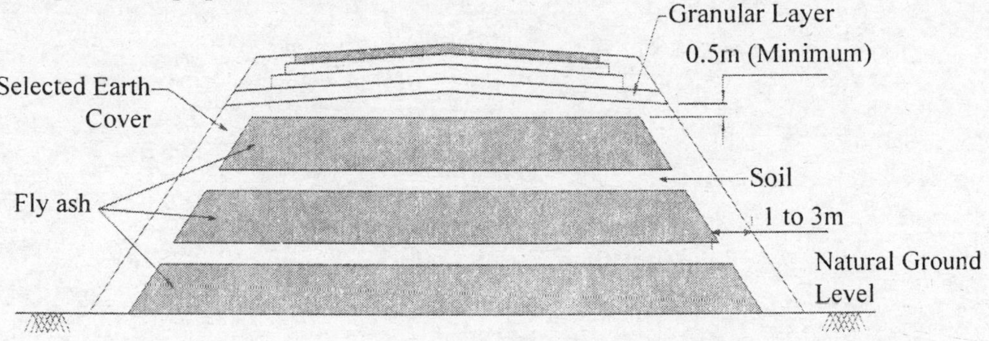

Fig. 9.7: Typical Cross-section of Embankment with Alternate Layer of Fly-ash and Soil

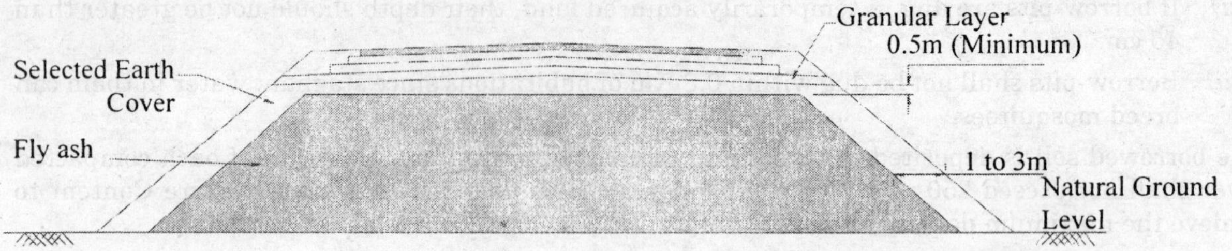

Fig. 9.8: Typical Cross-section of Embankment with Core of Fly-ash

REFERENCES

1. Guidelines for the Design of High Embankment, IRC:75-1979, Indian Roads Congress, New Delhi, 1979.
2. Terzaghi, K and R.B. Peck, Soil Mechanics in Engineering Practice, Asia Publishing House, New Delhi, 1962.
3. Specifications for Road and Bridge Works (Fifth Revision), Ministry of Road Transport and Highways, Published by the Indian Roads Congress, New Delhi, 2013.
4. Specifications for Rural Roads, Ministry of Rural Development, Published by the Indian Roads Congress, New Delhi, 2014.
5. Classification and Identification of Soils for General Engineering Purposes, IS:1498-1970, Bureau of Indian Standards, New Delhi, 1970.
6. Recommended Practice for Borrow-pits for Road Embankments constructed by Manual Operation, IRC:10-1961, Indian Roads Congress, New Delhi, 1961.
7. Guidelines for Use of Fly-ash in Road Embankments, IRC:SP:58-2001, Indian Roads Congress, New Delhi, 2001.

10

Chapter

Highway Drainage

10.1. Importance of Highway Drainage

Water is an important material needed for the construction of embankment, pavement layers and structures in a highway. But once the road is constructed, water is detrimental to the good performance of a road and must be drained off quickly. It is well said that cheap drainage is preferable to costly pavement.

10.2. Surface Drainage

The rain water that comes over a pavement should be led away to side ditches as quickly as possible. If allowed to remain on the pavement, it can percolate through the voids, cracks and joints in the pavement layers, reach the subgrade, soften it and reduce its supporting power. A film of water on a pavement can cause skidding. The surface drainage of a pavement is affected by adequate camber and longitudinal slope discussed in Chapter 7.

The Indian Roads Congress recommends that a return period of 10 years should be adopted for the estimation of run-off in rural areas and 25 years in urban areas (Ref 1).

The storm duration is chosen equal to the 'time of concentration' which is the time taken for the water from the remote part of the catchment to reach the point of exit (Ref 1, 2). The run-off from a catchment is given by the Rational Formula (Ref 3) given below:

$$Q = 0.028\, P\, I_c\, A \qquad (10.1)$$

where Q = Discharge in cu m per sec

P = Factor depending upon nature of surface (varying from 0.20 for sandy soils to 0.90 for rock, concrete or bituminous surface

I_c = Critical intensity of rainfall in cm per hour during the time of concentration

A = Catchment area in hectares

10.3. Design of Cross-Section of Drainage Channel

The discharge on the road is led to drainage channels on both sides of the road. The hydraulic design of the channel is carried out on the basis of the well-known Mannings formula:

$$V = \frac{R^{2/3} S^{1/2}}{n} \quad (10.2)$$

where V = Velocity in m/sec; which is 6 for lined channels and 1-2 in unlined earthen channels

S = Bed Slope

R = Hydraulic Mean Depth = $\frac{Area\ of\ Flow}{Wetted\ Perimeter}$

n = Manning's Rugosity Coefficient, which varies from 0.012 for paved drain to 0.025 for soils

A trapezoidal side ditch is generally provided for roads in open areas. A sketch is given in Fig. 10.1.

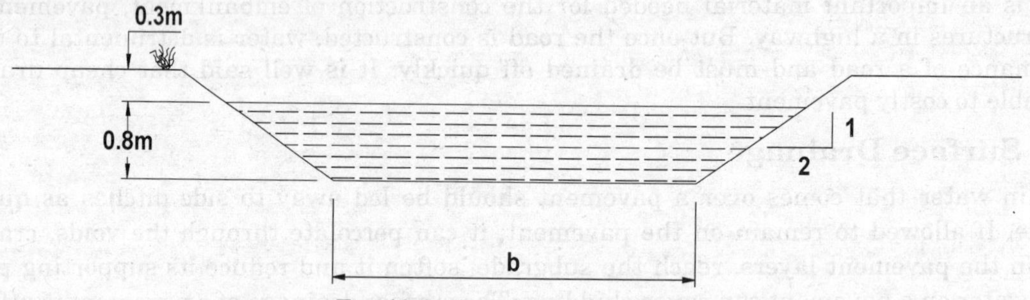

Fig. 10.1: Trapezoidal Side Ditch

10.4. Sub-Surface Drainage

The water percolating through the voids, cracks and joints in a pavement should be prevented from reaching the subgrade. This is done by providing a drainage layer, which runs from one edge of the embankment to the other. A typical drainage layer for a flexible pavement is given in Fig. 10.2.

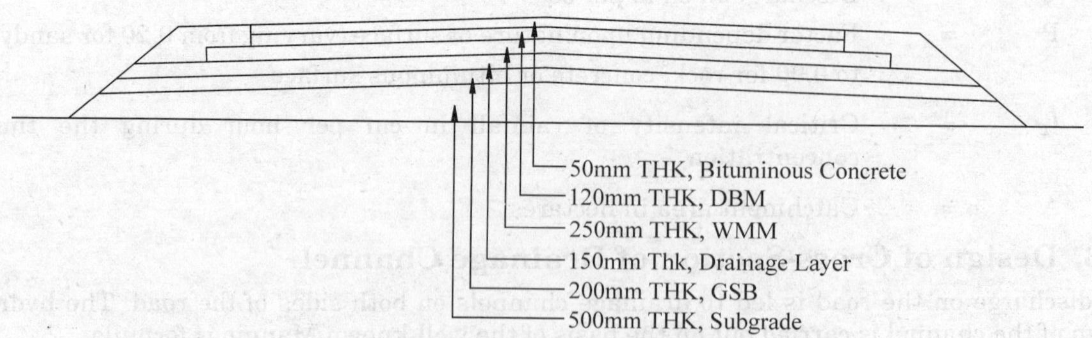

Fig. 10.2: Drainage Layer for Flexible Pavement

The suggested gradation for the drainage layer is given in Table 10.1 (Ref 4).

Table 10.1: Gradings for Drainage Layer (Ref 4)

IS SIEVE	PERCENT BY WEIGHT PASSING IS SIEVE	
75.0 mm	100	–
53.0 mm	80–100	100
26.5 mm	55–90	75–100
9.5 mm	35–65	55–75
4.75 mm	25–50	30–55
0.85 mm	2–10	–
0.425 mm	0–5	0–8
0.075 mm	–	0–3

10.5. Protection of Embankment Slopes Against Erosion

Embankment slopes are liable to get eroded by rain water. The most common method is to protect the slope by turfing. Pitching the slopes where there is a flow of water along the road is done as given in Fig. 10.3.

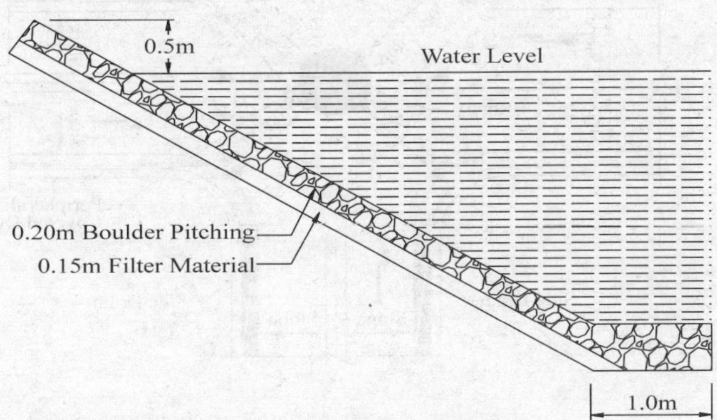

Fig. 10.3: Pitching Embankment Slopes

For protecting slopes of high embankment, a system of kerbed gutter and flumes at 5 m c/c is used (Fig. 10.4).

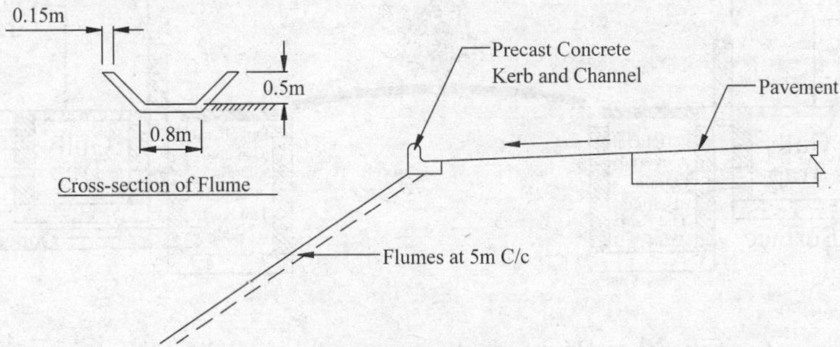

Fig. 10.4: Protection of High Embankment Slope by Kerb, Channel and Flumes

10.6. Drainage of Medians

Raised medians 4-5 m wide with kerbs are commonly adopted for dual carriageways in India. The median is then filled with soil between the kerbs. It is often turfed to prevent erosion. The earth filling should be sloped from the centre of the median to the kerb.

It is the common practice in India to plant flowering shrubs so that glare from the opposing traffic is prevented. Watering these plants is a regular requirement and this water can reach the pavement by seepage. To avoid this, it is a good practice to plant these shrubs in concrete pots.

If the median is wide and depressed, the water should be collected in a ditch at the centre and led away at suitable intervals by pipes laid below the carriageways and made to discharge eventually at the side ditches.

10.7. Drainage of Intersections

In roundabouts a part of the carriageway slopes towards the central island. This water needs to be collected in a drain around the island, and led away to an outlet by a pipe. Fig. 10.5 shows the arrangement.

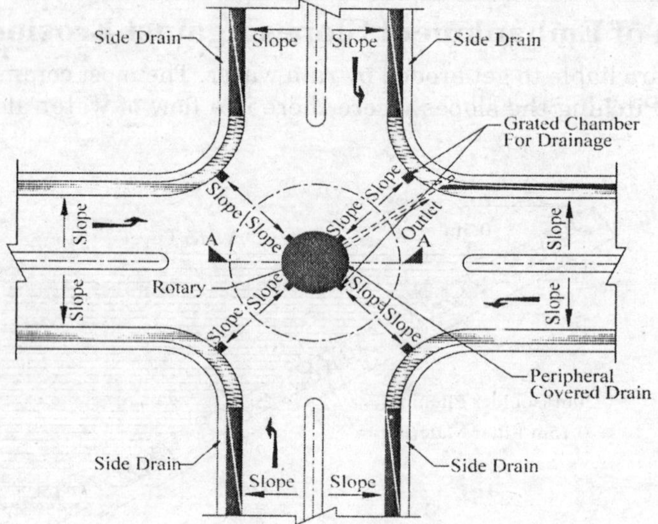

Plan

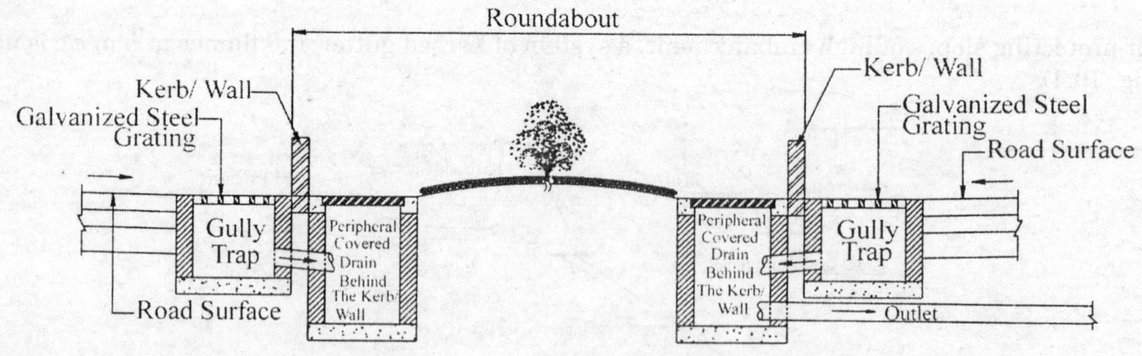

Section A-A

Fig. 10.5: Drainage Arrangement at a Roundabout

In other types of intersections, the surface water is collected in suitably located pits provided with gratings and eventually taken out at suitable locations by pipes laid beneath the pavement.

10.8. Drainage of Flyovers

Rain water which is received on the carriageway has to be led away. Providing outlets which discharge the water to the side carriageways below will fall on the pedestrians and other road users and should be avoided. The proper arrangement is to collect the water at the carriageway above at pier locations and provide pipes at the sides of the pier and discharge the water to the underground drainage system.

10.9. Drainage of Hill Roads

The water on the carriageway is led to the side drain at the cut face and is led to the nearest cross-drainage structure. Water falling on the cut slope is intercepted by catch water drains and is led to the nearest natural water course. Fig. 10.6 shows the arrangement.

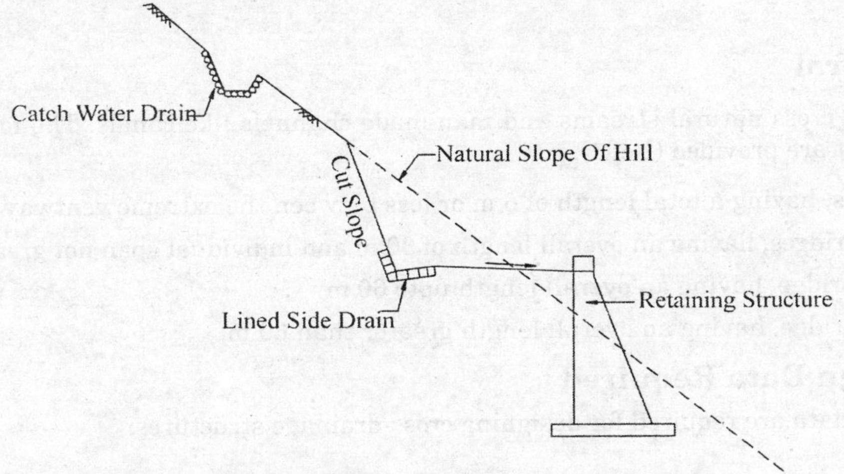

Fig. 10.6: Side Drain and Catch Water Drain in Hill Roads

10.10. Drainage of Urban Roads

Urban drainage is provided by the storm water drains below the carriageway. They are provided generally at the carriageway edges or beneath the medians. Inlets can be grated inlets which discharge the water directly into manholes at suitable intervals or grated inlets in the vertical face of kerbs. In the latter case, short pipes take the water from the inlets into manholes. The possibility of ground water harvesting should be investigated so that the recharging of the fast depleting ground water is facilitated (Ref 2).

REFERENCES

1. Guidelines of Road Drainage, IRC:SP:42-2014, Indian Roads Congress, New Delhi, 2014.
2. Guidelines on Urban Drainage, IRC:SP:50-2013, Indian Roads Congress, New Delhi, 2013.
3. Guidelines for the Design of Small Bridges and Culverts, IRC:34-1970, Indian Roads Congress, New Delhi, 1970.
4. Specifications for Road and Bridge Works (Fifth Revision), Ministry of Road Transport & Highways, Published by Indian Roads Congress, New Delhi, 2013.

11 Chapter

Cross Drainage Structures

11.1. General

Roads have to cross natural streams and man-made channels like canals. The following are the structures that are provided (Ref 1).

1. Culverts, having a total length of 6 m or less between the extreme ventway boundaries
2. Small bridges, having an overall length of 30 m and individual span not greater than 10 m
3. Minor bridge, having an overall length upto 60 m
4. Major bridge, having an overall length greater than 60 m.

11.2. Design Data Required

The following data are required for designing cross-drainage structures:

(i) Catchment Area, which can be measured by published topo sheets having contours or by traversing along the watershed

(ii) Cross-sections, one at the proposed location of the crossing, one upstream and one downstream at the following distances

Table 11.1: Distances Upstream and Downstream for Cross-Sections

	Catchment Area (Sq. Km)	Distance upstream and downstream from the proposed crossing
1.	Upto 3.0	100 m
2.	3.0 to 15	300 m
3.	Over 15	500 m

(iii) Highest Flood Level ascertained by local enquiry

(iv) Velocity of flow, observed during actual high flood

(v) Sub-soil borings or trial pits

IRC Codes of practice may be referred to for details (Ref 2, 3, 4).

11.3. Run-Off

Run-off from a catchment area is calculated in India on the basis of:

(i) Dicken's Formula

$$Q = CM^{3/4} \quad (11.1)$$

where Q = Discharge in cu m/sec

M = Catchment Area in sq. Km

C = 11-14 where rainfall is 60-120 cm/year

= 14-19 where rainfall is more than 120 cm/ year

= 22 in Western Ghats

(ii) Ryve's Formula (for Madras Presidency)

$$Q = CM^{2/3} \quad (11.2)$$

where Q and M have the same meaning as in (i) above,

C = 6.8 for areas within 25 Km of the coast

= 8.5 for areas between 25 Km and 160 Km of the coast

= 10.0 for limited areas near the hills

(iii) Ingli's Formula (for Bombay Presidency)

$$Q = \frac{125M}{\sqrt{M+10}} \quad (11.3)$$

where Q and M have the same meaning as in (i).

(iv) Rational Formula

$$Q = 0.028\,P\,f\,A\,I_c \quad (11.4)$$

which has already been discussed in Chapter 10 earlier.

(v) Manning's Formula which gives v, the velocity

$$V = \frac{R^{2/3} S^{1/2}}{n} \quad (11.5)$$

which has already been discussed in Chapter 10 earlier.

$$Q = A.V \quad (11.6)$$

where Q = Discharge in cu m/sec

A = Area of cross-section (sq. m)

V = Velocity in m/sec

The highest of the values obtained from the Empirical Formulae (Eqn. 11.1 to 11.13), Rational Formula (Eqn 11.4) and Manning's Formula (Eqn 11.5) is adopted, provided it does not exceed the next highest by more than 50 per cent. If it does, the discharge is restricted to that limit.

11.4. Linear Waterway

Where the banks are well defined and are non-erodible, the linear water is fixed as the width of the stream at the HFL. Where the banks are erodible as in alluvial streams, the Lacey's Equation is used.

$$W = 4.8 Q^{1/2} \quad (11.7)$$

where W = Regime Width in m

Q = Discharge in cu m/sec

11.5. Scour Depth

Erodible streams such as those in alluvial soil are subjected to scour during high floods. The scour depth is calculated from the formula

$$d_{sm} = 1.34 \left[\frac{Db^2}{K_{sf}} \right]^{1/3} \quad (11.8)$$

where d_{sm} = normal scour depth in m

d_b = Discharge in cu m per m width

K_{sf} = silt factor

= $1.76\sqrt{dm}$, dm being the weighted mean dia of the particles in mm

If the linear waterway provided in a bridge is less than the unobstructed width, the scour will be greater than that for the unobstructed flow. Equation 11.8 still holds good because d_b for the restricted waterway will be more than that for the unobstructed waterway.

11.6. Depth of Foundation

In rocky strata, or where the soil is not erodible at the calculated maximum velocity, the foundation shall be securely anchored into hard rocks to a depth of 0.6 m if the ultimate crushing strength of rock is 10 MPa or more. In all other cases, the depth of anchoring shall be 1.5 m.

In straight reaches of a single span bridge in alluvial strata, the scour depth is taken as 1.27 times the maximum scour depth calculated by the formula 11.8. The foundation are taken to a depth of not less than 2.0 m below the scour level.

In multi-span bridges and bad sites with curves or where diagonal current exit, the maximum scour depth is taken as 2 times the normal scour depth. Here also, the foundations are taken to 2.0 m below the scour level.

The above rules govern open foundations. For well foundations, the depth of wells is calculated as per IRC:78 (Ref 3).

11.7. Span

The following approximate relationship is used for fixing the economic span.

For masonry arch bridges S = 2 H

For RCC slab bridges S = 1.5 H

where S = Clear Span in metres

H = Height of abutment or pier from the bottom of its foundation to its top in metres

11.8. Types of Structures

In hill roads, dry stone masonry scuppers are provided as a temporary measure before permanent structures using RCC pipes or slab are constructed.

RCC Pipes using NP4 pipes are used for highway loadings, whereas NP3 pipes can be used for Rural Roads. The minimum diameter generally adopted is 1.0 m, which facilitates its cleaning. The type of bedding can be granular (minimum thickness 75 mm) or concrete (M-15) with minimum bedding thickness of 250 mm.

Granular bedding is adopted for fill height upto 4 m and concrete cradle bedding is adopted for fill heights greater than 4 m. If more than one pipe is adopted in a row, the minimum distance between pipes is half the diameter of the pipe subject to a minimum of 450 mm. Masonry (Brick or stone) arch bridges are adopted for culverts where good foundation conditions exist and for spans upto 15 m.

Simply supported RCC slabs are adopted for spans upto 10 m. RCC T-beam and slab bridges are adopted for spans in the range 10-15 m. They can be simply supported for any type of soil in the foundations and continuous for rocky foundation. RCC balanced cantilever bridges are adopted for deep foundations and spans in the range 25-60 m.

Simply supported prestressed T-beam and slab bridges are adopted for spans in the range 20-60 m for any type of foundation. Prestressed segmental balanced cantilever bridges are adopted for deep foundations and spans in the range of 60-100 m. Cable-stayed girders are adopted for deep foundations and spans in the range of 80-500 m. Submersible bridges, paved dips (Causeways) and vented causeways are adopted for low volume roads.

11.9. Vertical Clearance

The minimum vertical clearance between the HFL and bottom of superstructure shall be as under (Ref 1).

Table 11.2: Vertical Clearance for Bridges and Culverts

Discharge cm/sec	Vertical Clearance in mm
Upto 0.30	150
0.3–3.0	450
3.0–30.0	600
30.0–300.0	900
300.0–3000.0	1200
Above 3000.0	1500

11.10. Protection Works

When a bridge is built on a wide meandering river with a constricted opening, it is necessary to provide guide bunds to guide and confine the flow.

Culverts and bridges may be provided with bed floors, dispersing with deep foundations. The top of the bed floor is kept 0.3 m below the bed level.

The bed floor has an impervious apron. Upstream and downstream curtain walls and stone boulder talus are provided (Fig. 11.1).

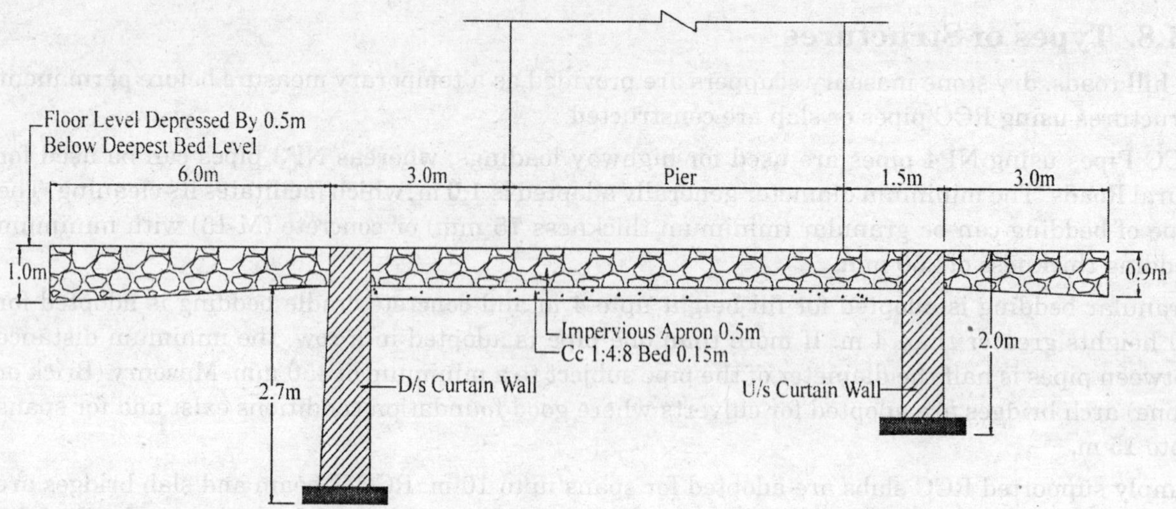

Fig. 11.1: Floor Protection for a Bridge with Shallow Foundations

REFERENCES

1. Guidelines for the Design of Small Bridges and Culverts, IRC:SP:13-2004, Indian Roads Congress, New Delhi, 2004.
2. Standard Specification and Code of Practice for Road Bridges, Section I, General Features of Design, IRC:5-1998, Indian Roads Congress, New Delhi, 1998.
3. Standard Specifications and Code of Practice for Road Bridges, Section VII, Foundations and Substructure, IRC:78-2000, Indian Roads Congress, New Delhi, 2000.

12 Chapter

Types of Pavements and Factors Governing Pavement Design

12.1. Types of Pavements

Pavements can be of the following types:

(i) Flexible Pavement which is a layered system consisting of several courses, and in which the wheel load is transmitted to the subgrade by the lateral distribution with increasing depth. The thickness is so selected that the stresses on the subgrade are kept within the safe bearing power.

(ii) Rigid Pavement, having high flexural strength, and transmitting the wheel load by beam action. Cement concrete pavement is a rigid pavement.

(iii) Semi-rigid pavement, which is an intermediate type between flexural and rigid, having a much lower flexural strength than a rigid and also deriving support by lateral distribution of load through the depth like a flexible pavement. Typical examples are soil-cement, lime-pozzolona concrete and lean concrete.

(iv) Some authorities identify a fourth type, composite pavement, which consists of layers of different composition. But strictly speaking all pavements are composite pavements since all layers have different composition.

12.2. Requirements of a Good Pavement

A pavement should have the following requirements:

(i) It should be structurally adequate to withstand the loads imposed on it.

(ii) It should be thick enough to distribute the loads to a safe value on the subgrade.

(iii) It should provide a hard wearing surface and not get abraded by the wheels of vehicles.

(iv) It should be skid-resistant

(v) It should have a good riding quality so that the vehicles can travel at a good speed without discomfort and at a low vehicle operating cost.

(vi) It should be dust-proof

(vii) It should not create high levels of noise

(viii) Its surface should be impervious so that water does not percolate through the voids and soften the subgrade.

12.3. Pavement Course

A pavement consists of various courses:

(i) At the top is the surfacing which provides a smooth, abrasion resistant, dust-free, noise-free surface, also resisting the ingress of water. In a concrete pavement, the Pavement Quality Concrete (PQC) slab is the top course.

(ii) The surfacing is supported by a base, which helps in distributing the loads in a flexible pavement.

(iii) Below the base is the sub-base, which further helps in distributing the loads. In a concrete pavement, the pavement rests on a sub-base, which is constructed as a Dry Lean Concrete (DLC) or soil-cement.

(iv) A drainage layer is needed to drain away the water percolating through the voids, cracks and joints in a pavement. In a flexible pavement, a part of the sub-base is made as the drainage layer.

12.4. Comparison Between Flexible and Concrete Pavement

The following comparison gives the advantages and disadvantages of the two types of pavements:

Table 12.1: Comparison of Flexible and Concrete Pavements

	Characteristics	Flexible Pavement	Concrete Pavement
1.	Life	Has a low life of 10-15 years	Has a long life of 30-40 years
2.	Maintenance	Requires heavy maintenance such as routine maintenance (filling up potholes, patch repairs etc.), renewal of surface once in 3-5 years and overlay once in 10-15 years.	Practically maintenance free. In a jointed plain concrete pavement, renewal of sealant in joints may be needed once in 5-10 years. A Continuously Reinforced Concrete Pavement does not require sealant renewal.
3.	Riding Quality	Though a good riding quality can be achieved after initial construction, the riding quality deteriorates with the passage of traffic, thus entailing higher vehicle operating costs.	A good riding quality is maintained throughout its life.
4.	Skid-resistance	Whenever the surface becomes smooth, a thin bituminous layer can be laid to restore anti-skid property.	If the surface texture wears out, regrooving can be done to restore anti-skid property.
5.	Stage Construction	Amenable to stage construction	Not amenable

(Contd...)

	Characteristics	Flexible Pavement	Concrete Pavement
6.	Binder Availability	Bitumen is derived from petroleum crude whose world-wide reserves are fast dwindling.	Cement is available in plenty in India and will continue to be available for many years in the future.
7.	Penetration of Water	Water percolates through voids, causing stripping of binder from aggregates and softening the subgrade.	Very little chances of water penetrating and hence it is ideal for high rainfall areas.
8.	Consumption of stone aggregates	High when the pavement is built initially and aggregates are needed for maintenance also.	Lower consumption than flexible pavement in initial construction, and no consumption at all for maintenance.
9.	Fuel Consumption of Trucks	14 per cent higher consumption of diesel than concrete pavement.	Lower consumption of diesel.
10.	Consumption of electricity for street lighting	10 per cent higher consumption than concrete roads because of dark colour.	Being grey or whitish, 10 per cent less consumption of electricity for street lighting.
11.	Initial Cost	May be slightly cheaper than concrete pavement in the initial cost.	May be marginally costlier (5-15 per cent) than flexible pavement.
12.	Life-cycle Cost	Life cycle cost is 2 to 3 times that of concrete pavement.	Life cycle cost is one-third to one-half that of flexible pavement.
13.	Management of traffic during construction	Can be thrown open to traffic immediately after construction.	The concrete has to be cured for about 14 days to gain strength and hence traffic cannot use a freshly laid concrete road.
14.	Utilities	If any repairs to utilities lying beneath the concrete pavement is to be done, it is easy to cut the pavement and reinstate it after repairs.	Concrete pavement cannot be cut and reinstated for repair of utilities. Special arrangements for relocating utilities outside the concrete pavement have to be done.
15.	Environmental pollution	Since bitumen and aggregates have to be heated, emissions from the hot-mix plant cause environmental pollution.	Since heating is not involved, environment pollution is less.

12.5. Factors Governing Pavement Design

The factors which govern pavement design are discussed below:

12.5.1. Wheel load

The design axle load in many countries, including India is 80 kN (18,000 lbs). Since vehicles are usually overloaded in India, it becomes necessary to express the damaging effect of higher axle loads in terms of the standard axle load. The fourth power rule is generally followed, which is given below:

$$F = \left(\frac{L}{L_s}\right)^4 \qquad (12.1)$$

where F = Equivalence Factor of a load L in terms of its damaging effect when compared to the standard axle load of 80 kN

Axle load spectrum studies are conducted and based on the fourth power rule, a multiplying factor known as VDF (Vehicle Damage Factor) is determined. If an axle load spectrum study is not conducted, the following values of VDF can be adopted (Ref 1).

Table 12.2: Indicative Values of VDF

Initial Traffic Volume in Terms of Commercial Vehicles per day (CVPD)	VDF Values for roads in	
	Rolling/ Plain Terrain	Roads in Hilly Terrain
0–150	1.5	0.5
150–1500	3.5	1.5
More than 1500	4.5	2.5

12.5.2. Distribution of Commercial Traffic over Carriageway

The number of lanes in a carriageway determines how much of the total two-way traffic should be considered as governing the traffic on the design lane. The following distribution is recommended in India (Ref 1):

(i) Single Lane Roads : Total number of CVPD

(ii) Two-lane Roads with bi-directional traffic : 50 per cent of the total two-way traffic (If the VDF in one direction is higher, the traffic in the direction of higher VDF is taken for design).

(iii) Four-lane Roads (undivided): 40 per cent of the total CVPD in both directions.

(iv) Dual Carriageway Roads: 75 per cent of the CVPD in each direction for dual two lane carriageway in each direction (*i.e.,* total four lanes). For a total six lane divided carriageway, the distribution factor is 60 per cent of the CVPD in each direction. For a total eight lane divided carriageway, the distribution factor is 45 per cent of CVPD in each direction.

12.5.3. Axle Load Repetitions for Design

The number of axle load repetitions during the design period governs the design. Since traffic grows every year, say at an annual growth rate of r (expressed in decimals), the number of axle load repetitions is given by the formula:

$$N = \frac{365 \times \left[(1+r)^n - 1\right]}{r} \times A \times D \times F \qquad (12.2)$$

where N = Number of repetitions of standard axle load of 80 kN

r = annual rate of growth of commercial vehicles in decimals (*i.e.,* for 5 per cent annual growth rate r = 0.05)

A = Initial commercial vehicle traffic per day in the year of completion of construction

D = Lane distribution (Refer para 12.5.3)

F = VDF (Refer para 12.5.2)

12.5.4. Subgrade

As per current Indian practice, the 500 mm thick subgrade should have a 4-days soaked CBR of minimum 8 for roads having traffic of 450 CVPD. For Low Volume Rural Roads, (Ref 2), the thickness recommended covers a wide range of CBR (ranging from 2 to 15).

12.5.5. Position of Wheel Load

For the design of jointed concrete pavements, the position of wheel loads in a slab panel is critical. These are discussed in Chapter 14.

12.5.6. Climatic Factors

Rainfall and frost affect pavement performance. Rainfall is taken into account of providing pavement drainage. Good pavement drainage also minimises the damages caused by frost action.

Variation of temperature causes warping stresses in concrete pavements. These are discussed in Chapter 14.

12.5.7. Properties of Materials

The sub-base, base and surfacing layers have strengths which are the basis for determining the stresses and strains in the layers. These are discussed in Chapter 13. The strength of concrete which is dependent on the properties of materials forming the concrete is discussed in Chapter 13.

REFERENCES

1. Guidelines for the Design of Flexible Pavements, IRC:37-2012, Indian Roads Congress, New Delhi, 2012.
2. Guidelines for the Design of Flexible Pavements for Low Volume Rural Roads, IRC:SP:72-2007, Indian Roads Congress, New Delhi, 2007.

13

Chapter

Design of Flexible Pavement

13.1. Possible Approaches

The following are the various approaches for designing a flexible pavement:

(1) Empirical Methods

(2) Methods based on Pavement Performance

(3) Analytical Methods

(4) Mechanistic-Empirical Method

13.2. Empirical Methods

The methods are based on observing the performance of pavements of certain thickness laid over subgrades of known supporting strength.

The California State Highway Department pioneered this method developing curves for thickness of pavement for wheel loads of 7000, 9000 and 12,000 lbs. TRRL UK used this method and developed similar curves for wheel loads ranging from 4,000 lbs to 16,000 lbs (Ref 1). They were subsequently revised in 1970, covering load repetitions in terms of million standard axles (Ref 2).

The IRC Guidelines adopted the UK

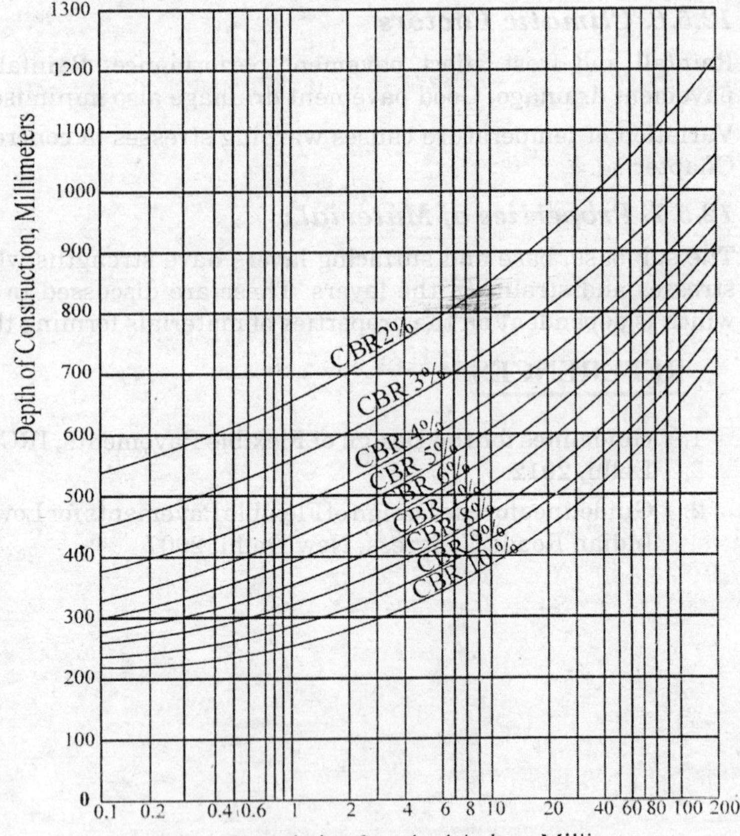

Fig. 13.1: Pavement Thickness Design Chart

Note: 1. Read total pavement thickness from continuous curves

2. Use dotted curves for proportioning sub-base thickness

practice and the 1984 version (Ref 3) considered the traffic in terms of cumulative number of equivalent standard axles of 80 kN in millions and provided charts for traffic upto 30 msa, and relating the thickness to the subgrade CBR values. Fig. 13.1 gives the charts.

13.3. Methods Based on Pavement Performance

In 1958, the AASHO undertook a very comprehensive highway research project with the objective of determining the relationships between the number of repetitions of specified loads and the performance of different thicknesses of flexible and rigid pavements (Ref 4). The test resulted in the development of relationships between the serviceability of pavements and load applications. The following relationships were evolved:

Rigid Pavement: $PSI = 5.41 - 1.80\,(1 + \overline{SV}) - 0.09\sqrt{C + P}$ (13.1)

Flexible Pavement: $PSI = 5.03 - 1.91\,(1 + \overline{SV}) - 0.01\sqrt{C + P} - 1.38(\overline{RD}^2)$ (13.2)

where PSI = Present Serviceability Index

$\overline{SV}$ = Average Slope Variance in both the wheel paths as measured by the AASHO profilometer

C = (i) for rigid pavement = major cracking in linear feet per 1000 sq ft of pavement area

(ii) for flexible pavement = area of cracking in sq. ft per 1000 sq. ft of pavement area

P (i) for rigid pavement = bituminous patching in sq. ft for 1000 sq. ft of pavement area

(ii) for flexible pavement = repair of surface either by skin patching or deep patching in sq. ft per 1000 sq. ft of pavement area $\overline{RD}$ (for flexible pavement) = rut depth measured in inches at the centre of the 4 ft. span in the deepest part of the rut, obtained by averaging 40-50 evenly spaced samples for a 1000 ft. section.

The following gives an approximate relationship between PSI and the rating done by visual inspection.

Table 13.1: PSI and Its Rating

PSI	Rating
4–5	Very Good
3–4	Good
2–3	Fair
1–2	Poor
0–1	Very Poor

The AASHTO Guide published in 1993 is based on the results of the AASHO Test and further research (Ref 5). The following equation has been proposed (all values in FPS units).

$$Log_{10} W_{18} = Z_R \times S_o + 9.36 \times \log_{10}(SN+1) - 0.20 + \frac{\log_{10}\left\{\frac{\Delta PSI}{4.2-1.5}\right\}}{0.40 + \frac{1094}{(SN+1)^{5.19}}}$$

$$= + 2.32 \times \log_{10}(M_R) - 8.07 \quad (13.3)$$

where W_{18} = Predicted number of 18,000 lb ESALS

Z_R = Standard normal deviate

S_o = Combined standard error of traffic prediction and performance prediction

ΔPSI = difference between initial design serviceability index(P_o) and the design terminal serviceability index.

M_R = Resilient Modulus (psi)

SN = $a_1D_1 + a_2D_2m_2 + a_3D_3m_3$ where a_1, a_2, a_3 are layer coefficients for the surface, base and sub-base respectively (these can be further modified based on drainage conditions)

The AASHTO design procedure also incorporates a new concept known as the "reliability" level, which is as under:

Table 13.2: Highway Class and Its Reliability

Highway Class	Recommended Reliability	
	Urban	**Rural**
Inter State and other Freeways	85–99.9	80–99.9
Principal Arterials	80–99	75–95
Collectors	80–95	75–95
Local	50–80	50–80

13.4. Analytical Methods

These are based on the determination of stresses and strains in a layered pavement structure. Burmister proposed an analysis of a two-layer system as early as in 1943 (Ref 6). Jones and Peattie extended this to three-layered analysis in 1962 (Ref 7 and 8). A rough formula relating E, the Modulus of Elasticity, and CBR was used.

$$E = 10 \text{ CBR } (MN/m^2) \quad (13.4)$$

The latest development is the Mechanistic-Empirical method by AASHTO and a software AASHTO Ware is available which incorporates all design variable like materials, local climate and traffic (Ref 9).

13.5. Current Indian Practice

In para 13.2, it was mentioned that the IRC Guidelines were revised in 1984, giving charts relating the pavement thickness to CBR and traffic upto 30 msa. In 2001, the guidelines were again revised using a semi-mechanistic approach and based on a research scheme implemented by IIT, Kharagpur (Ref 10), giving procedure for designing for traffic upto 150 msa, a value common

The guidelines have been further revised in 2012 incorporating the use of cementitious bases and sub-bases. The salient features are (Ref 11).

1. Traffic, which includes:

    commercial vehicles per day
    traffic growth rate
    spectrum of axle loads
    distribution of commercial traffic over the carriageway
    design life in years

The aspects of design life, vehicle damage factor, and distribution of commercial traffic over the carriageway have been covered in Chapter 12.

2. CBR—The earlier practice was to adopt the CBR value of the subgrade. The new guidelines propose the concept of "Effective CBR" which takes into the CBR of the soil below 500 mm of the subgrade as well. Fig. 13.2 gives how the effective CBR is arrived at.

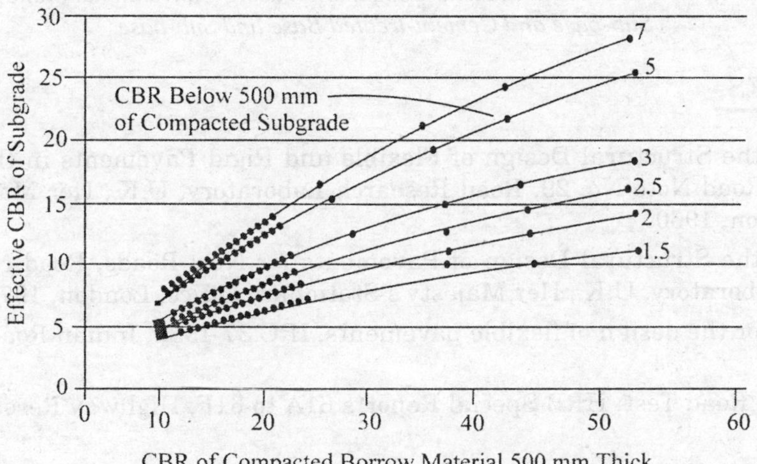

Fig. 13.2: Determination of Effective CBR

3. Resilient Modules of the subgrade soil

This is determined by laboratory tests. In the absence of testing facility, the following relationship between the Resilient Modulus (M_R) and CBR can be used. M_R (in MPa) = 10 x CBR for CBR of 5

M_R (in MPa) = 17.6 x CBR for CBR > 5

4. The flexible pavement is modelled as an elastic multi-layer structure and determining the stresses and strains at critical locations using a software IITPAVE developed by IIT (Kharagpur).

Fatigue resistance and rutting of bituminous layers has been incorporated in the design.

5. Design Charts are given for Pavement Thickness and Composition for various CBRs (3-15) and traffic (2-150 msa) using conventional stone aggregate bases and sub-bases as well as cement-treated bases and sub-bases. A comparison of the two for CBR of 10 and traffic of 150 msa is given in Fig. 13.3.

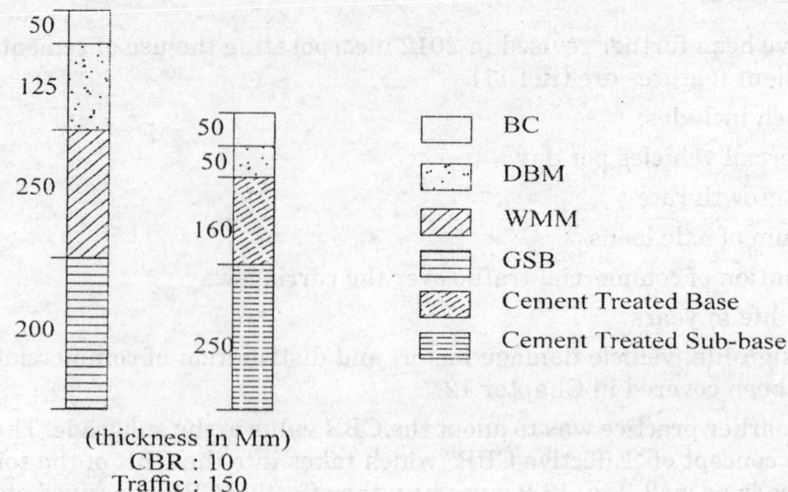

Fig. 13.3: Comparison of Pavement Thickness and Composition using Conventional Stone Aggregate base and Sub-base and Cement-treated Base and Sub-base

REFERENCES

1. A Guide to the Structural Design of Flexible and Rigid Pavements in the construction of new roads, Road Note No. 29, Road Research Laboratory, U.K., Her Majesty's Stationary Office, London, 1960.
2. A Guide to the Structural Design of Pavements for New Roads, Road Note No. 29, Road Research Laboratory, U.K., Her Majesty's Stationary Office, London, 1970.
3. Guidelines for the design of flexible pavements, IRC:37-1984, Indian Roads Congress, New Delhi, 1984.
4. The AASHO Road Test, HRB Special Reports 61A to 61F, Highway Research Board, 1961-1962.
5. AASHTO Guide for Design of Pavement Structures, American Association of State Highway and Transportation Officials, Washington D.C., 1993.
6. Burmister, D.M., The theory of stresses and displacements in layered systems and application to the design of airport pavements, Proceedings, Highway Research Board, Washington, 1943.
7 Jones A, Table of stresses for three-layered system, Highway Research Board Bulletin 342, Washington, 1962.
8. Peattie, K.R., Stress and Strain factors for three-layer elastic system, Highway Research Board Bulletin 342, Washington, 1962.
9. American Association of State Highway and Transportation Officials (AASHTO), 2008, Mechanistic-Empirical Pavement Design Guide, Interim Edition, A Manual of Practice, AASHTO, Washington DC, 2008.
10. Guidelines for the Design of Flexible pavements (Second Revision), IRC:37-2001, Indian Roads Congress, New Delhi, 2001.
11. Guidelines for the Design of Flexible Pavements, (Third Revision), IRC:37-2012, Indian Roads Congress, New Delhi, 2012.

14

Chapter

Design of Rigid Pavements

14.1. Factors Governing Design

The following factors govern the design of rigid pavements:

A. *Loading*

1. Wheel load and its repetitions
2. Area of contact of wheel load
3. Location of load with respect to panel of a slab

B. *Properties of the supporting medium*

1. Strength and properties of soil subgrade
2. Provision or non-provision of sub-base

C. *Properties of Concrete*

1. Strength
2. Modulus of Elasticity
3. Poisson's Ratio
4. Shrinkage Properties
5. Fatigue behaviour

D. *External Conditions*

1. Temperature variation
2. Friction between slab and sub-base

E. *Joints*

1. Whether joints are provided or not
2. Arrangement of joints
3. Dowel Bars and Tie Bars

F. *Reinforcement*

1. Quantity of Reinforcement in Jointed Pavement
2. Quantity of Reinforcement in Continuously Reinforced Concrete Pavements

14.2. Wheel Load, Area of Contact and Its Repetitions

Wheel load is the most important factor that affects design. In India, though the legal axle load is 10.2 tonnes for a single axle, 19.0 tonnes for a tandem axle and 24.0 tonnes for a tridem axle, overloading is common. In order to get over this phenomenon, an axle load spectrum study is done and the 98th percentile axle load is considered for design.

The tyre inflation pressure determines the area of contact. A tyre pressure of 0.8 MPa is adopted for design (Ref 1). The contact area is assumed to be circular.

The distribution of commercial traffic varies across the carriageway width, and has been discussed in Chapter 12.

The number of repetitions of the load affects the consumption of fatigue of the concrete and depends upon the design life. A design life of 30 years is commonly adopted. The rate of growth of traffic is determined as explained in Chapter 12. A rate of 7.5 per cent is considered in the Guidelines (Ref 1).

In jointed concrete pavements the location of the wheel load determines the stresses induced. The three critical locations are given in 14.1. Interior loading produces tensile stresses in the slab. Edge loading produces tensile stresses at the bottom of the slab and a small tensile stress at the top. Corner loading produces tensile stresses at the top of the slab due to cantilever action.

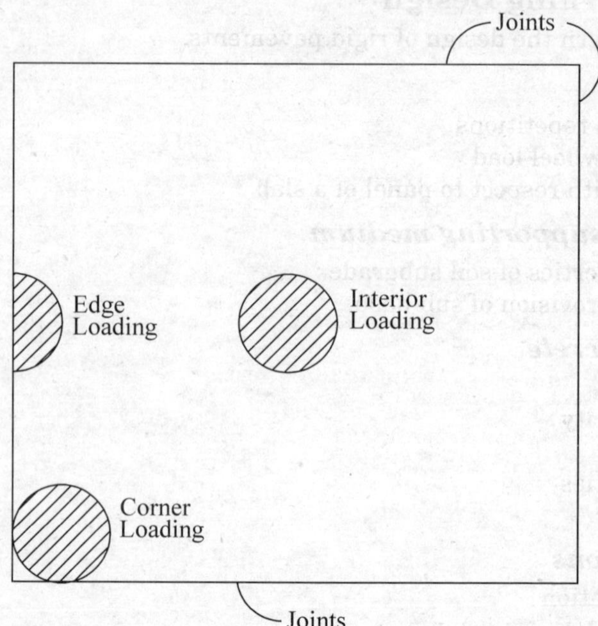

Fig. 14.1: Critical Loading Positions in a Concrete Slab Panel

14.3. Strength and Properties of Soil Subgrade

The strength of the soil subgrade determines how well it can support the slab. The strength is determined by the modulus of subgrade reaction, described in Chapter 12. The durability and susceptibility to frost action affect the design.

14.4. Provision or Non-Provision of Sub-Base

It is customary to provide a sub-base below the concrete slab and above the subgrade. The sub-base performs the following functions:

1. It provides a uniform and firm support to the pavement.
2. It prevents mud-pumping, which can occur due to water seeping through the joints, reaching

the subgrade, moistening it and when a load passes over the slab causing the water and mud to come up through the joints. This causes loss of support, which is the most important requirement for the design of a concrete slab.

3. It reduces frost action.
4. It provides a firm support to the pavers which lay the concrete slab.
5. It acts as a capillary cut-off in subgrades liable to capillary rise of water.

The materials used for sub-base are:

(i) Dry Lean Concrete having a 7-day average compressing strength of 10 MPa (Ref 1) and a thickness of 150 mm.

(ii) Cement-treated sub-bases having a minimum 28 day compressive strength of 7 MPa and thickness in the range of 100-200 mm.

(iii) Untreated granular sub-bases of thickness in the range of 150-300 mm.

It is customary to provide a debonding interlayer of polythene sheet of minimum thickness of 125 mm to allow smooth movement of the concrete slab caused by temperature changes.

14.5. Properties of Concrete

14.5.1. Strength

The design of concrete pavement slabs is based on the 28 day strength. A 90 day strength, which is 1.1 times the 28 day strength, can be permitted in view of the fact that during the initial 90 days, the number of repetitions of load is very small and has negligible effect on cumulative fatigue damage of concrete. For design of the concrete slab, the flexural strength is required, which can be obtained after conducting the beam test. Alternatively, it can be derived as per the following relationship.

$$F_{cr} = 0.7\sqrt{f_{ck}} \tag{14.1}$$

where

F_{cr} = Flexural strength (modulus of rupture), MPa

f_{ck} = cube strength, MPa

In no case the 28 day flexural strength of pavement quality concrete shall be less than 4.5 MPa (Ref 1). The target mean flexural strength to be achieved while designing the mix shall be such that there is 95 per cent portability that the characteristic strength would be achieved while the mix is produced in the field. Thus,

$$f'_{cr} = f'_{cr} + 1.96\,\sigma \tag{14.2}$$

where f'_{cr} = target mean flexural strength at 28 days, MPa

f_{cr} = characteristic flexural strength at 28 days, MPa

σ = standard deviation of field test samples, MPa

1.96 = factor compressing to the desired confidence level for 5 per cent tolerance level

14.5.2. Modulus of Elasticity and Poisson's Ratio

The value of Modulus of Elasticity, E, is generally taken as 30,000 MPa. The Poisson's Ratio is taken as 0.15.

14.5.3. Coefficient of Thermal Expansion

Since a concrete slab expands and contracts due to changes in temperature, stresses are induced which depend upon the coefficient of thermal expansion, which is taken as 10 x 10^{-6} per ^{0}C.

14.5.4. Shrinkage

Concrete expands slightly during setting due to hydration of cement, but on drying it shrinks. The stresses induced are small and are generally opposed to those of temperature and are normally not critical to design.

14.5.5. Fatigue Behaviour

Due to repetitive application of loads, progressive deterioration of the internal structure of concrete takes place. The ratio of the flexural stress and the flexural strength, known as the stress ratio (SR), is an indication of the ability of concrete to withstand the repetitive loads. The IRC guidelines (Ref 1) recommend that if SR is less than 0.45, the concrete is expected to sustain infinite number of repetitions. As the stress ratio increases, the number of load repetitions required to cause cracking decreases. The following relationships are given in Ref 1:

$$N = \text{Fatigue life} = \text{Unlimited for } SR<0.45 \quad (14.3)$$

$$N = \left[\frac{4.2577}{SR-0.4325}\right]^{3.268} \quad \text{for } 0.45 \le SR \le 0.55 \quad (14.4)$$

$$N = \frac{0.9718-SR}{0.0828} \quad \text{for } SR > 0.55 \quad (14.5)$$

14.6. External Conditions (Temperature Variation)

Charges in temperature cause stresses in the slab when they curl upwards or downwards. The curling stresses at the edge of the slab are to be added to the wheel load stress. The warping stress at the edge has been calculated by Bradbury (Ref 2) as per the following formula:

$$f_{te} = \frac{E\,\alpha\,\Delta t}{2}C \quad (14.6)$$

where f_{te} = temperature stress (in psi) at the edge

E = Modulus of Elasticity, psi

α = Coefficient of Thermal Expansion of concrete

Δt = change in temperature, ^{0}F

C = Bradbury's Coefficient, depending upon $\frac{W}{\lambda}$ or $\frac{L}{\lambda}$ where W = width of panel of slab

L = Length of panel

λ = radius of relative stiffness

$$= \sqrt[4]{\frac{1000 \text{ x } Eh^3}{12(1-\mu^2)R}} \quad (14.7)$$

The values of Bradbury's Coefficient C are given in Table 14.1 below:

Table 14.1 : Bradbury's Coefficient

$\frac{L}{l}$ or $\frac{W}{l}$	C
1	0.000
2	0.040
3	0.175
4	0.440
5	0.725
6	0.920
7	1.030
8	1.077
9	1.080
10	1.075
11	1.050
12	1.000

The temperature variation for various regions in India are given in Table 14.2

Table 14.2 : Recommended Temperature Differentials for Concrete Slabs

Zone	State/ Regions	Max. Temperature Differential ^{0}C in Slab of Thickness			
		150 mm	200 mm	250 mm	300 mm to 400 mm
(i)	Hilly region Uttaranchal, West Bengal, Jammu & Kashmir, Himachal Pradesh and Arunachal Pradesh	12.5	13.1	14.3	15.8
(ii)	Punjab, UP, Uttaranchal, Gujarat, Rajasthan, Haryana and North M.P., excluding hilly regions	12.5	13.1	14.3	15.8
(iii)	Bihar, Jharkhand, West Bengal, Assam and Eastern Orissa, excluding hilly regions and coastal areas.	15.6	16.4	16.6	16.8
(iv)	Maharashtra, Karnataka, South M.P., Chhattisgarh, Andhra Pradesh, Western Orissa and North Tamil Nadu, excluding hilly regions and coastal areas.	17.3	19.0	20.3	21.0
(v)	Kerala and South Tamil Nadu, excluding hilly regions and coastal areas.	15.0	16.4	17.6	18.1
(vi)	Coastal areas bounded by hills.	14.6	15.8	16.2	17.0
(vii)	Coastal areas unbounded by hills.	15.5	17.0	19.0	19.2

14.7. Friction Between Slab and Sub-Base

The friction between slab and the sub-base determines the restraint on expansion and contraction of the slab panel due to temperature changes. It is used in the design of longitudinal and transverse steel according to the following formula.

$$A = \frac{L f W}{2S} \tag{14.8}$$

where A = Area of steel in cm^2 per m width or length of slab

L = distance in m between transverse joints (for longitudinal steel) or free longitudinal joints (for transverse steel)

W = Weight of slab in Kg/m^2

f = coefficient of friction

S = Allowable working stress in steel in Kg/cm^2 (usually taken as 50 to 60 per cent of the maximum yield stress of steel)

14.8. Joints

The following are the types of joints which are provided in a Concrete Pavements (Ref 1, 3).

1. Expansion Joint
2. Contraction Joint
3. Longitudinal Joint
4. Construction Joint

Details of various types of joints are given in Fig. 14.2.

Expansion Joints are provided to allow for the expansion of slabs. They are full depth joints provided transversely only where the slab abuts with a bridge or culvert.

Contraction joints are provided in jointed concrete pavement for relieving the tensile stresses in the concrete due to drying shrinkage, for preventing the formation of irregular cracks due to restraint in free contraction of concrete and for relieving stresses due to warping. They are sawn to a depth of one-fourth to one-third the slab thickness, facilitating the formation of a natural crack extending to the full depth of the slab. Dowel bars are provided for load transfer.

Longitudinal joints are provided in multi-lane pavements and when the pavement is more than 5 m wide.

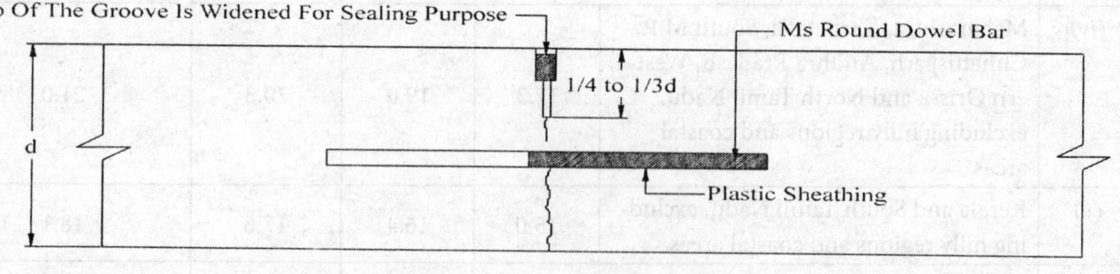

Contraction Joint With Dowel Bar

(a)

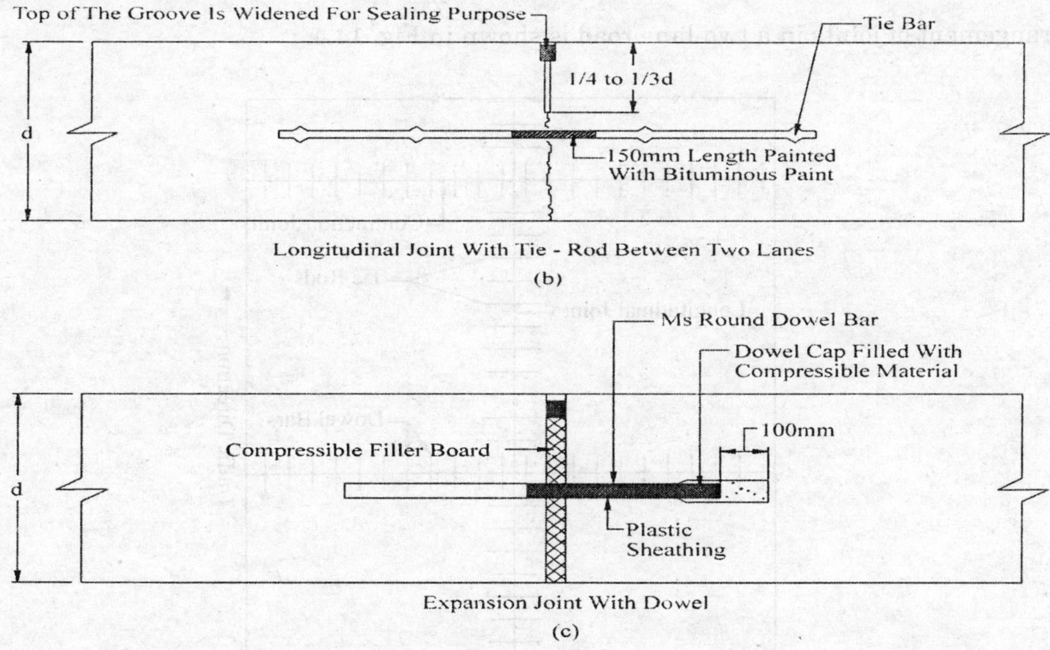

Fig. 14.2: Typical Cross-section of Joints

They are formed by sawing to a depth of one-third to one-half of the slab depth. Tie bars are provided to keep the adjoining slabs together.

Construction Joints are needed when the day's work is stopped at a location other than a contraction or expansion joint.

The joints are sealed with a sealant. Sealant can be (Ref 4):

(i) hot-poured (rubberised bitumen)

(ii) cold-poured (polysulphide or silicon)

(iii) preformed

The details of sealing are given in Fig. 14.3.

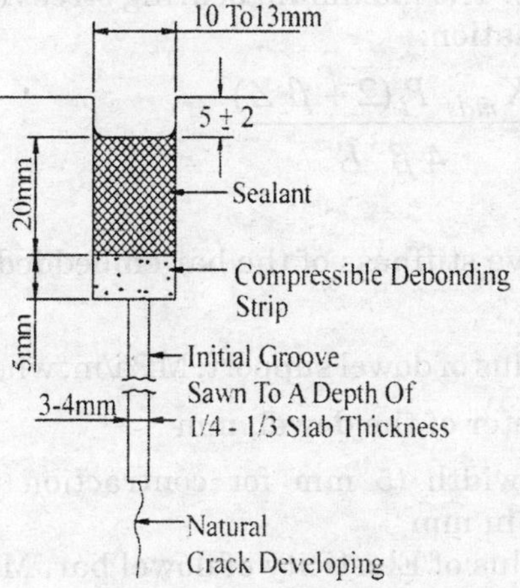

Fig. 14.3: Sealing of Joints

The arrangement of joints in a two-lane road is shown in Fig. 14.4.

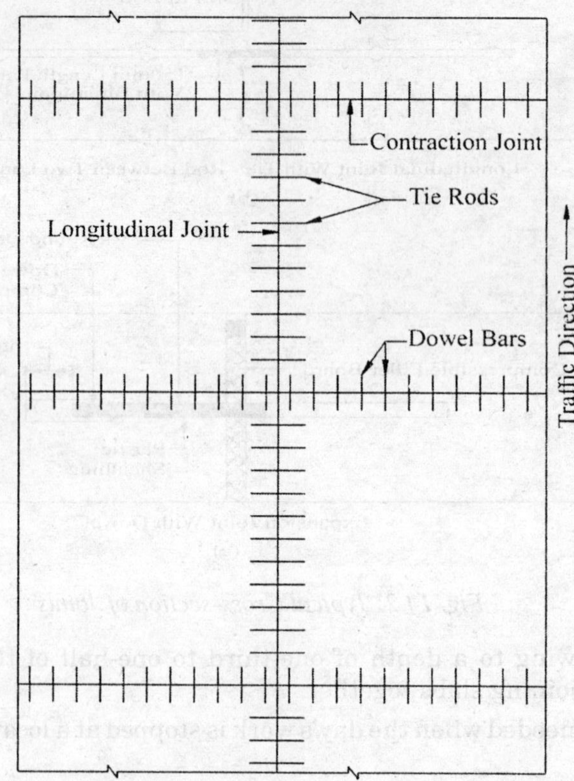

Fig. 14.4: Joints Configuration of Two Lane Road

14.9. Dowel Bars

Dowel Bars are provided in transverse joints for transfer of load. From the experience gained all over the world, it is found that it is only the bearing stress in concrete that is responsible for the performance of dowel bars (Ref 1). The maximum bearing stress ($F_{b\max}$) between the concrete and the dowel bar is given by the equation:

$$F_{b\max} = \frac{K_{mds}\ P_t (2 + \beta Z)}{4\beta^3 E} \qquad (14.9)$$

where β = relative stiffness of the bar embedded in concrete, mm = $\sqrt[4]{\dfrac{K_{mds}\ b_d}{4\,EI}}$

K_{mds} = modulus of dowel support, MPa/m, whose value may be taken as 4,15,000

b_d = diameter of the dowel, mm

Z = joint width (5 mm for contraction joint and 20 mm for expansion joint) in mm

E = Modulus of Elasticity of dowel bar, MPa

I	=	Moment of Inertia of the dowel, mm^4
P_t	=	Load transferred by the dowel bar, kN

The allowable bearing stress on concrete is calculated from the formula:

$$F_b = \frac{(101.6 - b_d) f_{ck}}{95.25} \qquad (14.10)$$

14.10. Tie Bars

The area of steel required per metre length of joint may be calculated from:

$$A_s = \frac{b\ f\ w}{S_{st}} \qquad (14.11)$$

where

A_s	=	Area of steel in mm^2 per metre length of joint
b	=	lane width in metres
f	=	coefficient of friction between slab and sub-base, which may be taken as 1.5
w	=	weight of slab in kN/m^2
S_{st}	=	Allowable working stress of steel in MPa (125 MPa for plain bars and 200 MPa for deformed bars)

The length of tie bar is calculated from the formula:

$$L = \frac{2 S_{st}\ A_{cs}}{B\ P_{ptb}} \qquad (14.12)$$

where

L	=	Length of tie bar in mm
S_{st}	=	Allowable working stress in steel (values as given in Formula 14.11)
A_{cs}	=	cross-sectional area of one tie bar, mm^2
B	=	Permissible bond stress MPa (1.75 MPa for plain bars and 2.46 MPa for deformed bars)
P_{ptb}	=	Perimeter of tie bar (mm)

14.11. IRC Recommendations for Dowel Bars and Tie Bars

The diameter and spacing of dowel bars and tie bars recommended by IRC (Ref 1) are given in Tables 14.3 and 14.4 respectively.

Table 14.3 : Dowel Bar Dimensions

Slab Thickness (mm)	Dowel Bar Dimensions		
	Diameter (mm)	Length (mm)	Spacing (mm)
200	25	360	300
230	30	400	300
250	32	450	300
280	36	450	300
300	38	500	300
350	38	500	300

Table 14.4 : Dimensions of Tie Bars

Slab Thickness (mm)	Dimensions (mm)				
	Diameter	Max. Spacing		Min. Length	
		Plain Bars	Deformed Bars	Plain Bars	Deformed Bars
150	8	330	530	440	480
	10	520	830	510	560
200	10	390	620	510	560
	12	560	900	580	640
250	12	450	720	580	640
300	12	370	600	580	640
	16	660	1060	720	800
350	12	320	510	580	640
	16	570	910	720	800

14.12. Reinforcement

14.12.1 Function of Reinforcement in Jointed Concrete Pavement

The reinforcement, if at all provided in jointed concrete pavement, is not intended to increase the flexural strength. Its role is to control cracking.

14.12.2. Quantity of Reinforcement in Jointed Concrete Pavement

The quantity of reinforcement (longitudinal and traverse) is calculated from the following formula:

$$A = \frac{L_d f W}{2S} \qquad (14.13)$$

where A = Area of steel in mm² required per m width or length of slab

L_d = Distance (m) between free transverse joints (for longitudinal steel) and free longitudinal joints (for transverse steel)

f = coefficient of friction between pavement and sub-base, taken as 1.5

W = Weight of slabs in kN/m²

S = Allowable working stress in steel in MPa (usually taken as 50 to 60 per cent of the minimum yield stress of steel)

The steel is placed 50-60 mm below the surface.

14.12.3. Quantity of Reinforced in Continuously Reinforced Concrete Pavements (CRCP)

As per Indian practice (Ref 5), the quantity of longitudinal steel for normal work is 0.7 per cent of the cross-sectional area of the pavement. The longitudinal steel is placed 100-115 mm below the surface for slab thickness of 250-330 mm. Transverse steel reinforcement is calculated from the following formula:

$$\text{\% of transverse steel, } P_t = \frac{\gamma_c W_s F}{2_{fs}} \qquad (14.14)$$

where P_t = transverse steel, %

γ_c = unit weight of concrete (kN/ m³)

W_s = total pavement width (m)

fs = allowable working stress in steel (taken as 75% of yield strength)

14.13. Analysis of Stresses in a Concrete Pavement

The pioneering work in the design of concrete pavement was done by Westergaard who presented his results in 1925 (Ref 6). The analysis is based on the following assumptions:

1. The concrete slab is homogeneous and isotropic and has uniform elastic properties.
2. The reaction of the subgrade is only vertical and is proportional to the deflection of the slab. In other words, the support provided by the subgrade is similar to that given by a dense fluid and the subgrade has no shear strength.
3. The reaction of the subgrade at a point is equal to k x deflection at that point, k being the Modulus of Subgrade Reaction (see Fig. 14.5).

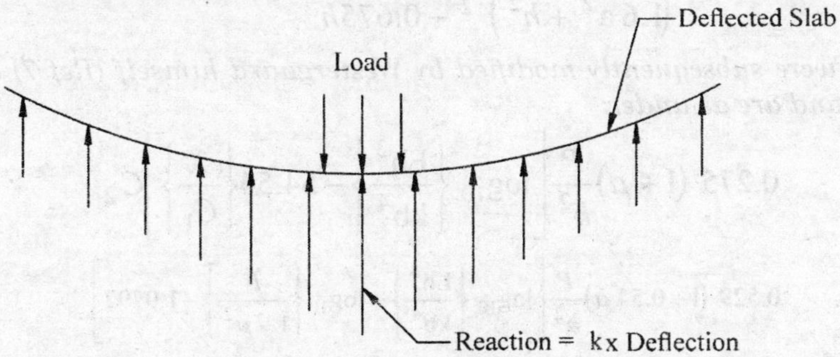

Fig. 14.5: Reaction of Subgrade under Loaded Concrete Slab

4. The slab is of uniform thickness
5. The loading positions are as given in Fig. 14.1.

The radius of relative stiffness (ℓ) which Westergaard used in his analysis is given by the formula:

$$\ell = \sqrt[4]{\frac{Eh^3}{12(1-\mu^2)k}} \qquad (14.15)$$

where ℓ = radius of relative stiffness (in)
E = Modulus of Elasticity (psi)
h = pavement thickness (in)
μ = Poisson's ratio, 0.15
k = Modulus of subgrade (lb/in³)

The three original equations proposed by Westergaard in 1925 (Ref 6) are:

$$f_i = 0.31625\ \frac{P}{h^2}\left[4\log_{10}\left(\frac{\lambda}{b}\right)+1.0693\right] \qquad (14.16)$$

$$f_e = 0.57185 \frac{P}{h^2}\left[4\log_{10}\left(\frac{\lambda}{b}\right) + 0.3593\right] \quad (14.17)$$

$$f_c = \frac{3P}{h^2}\left[1 - \left\{\frac{a\sqrt{2}}{\ell}\right\}^{0.6}\right] \quad (14.18)$$

where f_i = tensile stress at the bottom due to interior loading (psi)

f_e = tensile stress at the bottom due to edge loading (psi)

f_c = tensile stress at the top of the slab due to corner loading (psi)

P = Wheel load (ℓ b)

ℓ = radius of relative stiffness (Eq^n 14.15)

h = slab thickness in inches

a = radius of contact area of wheel in inches

b = $\left(1.6\,a^2 + h^2\right)^{1/2} - 0.675h$

The above formula were subsequently modified by Westergaard himself (Ref 7) and by Teller and Sutherland (Ref 8) and are as under:

$$f_i = 0.275\,(1+\mu)\frac{P}{h^2}\left[\log_{10}\left\{\frac{Eh^3}{kb^4}\right\} - 54.54\left\{\frac{\lambda}{C_1}\right\}C_2\right] \quad (14.19)$$

$$f_e = 0.529\,(1+0.54\,\mu)\frac{P}{h^2}\left[\log_{10}\left\{\frac{Eh^3}{kb^4}\right\} + \log_{10}\left\{\frac{b}{1-\mu^2}\right\} - 1.0792\right] \quad (14.20)$$

$$f_c = \frac{3P}{h^2}\left[1 - \left\{\frac{12\left(1-\mu^2\right)k}{Eh^3}\right\}^{0.3}\left(a\sqrt{2}\right)^{1.2}\right] \quad (14.21)$$

In the above formula, P, μ, h, E a and ℓ are as described earlier and

$$b = \left(1.6\,a^2 + h^2\right)^{1/2} - 0.675\,h \text{ where } a < 1.724\,h \quad (14.22)$$

and $$b = a \text{ where } a \geq 1.724\,h \quad (14.23)$$

C_1 = 5 ℓ and C_2 = 0.2

In the above equation, the radius of contact area is found by dividing the wheel load by the tyre pressure and treating the area as a circle.

Thus $$a = \sqrt{\frac{P}{Q}\frac{1}{\Pi}}$$

where Q = Tyre pressure in MPa

P = Load in N

a = radius of contact area in mm

For dual tyre, the total area of the two tyres plus the rectangular area between them is considered and a is obtained as above.

The above analysis was the procedure adopted in India till 2011, when the procedure was revised based on the Finite Element Method and a software IIT SLAB-II developed by the I.I.T. Kharagpur (Ref 1). The analysis has been done for (i) bottom-up cracking involving combination of load and positive non-linear temperature differential and (ii) top-down cracking involving combination of load and negative linear temperature differential. Single axles, tandem and tridem axles have been considered. Several regression equations have been presented making the designer's task easy. Charts have also been presented giving the flexural stress developed for various combination of axle loads, temperature differential and concrete slab thickness.

14.14. Design of Thickness of CRCP

The thickness of CRCP is desired firstly by treating it as a jointed pavements. In the USA, the thickness was reduced by 20 per cent as the performance of such pavements was not found to be satisfactory (Ref 5). But the recently finalised AASHTO guidelines using the Mechanistic-Empirical approach give a reduced thickness for CRCP (reduction being about 20 per cent).

REFERENCES

1. Guidelines for the Design of Plain Jointed Rigid Pavements (Third Revision), IRC:58-2011, Indian Roads Congress, New Delhi, 2011.
2. Bradbury, R.D., Reinforced Concrete Pavements, Wire Reinforcement Institute, Washington D.C., 1938.
3. Standard Specification and Code of Practice for Construction of Concrete Roads (Fourth Revision), Indian Roads Congress, New Delhi, 2011.
4. Recommended Practice for Sealing of Joints in Concrete Pavements, IRC:57-2006, Indian Roads Congress, New Delhi.
5. Guidelines for Design and Construction of Continuously Reinforced Concrete Pavement (CRCP), IRC:118-2015, Indian Roads Congress, New Delhi, 2015.
6. Westergaard, H.M., Computation of stresses in concrete roads, Proceedings of the Highway Research Board, Washington D.C., 1925.
7. Westergaard, H.M., Analytical tools for judging results of structural tests of concrete pavements, Public Roads 14(10), Washington D.C., 1933.
8. Teller, L.W., and E.C. Sutherland, the structural design of concrete pavements, Part 5, An experimental study of Westergaard analysis of stress conditions in concrete pavements of uniform thickness, Public Roads, 23(8), Washington D.C., 1943.

15 Chapter

Pavement Materials

15.1. Materials Used For Various Specifications

The following are the pavement materials in common use:

(A) Flexible Pavements

1. Stabilised materials
2. Granular Bases and Sub-bases
3. Brick, Stone and Concrete Blocks
4. Bituminous Materials

(B) Concrete Pavements

1. Cement
2. Coarse Aggregates
3. Sand
4. Admixtures and Additives
5. Steel
6. Sealants

15.2. Soil Stabilisation

Soil stabilization is a general term for any physical, chemical, mechanical, or combined method of changing a natural soil to meet an engineering purpose.

15.2.1. Need for Soil Stabilisation

As stone aggregates, which are the main materials used in pavement construction, are getting fastly depleted, alternatives have to be considered. One of the promising techniques is soil-stabilisation, where the local soil or marginal materials is blended with a stabilising agent to improve its strength, eliminate undesirable properties like excessive swelling and shrinkage, reduce compressibility and settlements and bring about economy in road construction.

15.2.2. Types of Soil-stabilisation techniques

The following are the broad soil-stabilisation techniques being adopted:

1. Mechanical Stabilisation
2. Stabilisation by additives

15.2.3. Mechanical Stabilisation

15.2.3.1. Soil-aggregate mixes

This technique involves the careful blending of locally available materials with soil to obtain a dense and stable mix. Soil-aggregate mixtures, sand-clay mixtures, sand-gravel mixtures and soil-overburnt brick ballast mixtures fall under this category. The particle-size distribution is so selected that a close-graded dense mixture results. The grain-size distribution should follow the Fuller's curve given by the equation:

$$P = 100\left(\frac{d}{D}\right)^n \qquad (15.1)$$

where

P = percentage passing any sieve

d = aperture of the sieve in question

D = the maximum size of the aggregate

n = exponential, whose value may be taken as 0.5.

Some typical gradation specifications are given in Table 15.1.

Table 15.1: Gradation Specifications for Soil-Aggregate Stabilised Mixes

Sieve Designation	Percent by weight passing the sieve for a nominal maximum size of		
	80 mm	40 mm	20 mm
80 mm	100	–	–
40 mm	10–100	100	–
20 mm	60–80	80–100	100
10 mm	45–65	55–80	80–100
4.75 mm	30–50	40–60	50–75
2.36 mm	–	30–50	35–60
600 micron	10–30	15–30	15–35
75 micron	0–10	0–10	0–10

In the above mixes, the percent passing 425 micron sieve shall have maximum Liquid Limit of 25 and maximum Plasticity Index of 6.

15.2.3.2. Sand-clay Mixtures

Clay is an undesirable soil in a pavement layer. If it is mixed with sand, the mix can be used as sub-base and base courses. For heavily trafficked roads, for use as a base course the mix should have a minimum soaked CBR of 80. The requirements of Liquid Limit and Plasticity Index given in para 15.2.3. apply to sand-clay mixes also.

15.2.3.3. Stabilisation of soil with soft aggregates

Soft aggregates such as kankar (lime-stone), over-burnt brick ballast (jhama metal) and murram can be used for pavement layers by embedding them in soil mortar. The volume of soil mortar is roughly about one-third of the total volume. This method was introduced in India and is known as the Mehra's method. To improve the layer and make it suitable for higher traffic, stone aggregates can be grafted to a thickness of 25 mm and surface with a bituminous treatment.

15.2.4. Soil-lime stabilisation

Large areas in India have clayey soil, popularly known as Black Cotton Soil. The properties of this soil can be vastly improved by adding lime (CaO). The following reactions take place:

(i) **Ion exchange**: Clay particles are negatively charged with ions of sodium, magnesium, potassium or hydrogen adsorbed on the surface. The strong positively charged ions of calcium present in lime replace the weaker ions of sodium, magnesium, potassium or hydrogen resulting in a preponderance of positively charged calcium ions. The plasticity of the soil is reduced, and flocculation takes place and pulverisation of the soil results.

(ii) **Pozzolonic reaction:** After ion exchange, the additional quantity of lime reacts with the aluminous and siliceous materials in the clayey soil and form cementitious gels in the presence of water. The pozzolonic reactions impart great strength to the mixture (Ref 2, 3, 4).

For effective stabilisation, the soil must have a fraction passing 425 micron sieve not less than 15 per cent and the Plasticity Index should be at least 10. The purity of lime (amount of Calcium Oxide) should be at least 50.

15.2.5. Soil-cement Stabilisation

Stabilisation of soil or granular material is a popular technique, and the following techniques are available:

(i) Soil-cement

(ii) Cement-modified soil

(iii) Cement-bound-granular material (CBGM)

Soil-cement is a mixture of pulverised soil, portland cement and water, compacted to a high density. The quantity of cement required to stabilise soils is given in Table 15.2 (Ref 5, 6).

Table 15.2: Quantity of Cement for Soil Stabilisation

Soil Type as per IS Classification	Usual range in cement requirement (% by weight)
GW, GP, GM	3–5
GM, GP	5–8
GM, GC, SM, SC	5–9
SP	7–11
ML	7–12
ML, MH	8–13
CL, CH	9–15
OH, CH	10 16

The fine grained soil should be pulverised by the addition of lime and at least 80 per cent of the soil should pass through 5.6 mm sieve and there should be no lumps larger than 26.5 mm.

The mix should be compacted to at least 95-100% of the standard Proctor density for low volume roads and heavy compaction (IS:270-Part VIII) for high volume roads. The water added shall be the optimum moisture content. The Unconfined Compressive Strength after 7 days moist curing shall be at least 1.75 MPa (Ref 7). After 12 cycles of wetting and drying, the loss in weight shall be as under (Ref 4).

1. **Base** : Less than 20 per cent
2. **Sub-base:** Less than 30 per cent
3. **Shoulder:** Less than 30 per cent

After mixing soil, gravel, cement and water, the layer (not more than 200 mm) is compacted and cured for 7 days.

15.2.6. Soil/ Gravel stabilised with Lime and Fly-ash

The soil/ gravel can be stabilised with a mixture of lime and fly-ash. The lime fly-ash ratio shall be 1:3, 1:4 or 1:5 depending upon the strength and durability desired. If necessary, 1 per cent of cement can also be added.

15.3. Granular Bases and Sub-bases

15.3.1. Types of Specifications

The various specifications used are:

1. Telford stone soling
2. Gravel bases and gravel sub-bases (GSB)
3. Water-bound-macadam (WBM)
4. Wet Mix Macadam (WMM)

15.3.2. Telford stone soling

This consists of laying single size stone boulders in thickness of 150-225 mm, and packing will spalls into the voids and rolling. This specification suffers from the disadvantages that it is purely labour-oriented and has not interlocking properties. It is thus not in use in India.

15.3.3. Gravel Base and Gravel Sub-base (GSB)

Gravel (popularly known as murram in India) is a naturally occurring material in India. The maximum size of particles should be one-half to one-third the thickness of the compacted layer. The maximum thickness of the compacted layer is 150 mm.

The grading requirements of granular sub-base are given in Table 15.3 (Ref 7).

Table 15.3: Grading for Granular Sub-Base

IS Sieve Designation	Percent By Weight Passing IS Sieve					
	Grading I	Grading II	Grading III	Grading IV	Grading V	Grading VI
75.0 mm	100	–	–	–	100	–
53.0 mm	80–100	100	100	100	80–100	100
26.5 mm	55–90	70–100	55–75	50–80	55–90	75–100
9.50 mm	35–65	50–80	–	–	35–65	55–75
4.75 mm	25–55	40–65	10–30	15–35	25–50	30–55
2.36 mm	20–40	30–50	–	–	10–20	10–25
0.85 mm	–	–	–	–	2–10	–
0.425 mm	10–15	10–15	–	–	0–5	0–8
0.075 mm	<5	<5	<5	<5	–	0–3

Gradings III and IV are preferably used in lower sub-base. Gradings V and VI are used as sub-base-cum-drainage layer. The thickness of each layer shall not be less than 150 mm.

The physical requirements of Materials for Granular Sub-base are given in Table 15.4 (Ref 7).

Table 15.4: Physical Requirements of Materials for Granular Sub-Base

Aggregate Impact Value	40 maximum
Liquid Limit	25 maximum
Plasticity Index	6 maximum
CBR at 98% Dry Density as per IS:2720-Part VIII	Minimum 30

15.4. Brick, Stone and Concrete Blocks

15.4.1. Brick Pavement

In the alluvial plains of North India, stone aggregates are to be obtained from long leads. Hence, bricks are used for:

(i) sub-base for important roads
(ii) base layer for middle category of roads
(iii) surfacing layer of unimportant roads like village roads
(iv) shoulders for single lane roads
(v) service lanes, footpaths and cycle tracks in towns and cities

Bricks, which should be over-burnt, should have a compressive strength of at least 27.5 MN/m^2. The water absorption should be in the range of 5-8 per cent.

Bricks are laid on a prepared sub-grade over which a cushion of 25-40 mm sand is introduced. The bricks can be laid as brick-on-edge or flat. The pattern of laying can be (i) header and stretcher bond (ii) herring bone bond or (iii) transverse bond.

15.4.2. Stone Block Pavement

Stones have a high hardness and stone blocks are used where heavy traffic is expected such as warehouses, godowns, harbours, railway yards etc. The blocks should have a minimum thickness of 150 mm and are laid over a layer of sand 50-63 mm thick. They can also be laid over a cement-sand mortar layer. The joints are filled with cement-sand mortar or sand-bitumen.

15.4.3. Interlocking Concrete Block Pavement

Interlocking concrete block pavements are used for:

1. Footpaths and side-walks
2. Cycle tracks
3. Residential streets
4. Car Parks
5. Fuel Stations
6. Rural Roads through villages
7. Roadside Rest Areas
8. Bus Depots
9. Approaches to Railway Level Crossings

10. City Streets
11. Truck Parking Areas
12. Industrial Floors
13. Warehouses and godowns
14. Container Depots
15. Port Wharf and Roads
16. Roads in high altitudes

They have a long life, are practically maintenance-free, unaffected by oil spillage, can be laid quickly and do not require curing at site and can be opened to traffic quickly. Since they are factory-made, their strength and quality are assured.

The compressive strength should be a minimum of 30 MPa (Ref 8). The suggested thickness of the blocks is as under:

1. Light Traffic, (footpaths, cycle tracks, residential streets) 60 mm
2. For Medium traffic 80 mm
3. For Heavy Traffic in Container Yards, Ports, Warehouses 100–120 mm

The blocks are laid over a cushion of sand 20-40 mm thick. The joints are filled with fine sand.

The base and sub-base thickness depends upon the CBR of the subgrade. Some guidance is given in Table 15.5.

Table 15.5: Suggested Thickness in mm of Sub-Base and Base for Interlocking Concrete Block Pavements

Type of Road	Layer	CBR %	
		Above 10	5–10
1. Cycle Tracks, Pedestrian Footpaths	Base	200	200
2. Residential Streets and Axle Load Repetitions less than 10 msa	Sub-base	200	250
	Base (WBM/WMM)	250	250
3. Collector Streets, Industrial Streets, Truck Parking Areas, Axle Load Repetitions 10–20 msa	Sub-base	200	250
	Base (WBM/WMM)	250	250
4. Arterial Streets, Axle Load Repetitions 20–50 msa Container Yards, Ports	Sub-base	200	250
	Base (WBM/WMM)	250	250
	or WBM/WMM	150	150
	and DLC over it	75	75

The blocks should have adequate edge restraint made with cement concrete strips or kerb or kerb and channel.

The water entering through the joints should be drained away. A typical arrangement is given in Fig. 15.1.

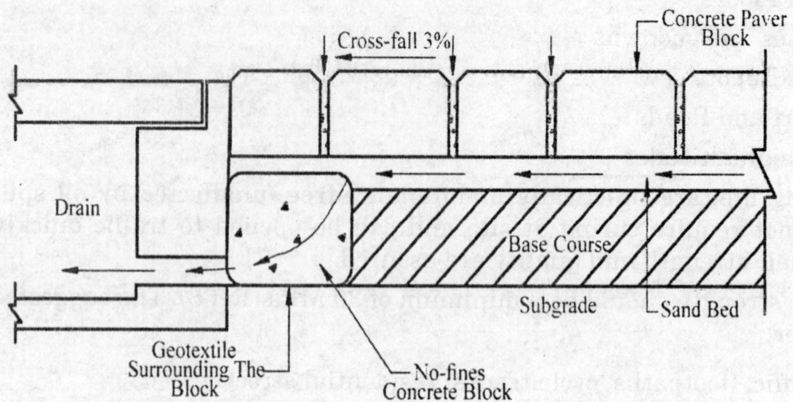

Fig. 15.1: Surface Drainage in a Block Pavement

15.5. Bituminous Materials

15.5.1. Terms used

Bitumen is the residue obtained after the fractional distillation of petroleum crude. The term "bitumen" is used in the British and Indian Specifications, whereas the term "asphalt" is used in the American Specifications.

A "straight-run-bitumen" is a bitumen whose viscosity composition has not been adjusted by blending or softening by fluxing, cutting back oil or other treatment.

A "penetration grade bitumen" is a bitumen whose degree of hardness can be measured by the standard penetration test.

A "viscosity grade (VG)" bitumen whose properties are defined by its viscosity rather than by standard penetration test.

A "cut-back" in a bitumen is one whose viscosity has been reduced by a volatile diluent like kerosene, so that it can be used without heating.

An "emulsion" is a bituminous binder which is in a liquid form rendered so by the addition of an emulsifier.

A "cationic emulsion" is one having positively charged particles.

An "anionic emulsion" is one having negatively charged particles.

15.5.2. Properties of Bitumen

The general properties of bitumen are given below:

1. They are predominantly hydrocarbons
2. They are soluble in carbon disulphide (CS_2)
3. They soften on heating
4. They are insoluble in water
5. They are impermeable to the passage of water
6. They are chemically inert
7. They oxidise slowly

15.5.3. Requirements of bitumen for use in roads

For use in roads, bitumen should possess the following properties:

1. The bitumen binder should be fluid enough at the time of mixing with aggregates by providing a thin film.
2. It should have low temperature susceptibility, *i.e.*, it should exhibit little change in viscosity with change in temperature.
3. It should have enough amounts of volatiles in it and it should not lose them excessively when subjected to higher temperature.
4. It should be ductile and not brittle.
5. It should have good affinity to the aggregates and should not cause stripping in the presence of water.
6. It should be capable of being heated to the design temperature without causing fire hazard.

15.5.4. Tests on Bitumen

15.5.4.1. Viscosity

Viscosity is the resistance offered by the fluid to its flow. It is thus opposite to fluidity. It is measured by viscometers. As per the current IS, paving bitumen are available in four grades, VG 10, VG 20, VG 30 and VG 40. The VG 10 is the softest, whereas VG 40 is the hardest. VG 10 corresponds to penetration grade 80-100, VG 20 corresponds to penetration grade 60-80, VG 30 corresponds to penetration grade 50-70 and VG 40 corresponds to penetration grade 40-60.

15.5.4.2. Penetration

An indirect measure of viscosity is the amount of penetration of a standard needle under the standard prescribed load (100 gms), temperature (25^{0}C) and time in which the penetration is recorded (5 secs) (Fig. 15.2).

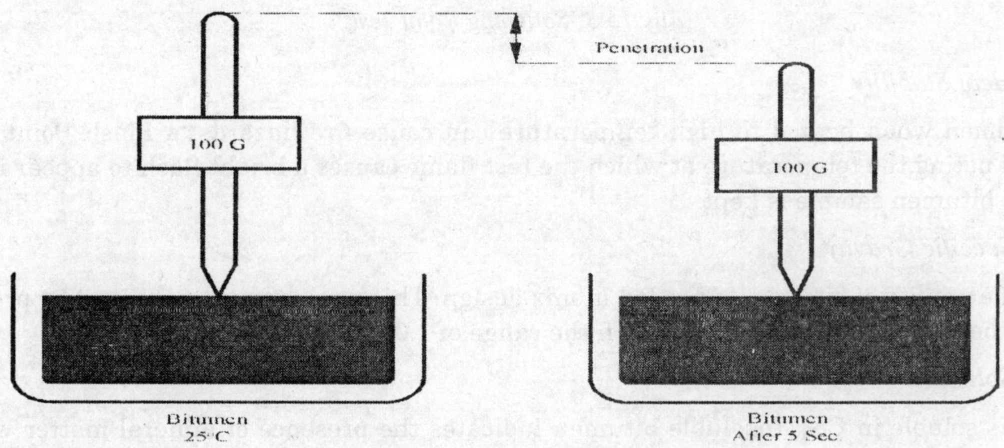

Fig. 15.2: Penetration Test

15.5.4.3. Softening Point

Bitumen loses its hardness on heating, but there is no defined softening point. The standard test to determine the softening point is the "Ring and Ball Test" under which a brass ring containing the bitumen is suspended in water or glycerine at a given temperature. A steel ball is placed on the disc of bitumen (Fig. 15.3). The liquid medium is then heated at 5^{0}C per minute. The temperature at which the softened bituminous material touches the bottom plate at a specified distance below the ring is recorded as the softening point. Bitumen VG 40 has a softening point of 50–65^{0}C and the softening point of VG 10 bitumen is in the range 30–45^{0}C.

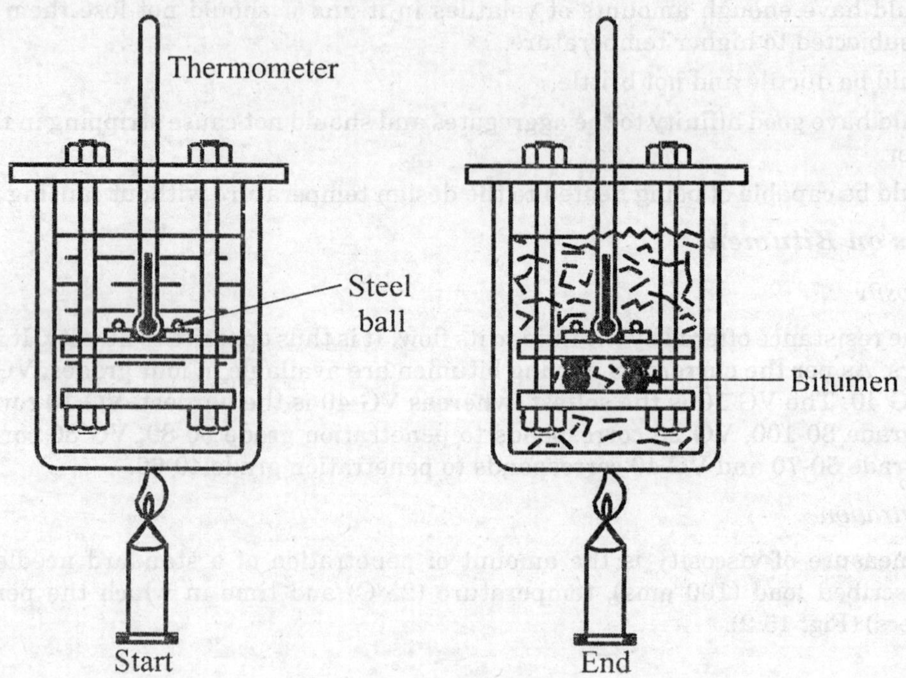

Fig. 15.3: Softening Point Test

15.5.4.4. Heat Stability

Since bitumen when heated to high temperature can cause fire hazards, a Flash Point Test is conducted noting the temperature at which the test flame causes a bright flash to appear in a cup where the bitumen sample is kept.

15.5.4.5. Specific Gravity

The specific gravity of bitumen is needed in mix design. This property is determined by preparing a cube of the same. The value is usually in the range of 1.02–1.04.

15.5.4.6. Solubility

Bitumen is soluble in CS_2. Insoluble bitumen indicates the presence of mineral matter which is undesirable. Indian specifications require 99 per cent solubility.

15.5.4.7. Thin Film Oven Test

When bitumen is mixed with aggregates in a hot-mixed plant it would tend to age harden This is determined by the Thin Film Oven Test and Rolling Thin Film Oven Test.

The requirements of Paving Bitumen in India are given in Table 15.6.

Table 15.6: Requirements of Paving Bitumen and IS Tests

Property	Viscosity Grade of Paving Bitumen				IS Test
	VG 10	VG 20	VG 30	VG 40	
1. Absolute Viscosity at 60°C, Poise (Minimum)	800	1600	2400	3200	IS 1206 (Part 2)
2. Kinematic Viscosity at 135°C, centistoke (Minimum)	250	300	350	400	IS 1206 (Part 3)
3. Flash Point	220	220	220	220	IS 1209
4. Solubility in CS_2 (Minimum %)	99.0	99.0	99.0	99.0	IS 1216
5. Penetration at 25°C	80–100	60–80	50–70	40–60	IS 1203
6. Softening Point °C (Minimum)	40	45	47	50	IS 1205
7. Tests on residue from Thin Film Oven Test/ Rolling Thin Film Oven Test					
(1) Viscosity ratio at 60°C (Max)	4.0	4.0	4.0	4.0	IS 1206 (Part 2)
(2) Ductility at 25°C (Min)	75	50	40	25	IS 1208

15.5.5. Cut-back

In order to avoid heating bitumen to bring down its viscosity, a volatile diluent is added to make it suitable to coat the aggregates with a thin film. There are three types of cut-backs

Rapid Curing (RC)	....	Bitumen fluxed with naptha type diluents
Medium Curing (MC)	....	Bitumen fluxed with Kerosene
Slow Curing (SC)	....	A liquid residue produced in the refining process, containing little or no volatile constituents

The use of cut-backs is not favoured because the diluent can find its way to water bodies and contaminate it.

15.5.6. Emulsions

Emulsions are widely used for cold-mix specifications which do not require a hot-mix plant and find application in repair works and for cold-weather work. But they are costly and contain 50-65 per cent bitumen, the rest being water. On application to the road, the emulsion breaks down, changing its brown colour to black when the water evaporates leaving the bitumen to provide a film over the aggregates.

The emulsions can be rapid setting (RS), medium setting (MS) or slow setting (SS). As mentioned earlier, emulsions can be anionic or cationic. Polymer modified emulsions are also available which enhance the properties of bitumen. The prefix C is added when the emulsions is cationic. Thus CRS indicates a rapid setting cationic emulsion.

15.5.7 Guide to the selection of bituminous binders

The following general guidelines give the correct bituminous binder for various types of work.

Table 15.7: Guidelines for Selection of Bituminous Binder

Type of Binder	Application
1. VG 40 (30/40 Penetration)	Hot mix bituminous macadam, bituminous concrete and penetration macadam in normal summer temperatures and in plains.
2. VG 30 (60/70 Penetration)	1. Hot mix bituminous macadam, bituminous concrete in normal summer temperatures and in plains. 2. Penetration macadam in winter and at high altitudes
3. VG 20 (80/100 Penetration)	1. Surface Dressing, seal coat, premixed chipping carpet in normal summer temperatures and in plains. 2. Penetration Macadam in summer temperature and in plains. 3. Hot mix bituminous macadam and bituminous concrete in summer temperatures and in plains. 4. Hot mix bituminous macadam and bituminous concrete in winter temperatures and at high altitudes.
4. Cut-back RC	1. Surface dressing and seal coats at normal summer temperatures and in plains. 2. Surface dressing at winter temperatures and at high altitudes. 3. Cold mix macadam and bituminous carpets at summer and winter temperatures, both in plains and at high altitudes.
5. Cut-back MC and SC	Priming
6. Emulsions (Ref 11)	1. Prime Coat and Tack Coat 2. Cold weather work 3. Wet condition of aggregates 4. Maintenance and Repair works 5. Sealing of fine cracks 6. Work in remote areas

15.5.8. Modified Bitumen

15.5.8.1. Advantages of Modified Bitumen

The properties of bitumen can be vastly enhanced by adding small quantities of modifiers. The advantages are (Ref 9):

1. Lower susceptibility to temperature variations
2. Higher resistance to cracking and reflective cracking
3. Higher resistance to deformation at higher temperature
4. Better age resistance properties

5. Better adhesion between aggregates and binder
6. Higher fatigue life
7. Overall improved performance

15.5.8.2. Types of Bitumen Modifiers

The bitumen modifiers can be classified as given in Table 15.8 (Ref 9).

Table 15.8: Classification of Rubber and Polymer-Based Bitumen Modifiers

Type of Modifiers	Examples	Indicative Dose (Per cent by weight of bitumen)
Plastomeric	Polyethylene (PE), Ethylene Vinyl Acetate (EVA), Ethylene Butyl Acrylate (EBA), Ethylene Methyl Acrylate Copolymers (EMA)	3–6
Elastomeric	Styrene Isoprene Styrene (SIS), Styrene-Butadiene-Styrene (SBS), Styrene-Butadiene Rubber, Ethlylene Ter Polymer (ETP)	3–5
Synthetic Rubber Latex	Styrene Butadiene Rubber Latex and any other synthetic rubber	3–5
Natural Rubber	Latex or Rubber Powder	2–4
Crumb Rubber or Treated Crumb Rubber	Crumb Rubber, Treated Crumb Rubber	10–12

15.5.8.3. Use of Waste Plastic

Waste Plastic in the form of polythene bags and other non-bio-degradable waste like bottles etc. can be advantageously used to modify the properties of bitumen and when used can enhance the properties of bitumen as described in para 15.5.7.1. The waste plastic is shredded and added to the mix in the hot-mix plant (Ref 10).

15.6. Bituminous Courses in Highway Pavements

15.6.1. Types of Bituminous Courses

Bituminous Courses is a flexible pavement and is of the following types:

1. Prime Coats and Tack Coats
2. Surface Dressing
3. Seal Coats
4. Thin open-graded premix carpets
5. Base Course
6. Dense surfacings

15.6.2. Prime Coat

A prime coat is an initial application of low viscosity bituminous material to an absorbent surface like granular base, Wet Mix Macadam, Water Bound Macadam etc. with the objective of plugging the voids and coat the loose mineral particles. The binder chosen is a slow setting emulsion. Medium setting (MC) or slow setting (SC) cut-backs can also be used with sub-zero temperature. The quantity of primer is 0.6–0.9 Kg per sq. m for surfaces with medium porosity and 1.2–1.5 Kg per sq. m for surfaces with high porosity.

15.6.3. Tack Coat

A tack coat is provided for bonding the new bituminous layer to an old one. There is a strong feeling that a primed surface does not need a tack coat. The quantity of binder is 0.2–0.25 Kg per sq. m for a normal bituminous surface and 0.25–0.30 Kg per sq. m for a dry and hungry bituminous surface. The binder used is a slow setting emulsion. Cut-backs should be used at sites at sub-zero temperatures.

15.6.4. Surface Dressing

Surface dressing is a process wherein a thin film of bituminous binder is sprayed on the road surface, covered with a coat of stone aggregates and rolled. It can be applied as one coat or two coats. The aggregates can be precoated with bitumen to reduce the whip-off due to traffic. The binder used is a low viscosity bitumen, VG-10. A rapid curing cut-back RC-3 and emulsion can also be used. The nominal sizes of aggregates normally specified are (Ref 12, 13).

First coat of surface dressing : 12 mm
Second coat of surface dressing: 10 mm

The quantities of aggregates recommended by IRC are (Ref 12, 13).

First coat : 0.14 – 0.15m^3 per 10 sq. m
Second coat : 0.09 – 0.11m^3 per 10 sq. m

The quantity of binder used in Kg per 10 sq. m is:

	Bitumen	**Cut-back**
First coat	17.2–19.5	19.5–22
Second coat	9.8–12.6	12.2–14.6

15.6.5 Open-graded Premix Carpet (PC)

This specification is used as a thin surfacing layer of 20 mm thick and is a popular specification in India for roads carrying low and medium traffic. The carpet is provided with a seal coat to protect water entering through the voids. The quantities of aggregates and binder are given below:

Table 15.9: Materials for 20 mm Carpet (Per 10 Sq M)

Material	Quantity
1. Aggregates for premixing	
12 mm size	0.18 cu m
10 mm size	0.09 cu m
2. Aggregates for Seal Coat	
Low rainfall area	0.06 cu m
High rainfall area	0.09 cu m
3. Bitumen for premixing	14.6 Kg
4. Bitumen for Premixed Seal Coat in low rainfall area	6.8 Kg
5. Bitumen for Liquid Seal Coat in high rainfall area	9.8 Kg

15.6.6. Built-up Spray Grout (BUSG)

This specification was popular in India till recently, but is not recommended any more now. The work consists of the following operations:

1. Apply a tack coat
2. Spread coarse aggregates 50 mm down at the rate of 0.5 cu m per 10 sq m and roll it.
3. Apply binder at 12.5 to 15.0 Kg per 10 sq m
4. Apply second layer of coarse aggregates at 0.5 cu m per 10 m^2 and compact it.
5. Apply binder at 12.5 to 15 Kg per 10 sq. m.
6. Spread Key aggregates (20 mm down) at 0.13 cu m per 10 sq. m and roll.
7. Provide a wearing course of 20 mm premix carpet.

15.6.7. Bituminous Macadam

Bituminous Macadam (BM) is an open-graded mix where aggregates are mixed with bitumen and is used as base and binder course.

The following two gradings are used (Ref 7).

Table 15.10: Grading of Aggregates and Bitumen Content for Bituminous Macadam

Grading	1	2
Nominal Maximum Aggregate Size (mm)	40	10
Layer Thickness (mm)	80–100	50–75
I.S. Sieve (mm)	Cumulative % by weight of total aggregates passing	
45	100	
37.5	90–100	
26.5	75–100	100
19	–	90–100
13.2	35–61	56–88
4.75	13–22	16–36
2.36	4–19	4–19
0.3	2–10	2–10
0.075	0–8	0–8
Bitumen Content (per cent by mass of total mix)	3.3	3.4

The mixing is done in a hot-mix plant. The mixing, laying and rolling temperatures are given in Table 15.11 (Ref 7).

Table 15.11: Mixing, Laying and Rolling Temperatures for Bituminous Mixes (Degree Celsius)

Bitumen Grade	Bitumen Temperature	Aggregate Temperature	Mixed Material Temperature	Laying Temperature	Rolling Temperature
VG 40	160–170	160–175	160–170	150 Min	100 Min
VG 30	150–165	150–170	150–165	140 Min	90 Min
VG 20	145–165	145–170	145–165	135 Min	85 Min
VG 10	140–160	140–165	140–160	130 Min	80 Min

The rolling is done in the following sequence:

(i) Initial or break-down rolling with 8–10 tonne static weight smooth wheel roller.

(ii) Intermediate rolling with 8–10 tonne static weight or vibratory roller or pneumatic tyre roller of 12 to 15 tonne weight with a tyre pressure of at least 0.56 MPa.

(iii) Finish rolling with 6–8 tonne tandem roller.

15.6.8. Dense Bituminous Macadam (DBM)

Dense Bituminous Macadam is used for Base Courses.

The following two gradings are commonly adopted for DBM.

Table 15.12: Composition of DBM

Grading	1	2
Nominal Size of Aggregate (mm)	37.5	26.5
Layer Thickness (mm)	75–100	50–75
I.S. Sieve (mm)	Cumulative % by Weight of Total Aggregate Passing	
45	100	
37.5	95–100	100
26.5	63–93	90–100
19	–	71–95
13.2	55–75	56–80
9. 5	–	–
4.75	38–54	38–54
2.36	28–42	28–42
1.18	–	–
0.6	–	–
0.3	7–21	7–21
0.15	–	–
0.075	2–8	2–8
Bitumen Content % by mass of total mix	Min 4.0	Min 4.5

The Bitumen Content is determined from the Marshall Mix Design procedure. The requirements of the mix are given in Table 15.13.

Table 15.13: Requirements of Dense Bituminous Macadam

Properties	Viscosity Grade Bitumen	Modified Bitumen	Cold Climate
Compaction Level	75 Blows on each face of specimen		
Min. Stability (kN at 60°C)	9.0	12.0	10.0
Marshall Flow (mm)	2–4	2.5–6	3.5–5
Marshall Quotient $\left(\frac{\text{Stability}}{\text{Flow}}\right)$	2–5	2.5–5	2.5–5
% Air Voids	3–5	3–5	3–5
% Voids Filled with Bitumen	65–75	65–75	65–75
Coating of Aggregates	95% Min	95% Min	95% Min
Tensile Strength Ratio	80% Min	80% Min	80% Min

The mixing and laying procedures are the same as given in Para 15.6.7.

15.6.9. Bituminous Concrete (BC)

Bituminous Concrete is used as the surface layer. The composition of two commonly adopted gradings are given in Table 15.14.

Table 15.14: Composition of Bituminous Concrete Layers

Grading	1	2
Nominal Aggregate Size (mm)	19	13.2
Layer Thickness (mm)	50	30–40
I.S. Sieve (mm)	Cumulative % by weight of total aggregate passing	
26.5	100	
19	90–100	100
13.2	59–79	90–100
9. 5	52–72	70–88
4.75	35–55	53–71
2.36	28–44	42–58
1.18	20–34	34–48
0.6	15–27	26–38
0.3	10–20	18–28
0.15	5–13	12–20
0.075	2–8	4–10
Bitumen Content % by mass of total mix	Min 5.2	Min 5.4

The mix is designed as per the Marshall method. The requirements are the same as given in Table 15.12 for DBM. The mixing and laying procedures are the same as given in para 15.6.7.

15.6.10. Penetration Macadam

This specification consists of a compacted layer of stone aggregates into which a bituminous binder is allowed to penetrate and fill the voids. After the bitumen is applied, a layer of key aggregates is spread on the surface and rolled. A seal coat is provided to make the surface impervious to water. The specification is used where hot-mix plant is not available. The quantity of bitumen is 50 Kg per 10 sq. m for 75 mm layer. The quantity of aggregates is 0.6 cu m per 10 sq. m of layer of thickness 50 mm, and 0.9 cu m per 10 sq m of layer of thickness 75 mm. The quantity of key aggregates (Ref) is 0.15 cu m and 0.18 cu m (per 10 sq m) for a layer of thickness 50 mm and 75 mm respectively.

15.6.11. Semi Dense Bituminous Concrete (SDBC)

The Semi-Dense Bituminous Concrete is an intermediate type of surfacing between the open textured premix carpet (PC) (para 15.6.5) and the Bituminous Concrete (BC) (para 15.6.9). The voids in the mixture is in the range of 5–10 per cent as against 3–5 per cent in the BC. The specification is not favoured for high traffic roads, and is adopted for low traffic roads. SDBC is laid in thickness of 20 mm or 25 mm. The gradings commonly adopted conform to Type III(a) and III(b) mixes of the Asphalt Institute (Ref 15) and are given below.

Table 15.15: Gradation of Aggregates for Semi-Dense Bituminous Concrete

Sieve Designation	Percent by weight passing	
	20 mm Thickness	25 mm Thickness
20 mm	–	100
12.5 mm	100	75–100
10 mm	75–100	60–85
4.75 mm	35–55	35–55
2.36 mm	20–35	20–35
600 micron	10–22	10–22
300 micron	6–16	6–16
150 micron	4–12	4–12
75 micron	2–8	2–8

The binder content is in the range of 4.5 to 6.0.

15.6.12. Mastic Asphalt

Mastic asphalt is a mixture of bitumen, miner filler, fine stone aggregates in such a proportion that a voidless compact mass is produced. The mixture is fluid at about 200^{0}C, but solidifies on cooling. Because of its toughness, it is used in city streets carrying high volume of traffic and in bridge decks. The bitumen is an industrial grade bitumen, and its quantity is in the range of 14–17 per cent by weight of total mix. Fine aggregates shall pass 2.36 mm sieve. The filler is usually limestone powder. The coarse aggregates shall conform to the following grading for thickness of 25–50 mm and its per cent in the total mix by weight shall be 30–50.

Table 15.16: Grading of Coarse Aggregates of Mastic Asphalt

I.S. Sieve	Per cent passing IS Sieve
19 mm	100
13.2 mm	88–96
2.36 mm	0–5

The mastic asphalt is prepared in a cooker at a temperature of 170–180°C, and laid in a width of 1 m confined between angle irons. To impart skid resistance, precoated chips of size 9.5 mm–13.2 mm are spread at a rate of 0.05 per 10 sq m and pressed into the surface when the mastic has a temperature of 80–100°C.

15.6.13 Stone Matrix Asphalt (SMA)

Stone Matrix Asphalt, also known as Stone Mastic Asphalt, is a gap graded mixture with about 70–80 per cent stone content, 6–7 per cent binder, 8–12 per cent of cellulose filler and 0.3 per cent of fibres. It is laid to a thickness of 30–75 mm. It is rut resistant and is used for heavily trafficked sections. The mixing is done in a hot-mix plant.

15.6.14. Slurry seal

Slurry Seal is a treatment used for maintenance of bituminous surface. It is a mixture of slow setting cationic emulsion, stone aggregates, filler (cement) and water (Ref 16). It is prepared by mixing the ingredients in a truck mounted equipment and laid by the spreader box which is a part of the tuck mounted machine. Rolling is done by pneumatic tyred roller.

15.6.15 Microsurfacing

Microsurfacing is a mixture of modified bitumen emulsion, stone aggregates, water and additives and is used for preventive maintenance as well as periodic renewals (Ref 16). The mixing and laying is done in a truck-mounted equipment.

15.6.16 Bituminous Cold Mix

Bituminous Cold Mixes are used in locations where hot-mix plants are not available, in remote areas and for maintenance operations. They can be used for preparing Bituminous Macadam (for base courses and pavement strengthening), Semi Dense Bituminous Concrete and storable mixes for maintenance operations (pot-hole filling and patch repairs) (Ref 17).

15.6.17 Mix-seal Surfacing

This specification is an open-graded surfacing about 20 mm thick. The gradation and bitumen contents are as under:

Table 15.17: Mix Seal Specifications

Gradation of Aggregates		
Sieve Size (mm)	Cumulative per cent by weight passing sieve	
	Type A	Type B
13.2	–	100
11.2	100	88–100
5.6	52–88	31–52
2.8	14–38	5–25
90 micron	0–5	0–5
Quantity of Aggregates (cu m per 10 sq m)	0.27	0.27
Quantity of Binder (Kg per 10 sq m)	22.0	19.0

15.7. Materials for Concrete Pavements

15.7.1. Materials used for Concrete Pavements

The following materials are used for concrete pavements (Ref 18).

1. Cement
2. Aggregates
   Coarse Aggregates
   Fine Aggregates
3. Admixtures and Additives
4. Steel
5. Sealants

15.7.2. Cement

The following varieties of cement are used:

1. Ordinary Portland Cement (OPC) 53 Grade, IS:12269
2. Ordinary Portland Cement (OPC) 43 Grade, IS:8112
3. Portland Pozzolana Cement (PPC), IS:1489 (Part 1) with flyash content not more than 20 per cent by weight of PPC
4. Portland Slag Cement (PSC), IS:455 (with Granulated Blast Furnace Slag Content not more than 50 per cent by weight of PSC)

15.7.3. Coarse Aggregates

The coarse aggregates shall be obtained by crushing rock. The water absorption shall not be more than 2 per cent. The maximum size shall be 26.5 mm for Dry Lean Concrete and 31.5 mm for pavement concrete. The Los Angeles Abrasion Value shall not exceed 35 per cent.

15.7.4. Fine Aggregates

Fine aggregates shall be natural sand or crushed stone sand, and shall conform to IS:383.

15.7.5. Combined Gradation of Coarse and Fine Aggregates

The coarse and fine aggregates shall be mixed to give a combined gradation given in Table 15.18.

Table 15.18: Combined Gradation of Coarse and Fine Aggregates for Concrete Pavements

I.S. Sieve	Percentage by weight passing the Sieve	Per cent passing IS Sieve
	For DLC	For Pavement Concrete
31.5 mm	-	100
26.5 mm	100	85–95
19.0 mm	75–95	68–88
9.5 mm	50–70	45–65
4.75 mm	30–55	30–55
2.36 mm	17–42	-
600 micron	8–22	8–30
300 micron	7–17	-
150 micron	2–12	5–15
75 micron	0–10	0–5

15.7.6. Admixtures and Additives

15.7.6.1. Admixtures

Plasticisers and Superplasticisers are used for improving the workability of concrete. The maximum quantity of these shall be 2 per cent by weight of cementitious materials. Silica Fume upto 10 per cent by weight of cementitious materials is added to increase the abrasion resistance of the concrete (Ref 19).

15.7.6.2. Additives

The most common additive used is flyash. The quantity of flyash shall be upto a maximum of 20 per cent by weight of the cementitious material. Site mixing may be allowed if proper facility is available. Granulated Blast Furnace Slag (GBFS) may be used upto 50 per cent by weight or Portland Slag Cement provided it is mixed in the factory.

15.7.7 Steel

Steel used for dowel bars shall be MS round bars with yield strength of 240 MPa. Steel used for tie bars shall be deformed/ plain bars and of Grade Fe 500. Steel for CRCP can be Grade Fe 415 or TMT Bars of Fe-500 (Ref 20).

15.7.8. Sealant

Sealants can be hot-poured, cold-poured or pre-formed. Hot-poured sealant can be rubberised bitumen. Cold-poured sealants commonly used are Polysulphide and Silicon (Ref 21). Preformed sealant strips are also used by inserting the strip into the groove.

REFERENCES

1. Mehra, S.R., A method of soil stabilisation for low cost roads, Paper No. 110, Journal of the Indian Roads Congress, Vol X-3, New Delhi, 1946.
2. Kadiyali, L.R., A critical review of the Theory and Practice of Lime-soil Stabilisation, Dissertation, University of Newcastle-Upon-Tyne, 1976.
3. Recommended Design Criteria for use of Lime-soil mixes in Road Construction, IRC:51-1973, Indian Roads Congress, New Delhi, 1973.
4. Guidelines for Soil and Granular Material Stabilisation using Cement, Lime and Fly-ash, IRC:SP:89-2010.
5. Use of Soil-Cement as Pavement Layer, Cement Manufacturers' Association, New Delhi, 1992.
6. Soil-cement Laboratory Handbook, Portland Cement-Association, Skokie, Illinois, 1956.
7. Specifications for Road and Bridge Works, Ministry of Road Transport & Highways, Published by the Indian Roads Congress, New Delhi, 2013.
8. Guidelines for the Use of Interlocking Concrete Block Pavements, IRC:SP:61-2004, Indian Roads Congress, New Delhi, 2004.
9. Guidelines on use of Modified Bitumen in Road Construction, IRC:53-2010, Indian Roads Congress, New Delhi, 2010.
10. Guidelines for the Use of Waste Plastic in Hot Bituminous Mixes (Dry Process) in Wearing Courses, IRC:SP:98-2013, Indian Roads Congress, New Delhi, 2013.
11. Use of Cold Mix Technology in Construction and Maintenance of Roads using Bitumen Emulsions, IRC:SP:100-2014, Indian Roads Congress, New Delhi, 2014.
12. Tentative Specifications for single coat bituminous surface dressing, IRC:17-1965, Indian Roads Congress, New Delhi, 1965.
13. Tentative Specifications for two coat bituminous surface dressing, IRC:23-1966, Indian Roads Congress, New Delhi, 1966.
14. Recommended Practice for Bituminous Penetration Macadam, IRC:20-1968, Indian Roads Congress, New Delhi, 1968.
15. Specifications and Construction Methods for Hot-Mix Asphalt Paving for Streets and Highways, Specification Series No. 1, The Asphalt Institute, Maryland, 1957.
16. Tentative Specifications for Slurry Seal and Microsurfacing, IRC:SP:81-2008, Indian Roads Congress, New Delhi, 2008.
17. Use of Cold Mix Technology in Construction and Maintenance of Roads Using Bitumen Emulsions, IRC:SP:100-2014, Indian Roads Congress, New Delhi, 2014.
18. Standard Specifications and Code of Practice for Construction of Concrete Roads, IRC:15-2011, Indian Roads Congress, New Delhi, 2011.
19. Guidelines for Use of Silica Fume in Rigid Pavement, IRC:114-2013, Indian Roads Congress, New Delhi, 2013.
20. Guidelines for Design and Construction of Continuously Reinforced Concrete Pavement, IRC:118-2015, Indian Roads Congress, New Delhi.
21. Recommended Practice for Sealing of joints in Concrete Pavements, IRC:57-2006, Indian Roads Congress, New Delhi, 2006.

16 Chapter

Construction of Pavements

16.1. Construction of Soil-Stabilised Roads

16.1.1 Technology adopted

The following types of technologies are adopted for soil-stabilised roads:

(i) Labour intensive

(ii) Machinery intensive

(iii) Intermediate technology

16.1.2. Labour Intensive Methods

If the soil is clayey, and stabilisation with lime, cement or flyash is adopted, it is very necessary to pulverise the clay lumps. Clods can be broken by using pick-axes or rammers. The additives like lime, flyash or cement are then placed in the required quantity and are spread by manual means. Water to the required quantity is added in stages and the materials are mixed. Compaction is done by rollers.

16.1.3 Machinery Intensive Methods

Three possibilities exist:

(i) Mix-in-place, using equipment such as rippers, rotary tillers, ploughs, scarifiers and disc harrows, watering by a spraying tanker, again mixing with the above equipment and compaction by rollers. A single pass stabiliser performs all these operations in one go.

(ii) Travelling plant method, which travels along the job, picks up the soil and stabiliser, mixes in a mixer, discharges the mixed material to a predetermined loose layer thickness. Compaction is separately done by rollers.

(iii) Stationary Plant where the soil, additives and water are mixed in a centrally located plant, and conveying the mix to the site, laying and compacting.

16.1.4. Intermediate or Appropriate Technology

This uses simple agricultural attachments such as disc harrows, disc ploughs, grader blades and rotillers which can be towed by an agricultural tractor. Watering is done by a water-sprinkler, also towed by an agricultural tractor. Rolling is done by a roller.

16.2. Gravel Courses

Naturally occurring gravel or material mixed with stone aggregates, sand and soil is laid and spread to obtain a maximum compacted thickness of 150 mm. Water in the quantity needed to achieve the Optimum Moisture Content is added and the mix is rolled with a 8-10 tonne smooth wheel roller to obtain the desired density with a number passes predetermined by a trial stretch.

16.3. Water Bound Macadam

The required quantity of aggregates is spread manually or using mechanical spreaders. The material is dry rolled with a 8-10 tonne static roller or a vibratory roller (if available). Wet rolling is then done adding copious amount of water. Screenings to the desired quantity are spread in thin layers and water is sprinkled. Hand brooms are used to sweep the material and facilitate the filling of voids. Binding material is then spread followed by watering. Rolling is then done with a 8-10 tonne smooth wheel roller or vibratory roller if available. A layer of coarse sand (0.06 cu m per 10 sq. m) is then spread followed by slight sprinkling with water and rolling.

16.4. Wet Mix Macadam

The coarse and fine aggregates and water in the required quantity are mixed in a mixing plant. The mix is then spread by a paver finisher and the layer is compacted by rolling with a smooth wheel roller 80-100 kN weight if the compacted layer thickness is upto 100 mm. If the layer thickness is more than 100 mm and upto 200 mm, the compaction is done with a vibratory roller of static weight 80-100 kN. The layer is allowed to dry for 24 hours.

16.5. Surface Dressing

The granular surface on which the surface dressing is to be done is primed. Over the primed surface, the bitumen binder is sprayed, and immediately thereafter the stone chips are spread. Rolling is then done with a static smooth wheel roller of weight 8-10 tonnes until the aggregates are firmly embedded in the binder and present a closed surface. Where two coats of surface dressing is to be done, the second coat is applied only after the first coat is open to traffic for about two weeks.

16.6. Premix Chipping Carpet

This specification results in a carpet of 20 mm thickness. The binder and aggregates are mixed in a hot mix plant, the mix is transported to the site and spread uniformly. Rolling is done by a smooth wheel static roller of 8-10 tonne weight. A seal coat (either liquid seal coat or premixed seal coat) is applied immediately after the carpet is rolled. Traffic is allowed only after the seal coat is laid.

16.7. Bituminous Macadam, Dense Bituminous Macadam and Bituminous Concrete

If the specification is to be laid on a gravel, WBM or WMM layer, the surface shall be primed. If it is to be laid over an existing bituminous layer, a tack coat is laid. The aggregates and binder are mixed in a hot-mix plant, the mix transported to the site and laid by a paver. Compaction is done by a 8-10 tonne smooth wheel roller for the break down rolling. The intermediate rolling is done with a 8-10 tonne static weight or vibratory rolling, or with a pneumatic tyred roller of 12-15 tonne weight having a tyre pressure of at least 0.56 MPa. The finish rolling is done with a smooth wheel tandem roller till the desired compaction is achieved.

16.8. Dry Lean Concrete Sub Base

The mixing of cement, stone aggregates and water is done in a batching plant. The material is transported to the site and placed with a paver having electronic sensors. The width to be laid shall have offsets (at least 750 mm) on either side so that the tracked wheels of the slip form paver can be accommodated. Compaction is done by a roller till a dry density of at least 98 per cent of that specified is achieved. Smooth wheel vibratory rollers of 8-10 tonne weight are used. Curing is done by sprinkling water after covering the surface with gunny bags or hessian. Alternatively resin based aluminised reflective curing compound can be sprayed. The surface is covered with hessian for three days and watered. A trial length of at least 100 m shall be first laid to evolve the sequence of operations.

16.9. Concrete Pavement

The mixing of cement, aggregates, water, additives and admixtures is done in a batching plant of adequate capacity. A separation layer of PVC sheet 125 micron in thickness is laid over the DLC sub-base. The concrete pavement is laid by a slip-form or fixed form paver, after the dowel bar assembly and tie bars are placed in position. Alternatively, a Dowel Bar Inserter, if available with the slip form paver, may be used. The concreting shall not be done when the ambient temperature is more than 35^{0}C. The paver shall spread, consolidate, screed and float finish and impart texture. The freshly laid concrete shall be protected from wind by portable tents. Immediately after texturing, the slab shall be cured by the application of curing compound followed by wet curing for 7-14 days. The contraction joints shall be sawn by a joint cutting saw in the sawing window which is generally 6-18 hrs after laying the concrete. After the joints are sawn, the cut is widened by further sawing, chamfering the edges, cleaning the joint groove by air-blasting, placing a compressible debonding strip, applying the primer of the sealant (if specified by the manufacturer) and applying the joint-sealant. The concrete pavement is opened to traffic after 28 days of curing if Ordinary Portland Cement is used. If Rapid Hardening Cement is used, the pavement can be opened to traffic after 7 days.

17

Chapter

Maintenance of Highways

17.1. Importance of Maintenance

The road, including the earthwork, pavement, traffic signs, traffic markings and cross-drainage structures, deteriorates due to the effect of traffic, rainfall, snowfall, temperature variations. Unless the road is maintained properly, the assets constructed with public money lose their value. Well maintained roads result in fewer accidents and lesser vehicle operating costs.

17.2. Maintenance of Earth Roads

Earth roads get damaged due to rain-cuts and rutting caused by traffic. They can be maintained by filling up the rain cuts in the embankment slopes after rainy season, maintaining turfing to protect the slopes and adding fresh earth to fill up ruts and rolling.

17.3. Maintenance of Gravel Roads

Gravel roads (known as murram roads in India) suffer damages due to loss of gravel, formation of ruts and local depressions. They can be repaired by adding fresh material to fill up depressions, ruts and potholes and rolling with sprinkling of water. Regravelling may be needed once in 2-5 years, by adding fresh gravel of 25-75 mm loose thickness and compacting with a roller after sprinkling with water.

17.4. Maintenance of Water-Bound-Macadam (WBM) Roads

WBM surface, without bituminous surface dressing or thin carpet, get ravelled due to the stones getting loosened and ripped off due to traffic. Local damages such as depressions, rutting and potholes can be repaired by adding fresh stone material, watering and compacting with a plate compactor. Resurfacing becomes necessary after 2-6 years depending upon the intensity of traffic. Resurfacing may be done in layers of 50-75 mm loose thickness adopting the same procedure as a new construction (para 16.3).

17.5. Maintenance of Bituminous Surfaces

17.5.1. The common defects noticed in a bituminous surfaced pavements are:

(a) Cracks
(b) Hungry Surface
(c) Fatty Surface
(d) Rutting

(e) Potholes
(f) Stripping
(g) Edge breaking

17.5.2. Pavement Distress Based Ratings

The Indian Roads Congress have formulated distress based ratings which are given in Tables 17.1, 17.2 and 17.3 (Ref 1).

Table 17.1: Pavement Distress Based Rating for National Highways and State Highways

Defects (Type)	Range of Distress		
Condition	**Poor**	**Fair**	**Good**
Cracking (%)	>10	5–10	<5
Ravelling (%)	>10	1–10	<1
Potholes (%)	>1	0.1–1	<0.1
Shoving (%)	>1	0.1–1	<0.1
Patching (%)	>10	1–10	<1
Settlement and Depression (%)	>5	1–5	<1
Rut Depth (mm) Using 3 m Straight Edge	>10	5–10	<5
Rating	1	1.1–2	2.1–3

Table 17.2: Pavement Distress Based Ratings for Major District Roads and Rural Roads

Defects (Type)	Range of Distress		
Condition	**Poor**	**Fair**	**Good**
Cracking (%)	>20	10–20	<10
Ravelling (%)	>20	10–20	<10
Potholes (%)	>1	0.5–1	<0.5
Patching (%)	>20	5–20	<5
Settlement and Depression (%)	>5	2–5	<2
Rating	1	1.1–2	2.1–3

Table 17.3: Pavement Distress Based Ratings for Urban Roads

Defects (Type)	Range of Distress		
Condition	**Poor**	**Fair**	**Good**
Cracking (%)	>15	5–15	<5
Ravelling (%)	>10	5–10	<5
Potholes (%)	0.5	0–0.5	Nil

(Contd...)

Defects (Type)	Range of Distress		
Settlement (%)	>5	1–5	<1
Rut depth (mm) using 3 m straight edge	>10	5–10	<5
Rating	1	1.1–2	2.1–3
Condition	Poor	Fair	Good

The serviceability indicators based on Roughness and Skid Resistance for Highways and Urban Roads are given in Table 17.4 and 17.5 (Ref 1).

Table 17.4: Serviceability Indicators for Highways

S.No.	Serviceability Indicators	Level 1 (Good)	Level 2 (Fair)	Level 3 (Poor)
1.	Roughness (Max. Permissible) (mm/ Km)	1800	2400	3200
2.	Skid Number	60	50	40

Table 17.5: Serviceability Indicators for Urban Roads

S.No.	Serviceability Indicators	Level 1 (Good)	Level 2 (Fair)	Level 3 (Poor)
1.	Roughness (Max. Permissible) (mm/ Km)	1800	2600	3400
2.	Skid Number	65	55	45

17.5.3. Treatment of Distress

Cracks may be treated by applying slurry seal and micro-surfacing.

Bleeding or fatty surface is treated by applying coarse sand, precoated chips or recycling.

Smooth surface may be roughened by applying surface dressing, premix carpet or micro-surfacing.

Hungry surface may be rectified by slurry seal or microsurfacing.

Rutting is repaired by applying fresh bituminous mixes after providing a tack coat.

Potholes are repaired by cutting the edges to a square/ rectangular shape, applying tack coat and filling the potholes with hot mix or cold mix.

Stripping is the loss of bond between the aggregates and the binder. The area affected is removed, and fresh bituminous material is laid and compacted, preferably adding an anti-stripping agent to the bitumen.

Edge breaking is repaired by removing the broken parts, placing fresh material and raising the shoulders to the level of the bituminous pavement and rolling the shoulder and the freshly laid bituminous material together.

17.5.4. Surface Renewal

The Ministry of Road Transport and Highways have given the norms for renewal which are given in Table 17.6.

Table 17.6: Norms for Surface Renewal

Category of Roads	Traffic Commercial Veh. Per day	Metal Renewal	Surface Dressing (2 Coats)	Premix Carpet plus Seal Coat	20 mm Mix Seal Surfacing	25 mm Semi Dense Carpet	25 mm Bit. Carpet	40 mm Bit. Carpet
NH/ SH (Normal)	>4500	–	–	–	–	–	5/4*	5/4*
	1500–4500	–	–	–	–	5/4*	5/4*	–
	450–1500	–	–	–	5/4*	5/4*	–	–
	<450	–	–	5/4*	5/4*	–	–	–
NH/ SH (Urban)	>4500	–	–	–	–	–	4/3*	4/3*
	1500–4500	–	–	–	–	4/3*	4/3*	–
	450–1500	–	–	4/3*	4/3*	4/3*	–	–
	<450	–	–	4/3*	4/3*	–	–	–
NH/ SH (Hills)	>1500	–	–	–	–	4/3*+	4/3*	4/3*+
	450–1500	–	–	4/3*	4/3*	4/3*	–	–
	<450	–	–	4/3*	4/3*	–	–	–
MDR/ ODR/VR (Normal)	>1500	–	–	–		5/4*	5/4*	
	450–1500	–	–	–	5/4*	5/4*	–	–
	150–450	–	–	–	5/4*	5/4*	–	–
	<150	5/4*	5/4*	5/4*	–			
MDR/ ODR/VR (Urban)	>1500	–	–	–	–	4/3*	4/3*	–
	450–1500	–	–	–	4/3*	4/3*	–	–
	150–450	–	3	4/3*	4/3*	–	–	–
	<150	3	4/3*	5/4*	5/4*	–	–	–
MDR/ ODR/VR (Hills)	>1500	–	–	–	–	4/3*+	4/3*+	
	450–1500	–	–	–	–	5/4*+	5/4*+	
	150–450	–	3	5/4*	5/4*+	–	–	–
	<150	4/3*	4/3*	5/4*	–	–		

* Indicates treatment for rainfall>3000 mm

\+ Indicates treatment for high altitude >2000 metres

17.5.5. Overlays

The design of overlays for flexible pavements in India is based on the Benkelman Beam Deflection Method (Ref 3). Fig. 17.1 gives the overlay thickness design curves, the thickness being in terms of thickness of Bituminous Macadam. For other material, the following conversion factors are given:

1 cm of bituminous macadam = 1.5 cm of WBM/ WMM/ BUSG

= 0.7 cm of DBM/ BC/ SDBC

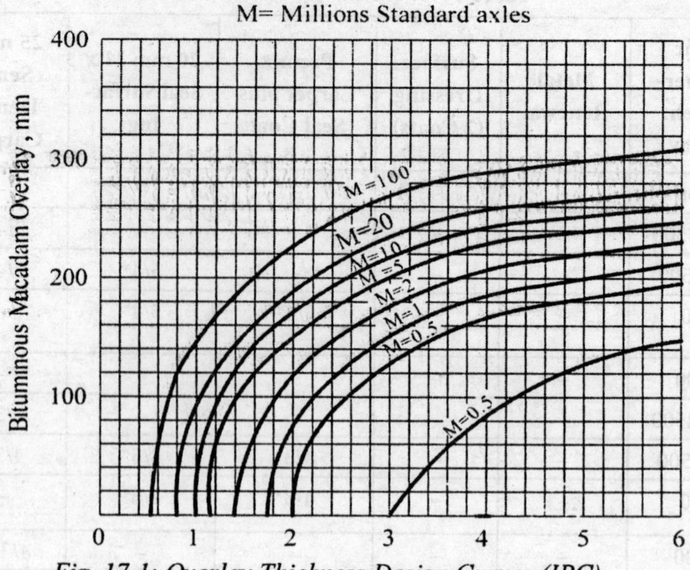

Fig. 17.1: Overlay Thickness Design Curves (IRC)

For rigid pavements, overlays can be of concrete or bituminous courses. The formulae of the Corps of Engineers (COE) and Federal Aviation Agency (FAA) are given below:

Unbonded Overlay (COE and FAA)

$$h_o = \left(h_m{}^2 - Ch_e{}^2\right)^{1/2} \quad (17.1)$$

Partially Bonded Overlay

$$h_o = \left(h_m{}^{1.87} - Ch_e{}^2\right)^{1/2} \quad (17.2)$$

(FAA)

$$h_o = \left(h_m{}^{1.4} - Ch_e{}^{1.4}\right)^{1/1.4} \quad (17.3)$$

(COE)

Fully Bonded Overlay

$$h_o = h_m - h_e$$

In the above formulae,

h_o = overlay thickness required in inches

h_m = thickness of monolithic slab required in inches

h_e = thickness of existing pavement slab in inches

C = Pavement condition factor
= 1.0 when the existing pavement is in good condition
= 0.75 when the existing pavement shows initial cracking
= 0.35 when the existing pavement is badly cracked

The Indian Roads Congress gives the following recommendation for flexible overlays (Ref 4).

For areas of medium rainfall (40-125 cm per year), favourable drainage conditions and low plasticity of soil (PI not exceeding 4)

(A) For very heavy traffic (exceeding 1500 commercial veh/ day)

(i) 4 cm Bituminous Concrete over 7.5 cm Bit. Macadam

or

4 cm Bituminous Concrete over 15 cm granular layer

(ii) For heavy and medium traffic (151-1500 commercial veh/day)

2 cm Premix Carpet over 7.5 cm Bit. Macadam

or

2 cm Premix Carpet over 7.5 cm Built-up-Spray Grout

or

2 cm Premix Carpet over 15 cm Granular Layer

or

4 cm Bituminous Concrete over 7.5 cm Granular Layer

(B) For areas with high intensity of rainfall (above 125 cm per year and upto 200 cm per year), unfavourable drainage conditions and subgrades of high plasticity (PI 20 and above)

(i) For very heavy traffic (exceeding 1500 commercial veh per day)

4 cm Bituminous Concrete over 10 cm Bituminous Macadam

(ii) For heavy and medium heavy traffic (151-1500 commercial veh. per day)

4 cm Bituminous Concrete over 7.5 cm Bituminous Macadam

(C) For areas of very heavy traffic and in areas of very high rainfall (above 200 cm per year), unfavourable drainage conditions and subgrades of high plasticity

(i) 4.0 cm Bituminous Concrete over 7.5 cm Bituminous Macadam over 5 cm Bituminous Macadam with 2.5-3.0 per cent bitumen

or

(ii) 4.0 cm Bituminous Concrete over 7.5 cm Bituminous Macadam

17.5.6. Rigid Overlays over Bituminous Surfaces (White-topping)

Rigid overlays over bituminous surfaces are popular all over the world, and are being followed in India also. The Indian Roads Congress recommendations (Ref 5) recognise three types of rigid overlays:

(i) Conventional concrete overlays of thickness 200 mm or more, without any consideration of any bond between the concrete overlay and underlying bituminous layer.

(ii) Thin White-topping of thickness 100-200 mm relying upon bond between the concrete overlay and the bituminous layer below. The joints are at a spacing of 0.6 to 1.25 m.

(iii) Ultra Thin White-topping, also relying upon the bond between the overlay and the bituminous layer below, and of a thickness less than 100 mm. The joints are spaced at 0.6 to 1.25 m.

When bonding between the overlay and the bituminous layer is taken into account in the design, the existing bituminous layer is milled and roughened to receive the overlay. The thickness of the bituminous layer after milling should be a minimum of 75 mm.

For Thin and Ultra Thin White-topping, the concrete should be of high strength (minimum 40 MPa) and fibres are added to achieve high strength and prevent cracks.

REFERENCES

1. Code of Practice for Maintenance of Bituminous Road Surfaces, IRC:82-2015, Indian Roads Congress, New Delhi, 2015.
2. Report of the Committee on Norms for Maintenance of Roads in India, Ministry of Road Transport & Highways, published by Indian Roads Congress, New Delhi, 2001.
3. Tentative Guidelines for strengthening of flexible pavements using Benkelman beam deflection technique, IRC:81-1984, Indian Roads Congress, New Delhi, 1984.
4. Recommendations about overlays on cement concrete pavement, IRC:SP:17-1977, Indian Roads Congress, New Delhi, 1977.
5. Tentative Guidelines for Conventional, Thin and Ultra Thin White-topping, IRC:SP:76-2008, Indian Roads Congress, New Delhi, 2008.

18

Chapter

Use of Geosynthetics in Highway Engineering

18.1. Terms Related to Geosynthetics

Geosynthetics cover a wide range of applications in Highway Engineering as under:

1. Geotextiles, which are synthetic woven or non-woven fabrics; 2. Geogrids, which are net like tensile elements with apertures of sufficient size to allow interlocking with soil and aggregates; 3. Geomembrane, which is an impermeable membrane acting as a barrier or liner; 4. Geonets; 5. Geocomposite 6. Geocell, used for erosion control of slopes 7. Paving Fabric, used for strengthening bituminous pavements and acting as stress relieving membranes and water barrier; and 8. Glass grids used for strengthening bituminous pavements.

18.2. Geotextile as Separation Layer

The geotextile acts to prevent intermixing of two materials. For example, if provided below a granular layer and above the earth subgrade, it prevents the soil mixing with the granular layer. It can also be used in railway track between embankment and ballast.

18.3. Geotextile for Filtration

Geotextile can be laid over soft soils over which the pavement is built so as to improve the load bearing property. Geotextile and geogrids are used as the reinforcing element in Reinforced Earth Walls.

18.4, Geosynthetics in Drainage

Geosynthetics can be used in subsurface drainage of roads. A Fin Drain consisting of a geonet sandwiched between two layers of geotextile and Prefabricated Vertical Drains are used to drain water in roads constructed over swampy or marshy soil.

18.5. Geosynthetics for Erosion Control and Slope

Geosynthetics are used to resist erosion of embankment slopes and allowing vegetation to grow.

REFERENCES

1. State of Art: Application of Geotextiles in Highway Engineering, Special Report 12, Highway Research Board, Indian Roads Congress, New Delhi-1994.
2. Specifications for Road and Bridge Works, Ministry of Road Transport and Highways.

19

Chapter

Urban Roads

19.1. Functional Classification and Design Speeds

Urban streets are classified into (i) Expressways (ii) Arterial Streets (iii) Sub-Arterial Streets (iv) Collector Streets and (v) Local Streets (Refer para 7.2. The design speeds are given in para 7.4. The geometric design aspects are dealt with in Chapter 7.

19.2. Pedestrian Facilities and Cycle Tracks

Pedestrian facilities are an important part of urban street design. They are dealt with in para 7.19. provision of dedicated cycle tracks is also equally important and their design is dealt with in para 7.20.

19.3. Bus Facilities

The layout of bus-bays is given in para 7.17. The provision of Bus Rapid Transit (BRT) is considered to be an essential requirement to reduce the use of personal vehicles in cities. Cities like Ahmedabad, Surat, Hubli-Dharwad and Pune have built BRTs and have benefitted from them. The cross-section of the Hubli-Dharwad road incorporating BRT is given in Fig. 19.1 (Ref 1).

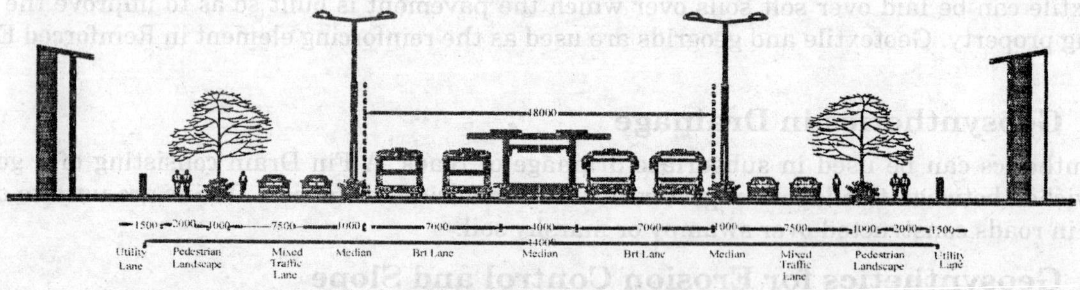

Fig. 19.1: Cross Section for Hubli Dharwad BRTs for 44 m RoW

The experience of the BRT in Delhi has not been encouraging, but more BRTs are being planned profiting from the earlier experience.

REFERENCES

1. BRTs, India Experience: Ahmedabad, Surat, Hubli-Dharwad, Pune, Traffic Infra Tech, Oct-Nov, 2015, Mumbai, 2015.

20 Chapter

Hill Roads

20.1. Hilly Terrain

Hilly terrain in India is classified under two groups as under:

	Percent Cross-slope of Country
Mountainous	25–60
Steep	greater than 60

The Himalayan region of the north and the Western Ghats of the south come under this category.

20.2. Design Speed

Table 20.1 gives the Design Speeds suggested for adoption in India (Ref 1, 2):

Table 20.1: Suggested Design Speeds (Kmph) for Hill Roads in India

Road Class		Mountainous Terrain		Steep Terrain	
Ruling		Minimum	Ruling	Minimum	Ruling
1.	National Highways and State Highways	50	40	40	30
2.	Major District Roads	40	30	30	20
3.	Other District Roads	30	25	25	20
4.	Village Roads	25	20	25	20

20.3. Minimum Radii of Horizontal Curves

The basic equation for the design of horizontal curves in a road is:

$$\frac{v^2}{gR} = e + f \qquad (20.1)$$

$$R = \frac{V^2}{127(e+f)} \qquad (20.2)$$

where v = speed in metres per second

V = speed in Km/hr

e = super-elevation in decimals (maximum 0.07 for snow bound areas and 0.10 for areas not affected by snow)

f = coefficient of side friction between tyre and road surface, taken as 0.15

R = Radius in metres

Table 20.2 gives the minimum radii for various classes of roads on the basis of the above formula.

Table 20.2: Minimum Radii of Horizontal Curves for Various Classes of Hill Roads

Road Class	Minimum Radius (m)							
	Mountainous Terrain				Steep Terrain			
	Areas not affected by snow		Snow-bound areas		Areas not affected by snow		Snow-bound areas	
	Ruling Min	Absolute Min	Ruling Min	Absolute Min	Ruling Min	Absolute Min	Ruling Min	Absolute Min
National Highways and State Highways	50	50	90	60	50	30	60	33
Major District Roads	50	30	60	33	30	14	33	15
Other District Roads	30	20	33	23	20	14	23	15
Village Roads	20	14	23	15	20	14	23	15

20.4. Gradients

The gradients recommended for Hill Roads are given in Table 20.3.

Table 20.3: Gradients for Hill Roads

Classification of Gradient	Mountainous Terrain and Steep Terrain more than 3000 m above Mean Sea Level	Steep Terrain upto 3000 m above Mean Sea Level
Ruling	1 in 20 (5%)	1 in 16.7 (6%)
Limiting	1 in 16.7 (6%)	1 in 14.3 (7%)
Exceptional	1 in 14.3 (7%)	1 in 12.5 (8%)

Ruling gradients are used as a matter of course. Limiting gradients are used where the adoption of gentler gradients would add enormously to the cost. Exceptional gradients are to be adopted only in very difficult situations and in short lengths not exceeding 100 m.

The cumulative rise/ fall in elevation over 2 Km length shall not exceed 100 m in mountainous terrain and 120 m in steep terrain.

Grades shall be eased at horizontal curves by an amount known as "grade compensation" calculated by the formula:

$$\text{Grade compensation (per cent)} = \frac{30+R}{R} \tag{20.3}$$

(Subject to a maximum of $\frac{75}{R}$)

where R = Radius in metres

20.5. Width

The width of carriageway is dependent on the projected traffic volume in the design year. The following Table gives the Design Service Volume.

Table 20.4: Carriageway Width and Capacity of Hill Roads

Lane Class	Design Service Volume PCU/day		
	Carriageway width (m)	For low curvature (0-200 degrees/Km)	For high curvature (above 200 degrees/Km)
Single Lane	3.75	1,600	1,400
Intermediate Lane	5.5	5,200	4,500
Two Lane	7	7,000	5,000

The Roadway Width including carriageway and shoulders are given in Table 20.5.

Table 20.5: Widths of Carriageway and Shoulders for Hill Roads

Road Class	Carriageway width (m)	Shoulders (m)	Roadway Width (m)
1. National Highways and State Highways			
(i) Single Lane	3.75	2 x 1.25	6.25
(ii) Double Lane	7.00	2 x 0.9	8.8
2. Major District Roads and Other District Roads	3.75	2 x 0.5	4.75
3. Village Roads	3.00	2 x 0.5	4.00

20.6. Hairpin Bends

Hairpin bends are unavoidable in certain locations where a sudden rise in elevation becomes necessary and the hill does not permit the provision of the required horizontal distance. The standards for the design of hair-pin bends are:

1. Minimum Design Speed — 20 Km/hr
2. Minimum Width at Apex
   - (i) NH, SH Single Lane — 9.0 m
   - (ii) NH, SH Double Lane — 11.5 m
   - (iii) MDR/ ODR — 7.5 m
   - (iv) Village Roads — 6.5 m
3. Minimum Radius of inner curve — 14.0 m
4. Minimum Transition length — 15.0 m
5. Gradient
   - Maximum — 1 in 40
   - Minimum — 1 in 200
6. Super-elevation — 1 in 10
7. Distance between two consecutive hairpin bends — 60 m

20.7. Protection Structures

Retaining Walls, Parapets and Breast Walls are needed on hill roads. Fig. 20.1 gives the details of dry stone masonry retaining wall with bands of stone masonry with cement mortar.

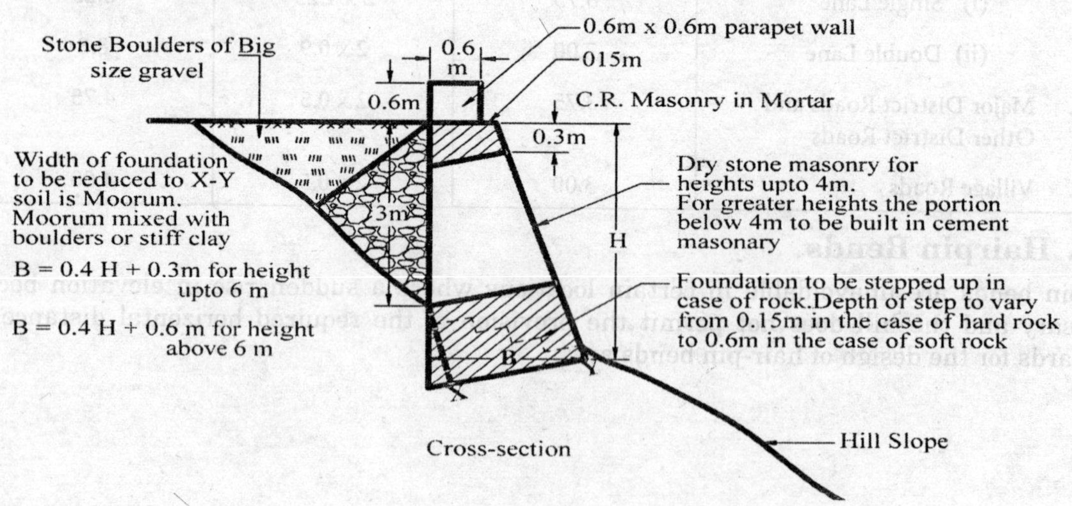

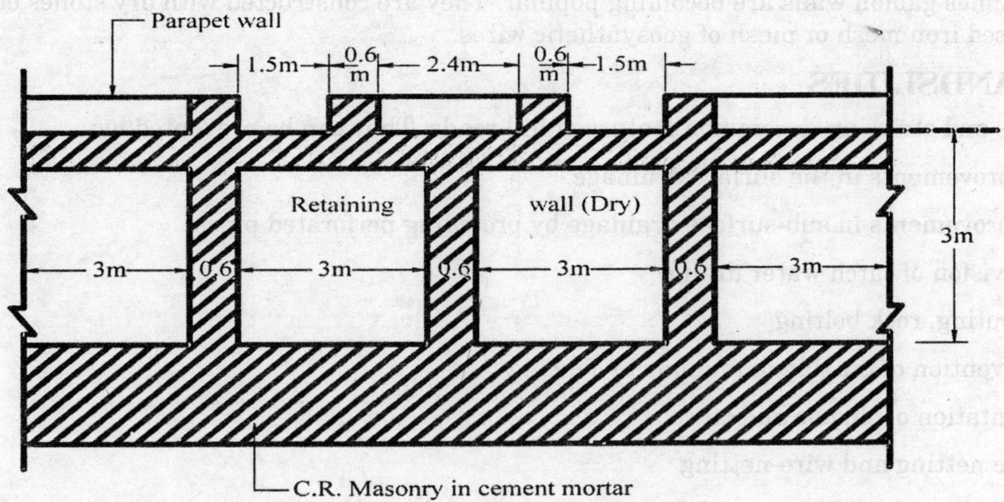

Note 1· The gap in parapet may be reduced or closed entirely, at the discretion of Engineer in charge. Suitable weep holes shall be provided if the parapet wall is built without gap

2. Top surface of parapet shall be given a weathered slope of one inch

Fig. 20.1: Typical Design for a Retaining Wall

The section of a breast wall needed to buttress the uphill slopes is given in Fig. 20.2.

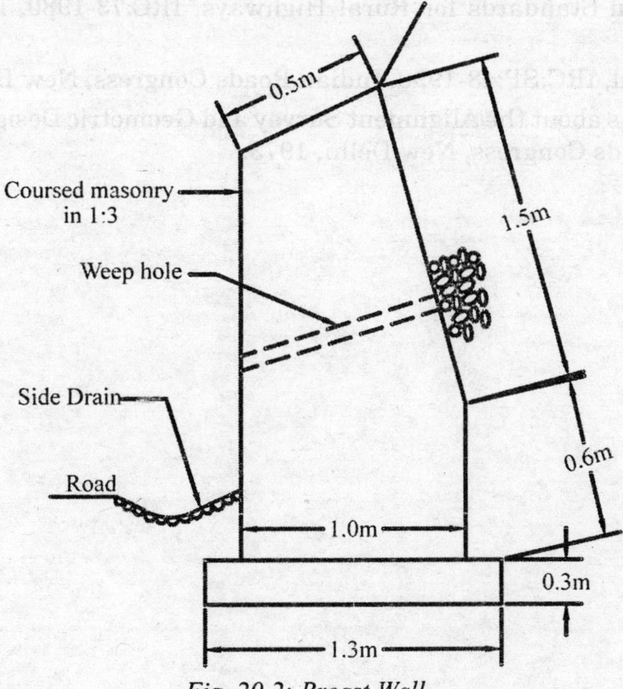

Fig. 20.2: Breast Wall

In recent times gabion walls are becoming popular. They are constructed with dry stones encased in galvanised iron mesh or mesh of geosynthetic wires.

20.8. LANDSLIDES

Landslides and strips are a common feature of hill roads. They can be prevented by:

(1) Improvements in the surface drainage

(2) Improvements in sub-surface drainage by providing perforated pipes

(3) Provision of catch water drains

(4) Grouting, rock bolting

(5) Prevention of grazing of animals by fencing

(6) Plantation of shrubs and trees

(7) Jute netting and wire-netting

(8) Asphalt mulch treatment of slopes to foster growth of vegetation

20.9. Snow Removal

Accumulation of snow causes serious disruption of traffic. Snow can be cleared by snow ploughs or by bull-dozing using dozers and graders. Application of salt (sodium chloride or calcium chloride) melts the ice.

REFERENCES

1. Geometric Design Standards for Rural Highways, IRC:73-1980, Indian Roads Congress, New Delhi, 1980.
2. Hill Road Manual, IRC:SP:48-1998, Indian Roads Congress, New Delhi, 1998.
3. Recommendations about the Alignment Survey and Geometric Design of Hill Roads, IRC:52-1973, Indian Roads Congress, New Delhi, 1973.

21

Chapter

Road in Desert Areas

21.1. Desert Area in India

Parts of Rajasthan fall under the Thar desert. The area is devoid of vegetation, has scanty rainfall and has loose fine sand. The fine sand gets shifted due to strong winds that sweep the desert in the hot summer months. As a result sand dunes are formed which can be at places as high as 100 m above the ground. Since the predominant wind direction is in the South-West to North-East direction, longitudinal sand dunes are formed in the same direction. When the prevailing wind direction changes, transverse sand dunes are formed. Barchan sand dunes are crescent shaped. Fig. 21.1 gives the sketches of these sand dunes.

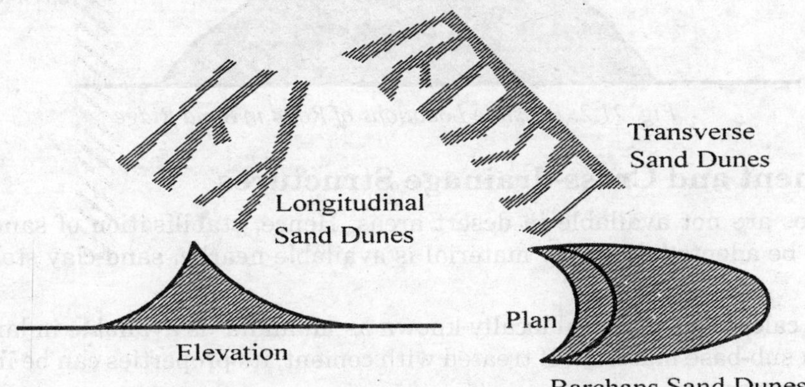

Fig. 21.1: Types of Sand-dunes

21.2. Principles of Road Location in Sand Dunes

The following principles govern the location of roads in sand dune area:

1. Avoid areas where sand is loose and unstable. Locate roads in areas where coarse sand is present.
2. If some vegetation is present, the road shall be located in such areas.
3. In areas having longitudinal sand dunes, the road can be located in the inter-dunal space (Fig. 21.2).
4. Avoid cutting slopes for locating the road, and avoid part cut and part fill.

5. Road can be located on top of the ridge of the longitudinal dune.
6. If the road has to be located cutting across transverse sand dunes, the gradients have to be steep, rising before a dune and falling after a dune is crossed. Such situations can be costly. The appropriate solution is to avoid locating the road cutting across a series of sand dunes.

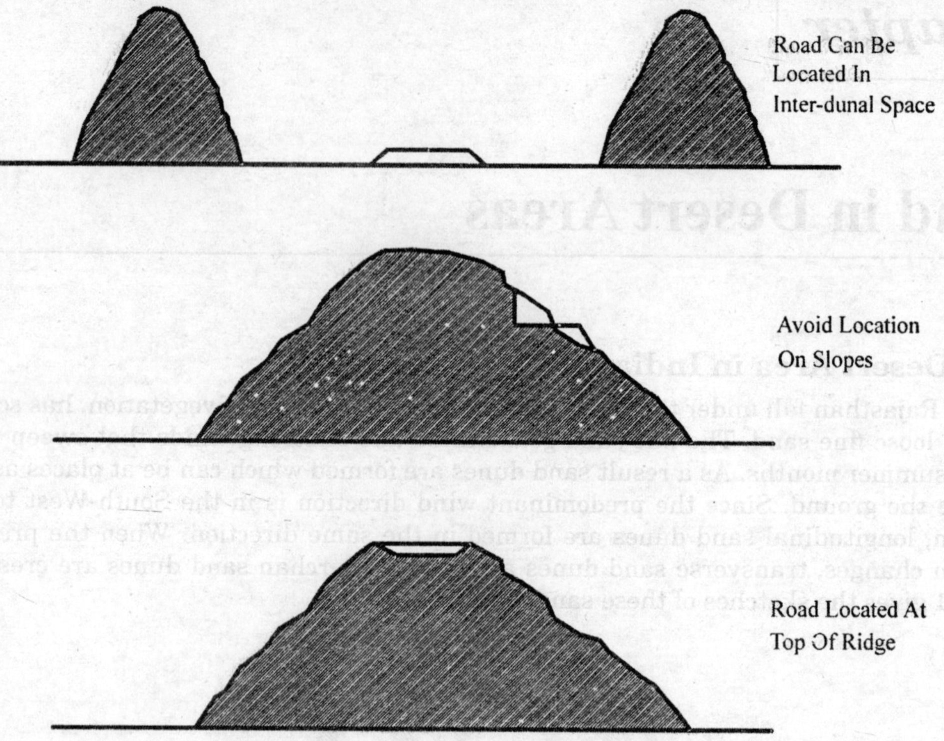

Fig. 21.2: Possible Locations of Road in Sand Ridge

21.3. Pavement and Cross-drainage Structures

Stone aggregates are not available in desert areas. Hence, stabilisation of sand using bitumen or cement shall be adopted. If clayey material is available nearby, sand-clay stabilisation can be adopted.

In Rajasthan, a calcareous material locally known as "dhandla" is available in large quantities. It can be used as a sub-base material. If treated with cement, its properties can be improved and can be used in base course.

Since water is scarce in desert areas, Water-bound-macadam and Wet Mix Macadam are difficult and can be substituted by Dry Bound Macadam. Precast cement blocks with arrangement for interlocking and load transfer have been developed by C.R.R.I. for use in deserts (Ref 1).

Since rainfall is scanty, natural drainage channels crossing the road alignment can be managed with paved dips and causeways.

REFERENCES

1. Central Road Research Institute, New Delhi, Development of Special Purpose Paving Blocks in India, 2nd International Conference on Concrete Paving, Delft, Holland, 1984.

22
Chapter

Toll Roads

22.1. Background

The requirement of funds for development and maintenance of roads in India is very large and the allocation of funds from the government alone cannot be adequate. Alternate sources of funds have to be explored. One promising source is Toll Roads.

22.2. International Experience on Toll Roads

Several countries have implemented Toll Roads and some of the experiences gained are given below (Ref 1):

1. **France:** About 6000 Kms of Autoroutes in France are toll-based.
2. **USA:** Though USA pioneered the toll roads, which were known as Turnpikes, the entire Inter-State Highway Programme was funded by the Government and no tolls are charged on them.
3. **Australia:** Australia has taken up a few projects through Toll financing. These are the Sydney Harbour Tunnel, M4 Motorway, M5 Motorway and M2 Motorway (Ref 2).
4. **Spain:** Spain has successfully constructed a network of Inter-city Expressways through Public-Private-Partnership.
5. **Malaysia:** The North-South Highway, 840 Km long, from Thai-Malaysian Border to Singapore, is an ambitious toll road.
6. **China:** The Guangzhou-Shenzhen Expressway is dependent on income from property development and user fee from vehicles.
7. **Pakistan:** The 340 Km long Lahore-Islamabad Expressway is partly financed by the government and partly by private concessionaire.
8. **U.K.:** The DBFO (Design, Build, Finance and Operate) is unique to UK (Ref 3). The government delegates the tasks of designing, building, funding and operation to the private entrepreneur.
9. **India:** India has implemented several projects on which toll is collected. Some of them are:

   (i) Durg Bypass (16 Km long)

   (ii) Hubli-Dharwar Bypass (30 Km long)

   (iii) Mumbai-Pune Expressway (120 Km long)

   (iv) Taj Expressway

(v) Several projects of NHs for widening to four/ six lanes

(vi) Rao-Pithampur Toll Road Project (Ref 4)

(vii) Bangalore Peripheral Road

22.3. Advantages of Toll Roads

The following are the advantages of Toll Roads:

1. They are an additional source of financing.
2. Tolling is an efficient method of charging the user for the facility.
3. The Private Sector invests money and hence they strive to complete the project in a short time.
4. Toll roads are designed and maintained to high standards so as to attract the vehicle users.
5. The wayside facilities like Rest Areas are very good and are appreciated by the users. Similarly, the Incidence Management system provides quick relief to the accident victims.

22.4. Disadvantages

The following are the disadvantages:

1. Road transport is already heavily taxed in India and only a fraction of the revenue earned is used for road development. Thus, toll is an additional burden. Already voices are being raised to discontinue collection of toll.
2. Since the private entrepreneur borrows money from the market, the interest rates are higher than what would be if the government borrows the capital.
3. Delays are experienced at Toll Plazas.
4. Cost of Toll collection is an additional cost component.
5. When a toll-free road exists as an alternative, the full potential of toll road is not utilised.
6. There are several uncertainties and risks since the concession period is long (20-30 years). Traffic projections over the further years can be very risky. Political uncertainties and force majeure also involve high risks.

22.5. Form of Public-Private Partnership in Toll Road Projects

BOT

Build Operate Transfer under which the entrepreneur builds, operates the facility and transfers it at the end of the concession period to the government. This was the most used format in India till recently.

BOLT

Build Operate Lease and Transfer, under which in addition to build, operate and transfer, the Concessionaire obtains lease of the facility.

BOO

Build Own Operate is similar to BOT, but without transfer of ownership to the private entrepreneur.

DBFO

Design Build Finance and Operate, under which the entrepreneur has been given additional task of designing.

EPC

Engineering, Procurement and Construction, under which the entrepreneur has to design, procure a suitable contractor for construction and thus completes the construction.

22.6. PPP Financing Pattern

Not all projects are viable as pure Toll Roads. In such cases, the government still pursue Public-Private Partnership through the following options:

1. **Annuity Payments:** The Concessionaire meets all the up-front cost and constructs the facility. The government makes a fixed payment at regular intervals (say half-yearly or yearly). The government may or may not collect toll, and if toll is collected the money goes to the government.

2. **Viability Gap Funding (VGF):** If a project is not found to be financially viable the basis of toll collection, the government agrees to give a grant to the Concessionaire to cover the viability gap. Currently, for NH projects it is restricted to 40 per cent of the project cost.

   In some rare cases, where the cost of construction is low and the revenue expected to be earned from tolls is high (because of high volume of traffic), the Concessionaire does not seek VGF, but, on the contrary, offers to give a premium (negative grant) to the government. The project is then awarded to the party which offers to give the maximum negative grant.

3. **Asset Securitisation:** Another form of raising resources is "asset securitisation" This concept was pioneered by China and is used to raise funds from lending institutions like the World Bank and the Asian Development Bank. Under this scheme, the existing highway assets are leveraged to raise fresh funds.

4. **Special Purpose Vehicle:** Special Purpose Vehicle (SPV) is a financial and construction company, specially created with equity participation by the government, financial institutions and the construction company. The Moradabad Bypass project in India is an example where the NHAI and ILFS have formed an SPV.

22.7. Government of India's policy on private sector participation on NHs

The Government of India have formulated a policy on private sector participation on NHs, the salient features of which are given below (Ref 5).

1. The policy of privatisation will be implemented by the NHAI and in exceptional cases by the State Public Works Departments.
2. The basic principles in identification of NH projects for private investments would be that the project is an approved project of the Ministry of Road Transport & Highways and it is capable of yielding adequate financial return.
3. The Government will provide support by carrying out all preparatory works, such as:
   - Feasibility study;
   - Land for right-of-way and roadside facilities;
   - Relocation of utility facilities;
   - Cutting of trees;
   - Removal of encroachments and resettlements and rehabilitation of persons affected from the projects;
   - Depending upon financial viability, the cost of the above activities may be recouped from the project;
4. Suitable traffic support/ guarantee will be provided on a case to case basis.
5. Exemption will be given from import duty on identified high quality construction plant and equipment.

6. Foreign Direct Investments upto 74 per cent would be permitted automatically. Beyond that, proposals would need clearance of the Foreign Investment Promotion Board.
7. A five year tax holiday is available followed by a further five year period of 30 per cent exemption and this concession can be availed of in any 10 consecutive years during the first 20 years of operation (Section 80 IA (12) (ca) of the Income Tax Act, 1961).
8. Upto 40 per cent may be excluded from Income Tax of the income derived by financial institutions from finance they provide for infrastructure projects.
9. Exemption is available from Income Tax on the income from dividend and interest on long term capital gain derived from investments in the form of shares or long term finance to any enterprise set up to develop, maintain and operate an infrastructure facility.
10. Subscription to equity shares and debentures are eligible for deductions under Section 99 of the Income Tax Act.
11. Real estate development can be made integral part of BOT projects to enhance their financial viability.
12. NHAI can provide capital grants to the developers of a road project for project cost on a case to case basis.
13. NHAI can participate upto 30 percent of total equity of a company floated to develop a road project.
14. The ownership of the land for the highway construction and roadside facilities would continue to vest in Government. Mortgaging and subleasing of this land for raising finances is not allowed. However, land will be given on lease to entrepreneurs.
15. There will be provision of unified check barriers at the inter-state borders for the use of all government authorities. Such barriers would be located outside the right-of-way with proper entry/exit layout.
16. Disputes resolution and arbitration would be under the Indian arbitration and Conciliation Act, 1996. (it incorporates UNCITRAL provisions).
17. Entrepreneurs would be protected against force majeure situations including political, non-political and legislative changes.
18. There will be a standard concession agreement, which addresses the concerns of various stake holders in the project.

REFERENCES

1. Approach Paper on Road Development Plan 2001-2021, (Draft), Indian Roads Congress, New Delhi, 1999.
2. Ford, K.B., Toll Road Developments in New South Wales, Australia, Seminar on Private Sector Participation in Highway Projects, Indian Roads Congress, New Delhi, 1995.
3. Jenkinson, M, DBFO, The U.K. Experience, Seminar on Private Sector Participation in Highway Projects, Indian Roads Congress, New Delhi, 1995.
4. Sidharth Sinha, Toll Financing for Roads : MP Tolls Limited (Rao-Pithampur Toll Road Project)- A case study, Infrastructure Development and Financing, Macmillan India, New Delhi, 1999.
5. Road Development Plan, Vision: 2021, Ministry of Road Transport and Highways, published by the Indian Roads Congress, New Delhi, 2001.

23
Chapter

Quality Assurance

23.1. Definitions

The following terms are defined as under (Ref 1):

1. **Quality Assurance:** Quality Assurance is defined as all the planned and systematic activities implemented with Quality System and demonstrated as needed to provide adequate confidence that an entity will fulfil the requirements.
2. **Quality System:** The organisation structure, responsibilities, procedures, processes and resources needed for implementing quality management.
3. **Quality Control:** The operational techniques and activities that are used to fulfil the requirements for quality. All those planned and systematic actions necessary to provide confidence that a product or service will satisfy given requirements for quality.
4. **Quality Audit:** A systematic and an independent examination to determine whether quality activities and related results comply with planned arrangement, are implemented effectively and are suitable to achieve objectives.

23.2. Quality Assurance Manual

In order to fulfil the requirements of quality of a project, a Quality Assurance Manual (QAM) is to be prepared for each of the following activities.

(a) Design and Project Preparation
(b) Construction
(c) Operation and Maintenance
(d) Production of Manufactured Items

The QAM has to be prepared by:

(a) The Client
(b) The Design Consultant
(c) The Contractor
(d) The Supervision Consultant
(e) Operation and Maintenance Agency
(f) The Independent Engineer who supervises the work of a private operator

23.3. Quality Control

Control of the quality of materials used in the construction and the finished work is an essential part of the highway engineering work. The control is exercised in terms of the specifications prescribed and the acceptance criteria laid down. To exercise proper control, field laboratory for conducting routine tests and a central laboratory for conducting special tests are to be set up.

Statistical Quality Control is to be exercised since the materials and finished work display some dispersal, and generally follow the Normal Distribution (Ref 1, 2). The definition of some of the terms used in this process are:

(i) Mean, $\overline{x}$, (also referred to as the average) is the sum of the individual observations (Σx) divided by the number of observations (n).

Thus

$$\overline{x} = \frac{\sum x}{n} \qquad (23.1)$$

(ii) Standard Deviation j is given by the equation:

$$j = \sqrt{\frac{\sum\left(x-\overline{x}\right)^2}{n-1}} \qquad (23.2)$$

(iii) Tolerance Level is the number of (1 in N) or percentage (P%) of test data falling below or above it and is given by

$$x \text{ min} = \overline{x} - r\,j \qquad (23.3)$$

$$x \text{ max} = \overline{x} + r\,j \qquad (23.4)$$

The values of r are given in Table 23.1.

Table 23.1: Values of r for Different Levels of Confidence

Confidence level in terms of extent of test values that can be tolerated below or above specified value	r
1 in 10 (10%)	1.28
1 in 20 (5%)	1.64
1 in 40 (2.5%)	1.96
1 in 100 (1%)	2.326

The typical normal distribution curve illustrating the above principles is given in Fig. 23.1.

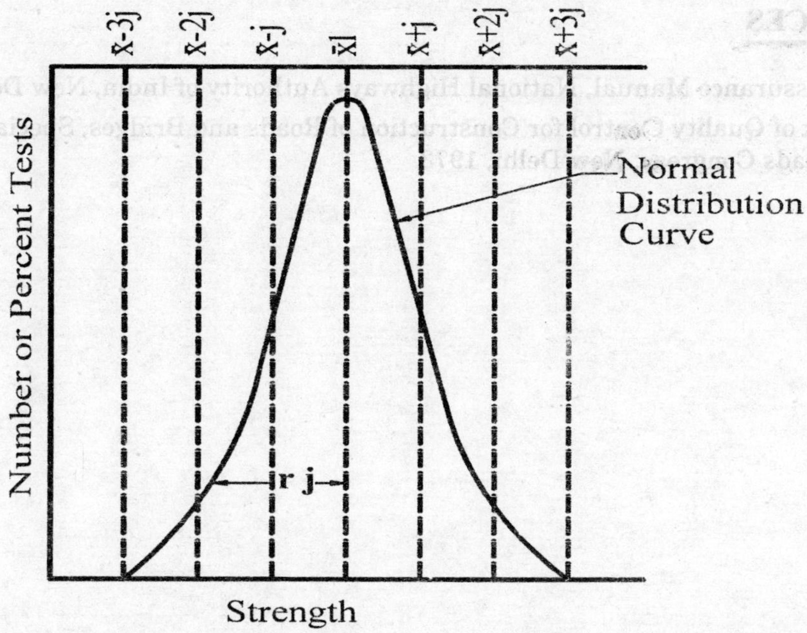

$\bar{x}$ = MEAN VALUE, j = STANDARD DEVIATION

r = FACTOR FOR TOLERANCE LEVEL

Fig. 23.1: Normal Distribution Curve

23.4. Frequency of Tests

Since the number of tests may not be many, the mean of results of n samples can be assumed to be $\bar{x}$ with a standard deviation of $\frac{j}{\sqrt{n}}$. Thus for controlling the density of compaction of soil, the acceptance criteria shall be subject to the condition that the mean density shall not be less than the specified density plus $\left[1.64-\frac{1.64}{\sqrt{n}}\right]$ times the standard deviations, with the stipulation that the tolerance level is 1.64 (*i.e.*, 1 in 20 values can be below the specified value).

The following example illustrates the procedure.

The mean of density of compacted soil is 1.90 gm/cc and the standard deviation is 0.30. 6 samples were tested and the mean density observed was 2.0 gm/cc. If the acceptance criteria is that 1 in 20 values can be below the specified value, are the results acceptable?

$$\bar{x} = 1.90$$

$$j = 0.30$$

$$n = 6$$

$$= \left[1.64-\frac{1.64}{\sqrt{6}}\right] \times 0.30 = 0.201$$

$$= 1.9 + 0.201 = 2.101 < 2.0$$

Hence results are not acceptable.

REFERENCES

1. Quality Assurance Manual, National Highways Authority of India, New Delhi, 2006.
2. Handbook of Quality Control for Construction of Roads and Bridges, Special Publication II, Indian Roads Congress, New Delhi, 1973.

24 Chapter

Highway Administration and Finance

24.1. Classification of Roads

Chapters 6 and 7 give the classification of urban and non-urban roads in India. Roads are an important constituent of an infrastructure and hence roads are built, maintained and administered by government agencies, though of late the private sector was being involved in building and maintaining roads.

24.2. Administration of National Highways

National Highways (NHs) are owned and administered by the Central Government in India under the National Highway Act, 1956. The Ministry of Road Transport and Highways (MORTH) currently are in charge of the National Highways.

The MORTH delegates some of its functions to the National Highway Authority of India (NHAI) or the State Public Works Departments (PWDs), or the Central Public Works Department (CPWD) or the Border Roads Organisation (BRO). When the task of designing, construction and maintenance of NHs is delegated to the State PWDs, agency charges (currently 9%) are paid to them.

The planning of the development of NHs is done by the MORTH. The Roads Wing in the MORTH handle all matters pertaining to NHs. It is headed by a Director General (Road Development) (DGRD), assisted by Additional Director Generals (ADGs), Chief Engineers and other supporting engineers. The Roads Wing have Regional Offices in every State. Specialist functions like Research and Mechanisation are looked after by Chief Engineers.

The DG(RD) is also the ex-officio Treasurer of the Indian Roads Congress (IRC). In addition to NHs, the Roads Wing also administer the Central Road Fund (CRF) and projects funded by CRF and Roads of Inter-State and Economic Importance and Strategic Roads in the sensitive national borders.

24.3. National Highways Authority of India (NHAI)

The NHAI was established under the National Highways Authority of India Act, 1988 and was operationalised in 1995. It is an autonomous body with executive responsibility of the development, maintenance and operation of NHs transferred to it by the MORTH.

24.4. Border Roads Organisation (BRO)

The BRO was set up in 1960 with the objective of construction and maintenance of roads in the sensitive border areas of the country. The organisation has undertaken several challenging jobs including roads, bridges and tunnels.

24.5. Administration of Rural Roads

The Ministry of Rural Development (MORD) looks after the development of Rural Roads. The National Rural Road Development Agency (NRRDA) under the MORD looks after the important programme known as the Pradhan Mantri Gram Sadak Yojana (PMGSY). The actual implementation of the programme is done by the State Rural Road Development Agencies (SRRDA).

24.6. Administration of Roads in the Military Areas

The construction and maintenance of roads in the military areas are looked after by the Military Engineering Service (MES) under the Ministry of Defence. Roads in the cantonment areas, serving civil needs, are looked after by the Cantonment Boards under the Ministry of Defence.

24.7. Road Research

Road Research in India is looked after by the Central Road Research Institute (CRRI) which is one of the research institutes under the Council of Scientific and Industrial Research (CSIR). The CRRI was set up in 1950 in Delhi. It has several divisions looking after various branches of research on roads, bridges and traffic. It is also running post-graduate courses.

Several States have their own Road Research Institutes. Mention may be made of the following which have contributed considerably to Road Research.

1. Highway Research Station (HRS), Chennai
2. Maharashtra Engineering Research Institute, (MERI), Nasik
3. Gujarat Engineering Research Institute, (GERI), Vadodara
4. U.P. Research Institute, Lucknow

24.8. Indian Roads Congress (IRC)

The IRC was established in 1934 and is a body of professional highway engineers. It has produced several standards, codes of practice, Manuals and Special Publications dealing with various aspects of roads, bridges, machinery and equipment (see also Chapter 5 and 6).

The Highway Research Board (HRB) under the IRC looks after the identification of research needs, evaluates the findings from specific research studies and disseminates the findings.

24.9. Training

In order to impart training for engineers at-entry or in-service, the National Institute for Training Highway Engineers (NITHE) has been established at NOIDA. It is now named as Indian Academy of Highway Engineers (IAHE). It is under the administrative control of the MORTH.

Similar training institutes also exist in some of the States.

24.10. Highway Financing and Taxation in India

The main source of financing road construction and maintenance is the taxes levied on road transport. Road transport taxes are in the following forms:

(1) Taxes on vehicles, spare parts, tyres

(2) Taxes on fuel and lubricants

(3) Import duty on petroleum crude, vehicles, spare parts, tyres

(4) Permit fees

(5) Toll levied by the State (other than toll levied by private operators for projects financed by them).

It is the general belief that road transport is highly taxed in India and the taxes raised are not spent exclusively on roads and are a means to generate general revenues. An estimate (Ref 1) indicates that only about 50 per cent of the road transport taxes is spent on construction and maintenance of roads in India. In USA the taxes raised are fully spent on road construction and maintenance.

The Central Road Fund (CRF) derives its funds from the levy of 5 per cent of the basic price out of duty of customs and excise on petrol and diesel. About one-third is set apart for utilisation by the Central Government on NHs, and the balance is distributed to the States on the basis of petrol and diesel consumption in each State. Recently, levy of cess on petrol and diesel was imposed, and portions of the accruals were earmarked for maintenance and development of NHs, SHs, MDRs, Rural Roads and construction of Road Overbridges at unmanned rail crossings.

Toll is a big source of revenue, and the subject of Toll Roads is dealt with in Chapter 22.

REFERENCES

1. Road Development Plan, Vision:2021, Ministry of Road Transport & Highways, published by the Indian Roads Congress, New Delhi, 2001.

25

Chapter

Traffic Engineering

25.1. Scope of Traffic Engineering

Highway engineering deals with the provision of the road infrastructure on which traffic moves. Traffic Engineering deals with how the safe and efficient movement of traffic is effected.

25.2. Traffic Control

25.2.1 Traffic control deals with various devices that promote safe movement of traffic. The following devices are important part of traffic control:

(i) Traffic Signs
(ii) Road Markings
(iii) Traffic Signals

25.2.2. Traffic Signs

Traffic signs are of three types:

(i) Danger Signs (also known as Warning Signs or Cautionery Signs)
(ii) Regulatory Signs
(iii) Information Signs

The Indian Roads Congress standards for Road Signs are given in Figs. 25.1, 25.2, 25.3, 25.4 and 25.5 (Ref 1).

Fig. 25.1: Stop Sign

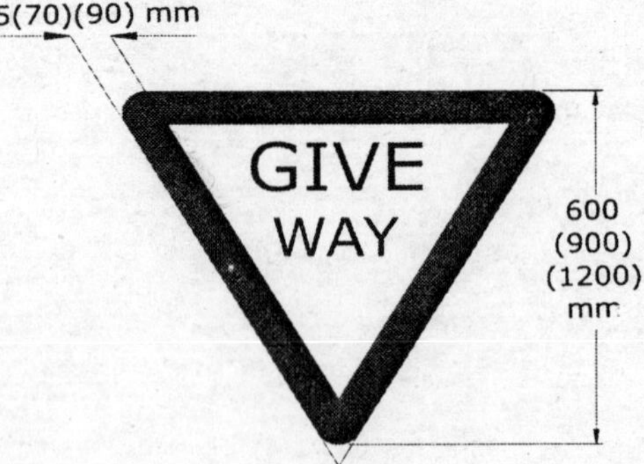

Fig. 25.2: Give Way Sign

Fig. 25.3: Mandatory Signs

Fig. 25.4: Warning Signs

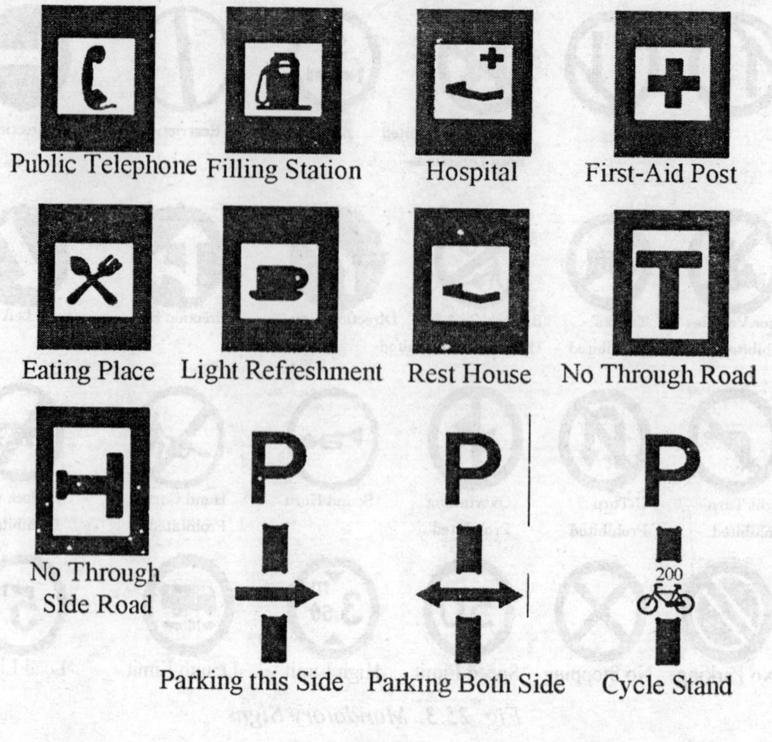

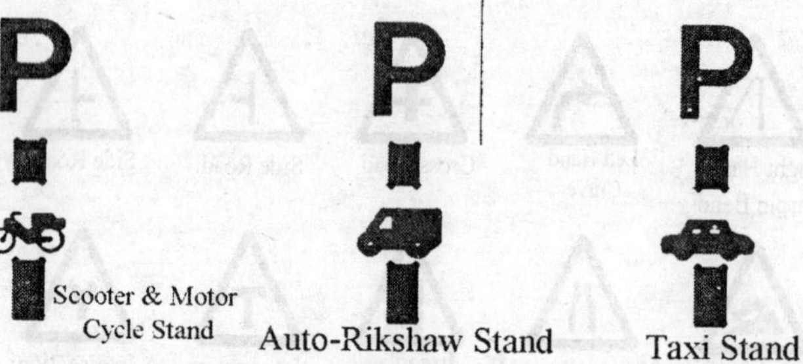

Fig. 25.5: Information Signs

The signs shall have retro-reflective sheeting to ensure good visibility. The signs are normally ground mounted for normal roads, but overhead signs are used at complex interchanges, roads with three or more lanes in each direction and expressways.

Variable Message Signs are provided (Ref 2) to provide information on congestion, speed restrictions, weather related problems, occurrence of accidents ahead to warn the drivers to drive carefully. They are operated with the assistance of a central controller.

25.2.3 Road Marking

Road markings are provided on the road pavement (carriageway markings) or on objects which are part of the road.

Carriageway markings are of the following categories:

(i) Centreline of carriageway

(ii) Traffic lane markings

(iii) Stop lines

(iv) Pedestrian crossings

(v) Bus Stops

(vi) Pavement edge lines

(vii) Direction arrows

(viii) Word Messages

(ix) Parking-related markings

The markings are painted with hot applied thermoplastic paints. The colour of the markings as per Indian practice is (Ref 3):

- **White Colour :** For all carriageway markings except those intended for parking restrictions
- **Yellow Colour:** For markings intended for parking restrictions, continuous centre line and barrier line markings
- **Alternate Bands of White and Black:** Kerb and object markings

Figs. 25.6, 25.7, 25.8, 25.9 and 25.10 give the road markings as per Indian practice.

On the kerbs, alternate black and white strips are painted, the width of the strips being 50 cm.

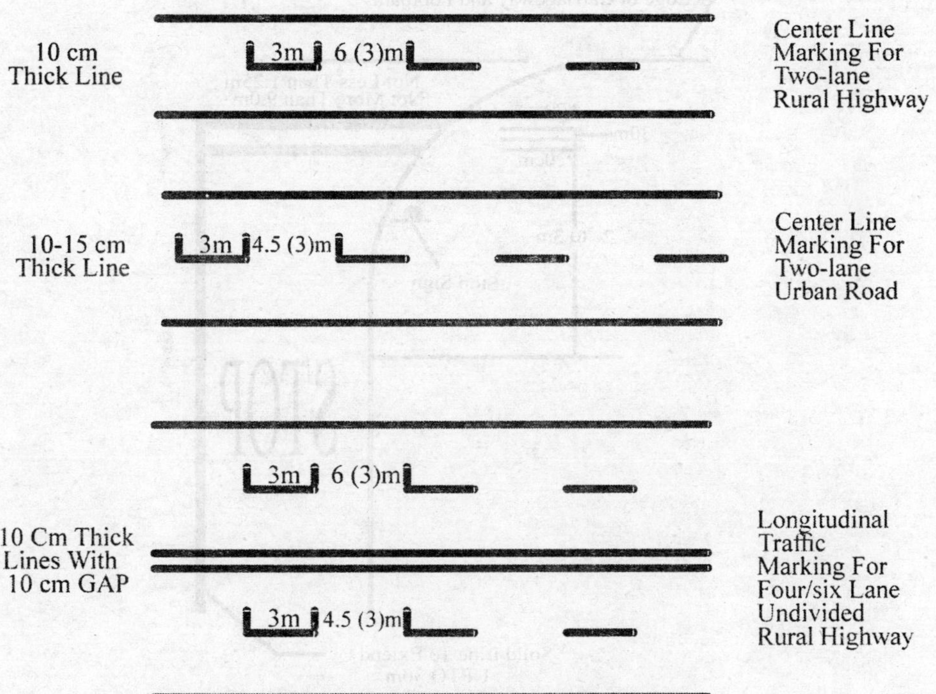

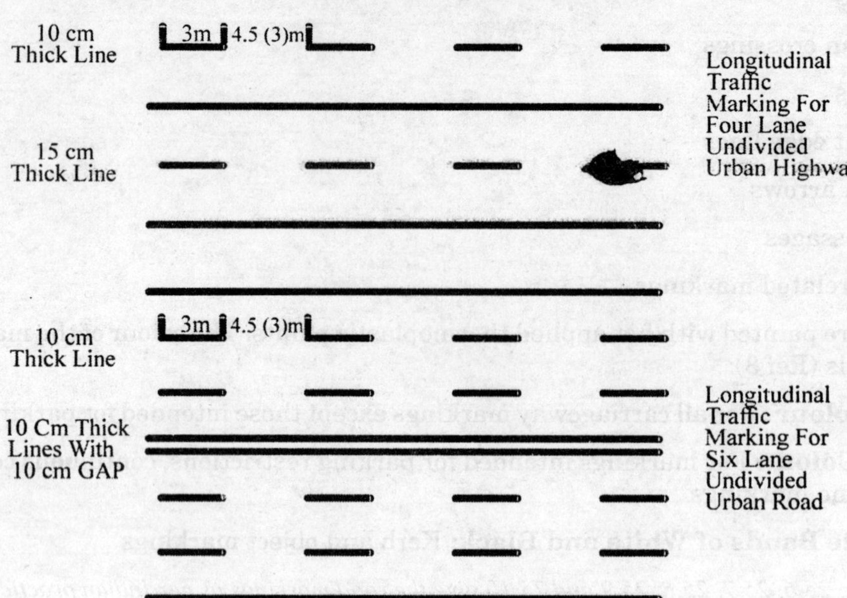

Fig. 25.6: Centre Line and Lane Markings

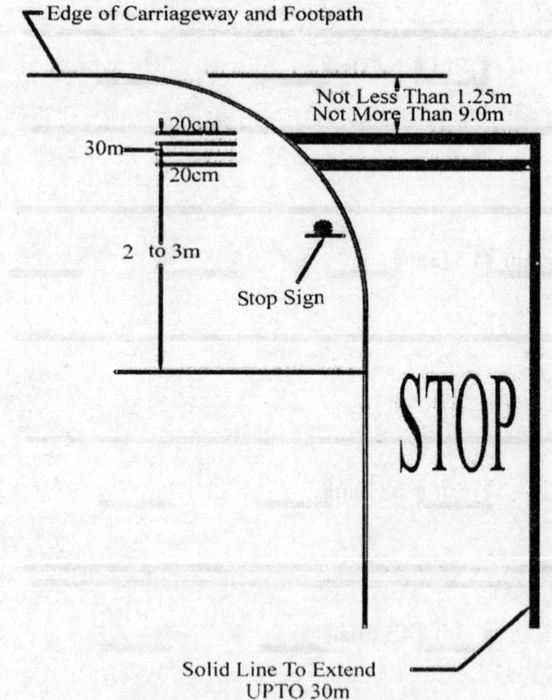

Fig. 25.7: Details of Stop Line

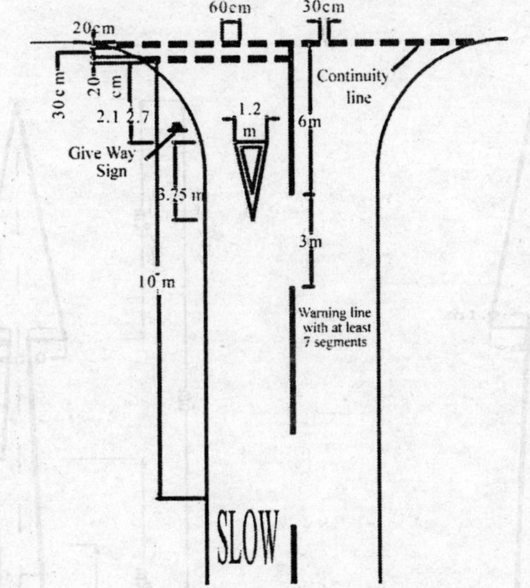

Fig. 25.8: Markings for GIVE WAY Indication

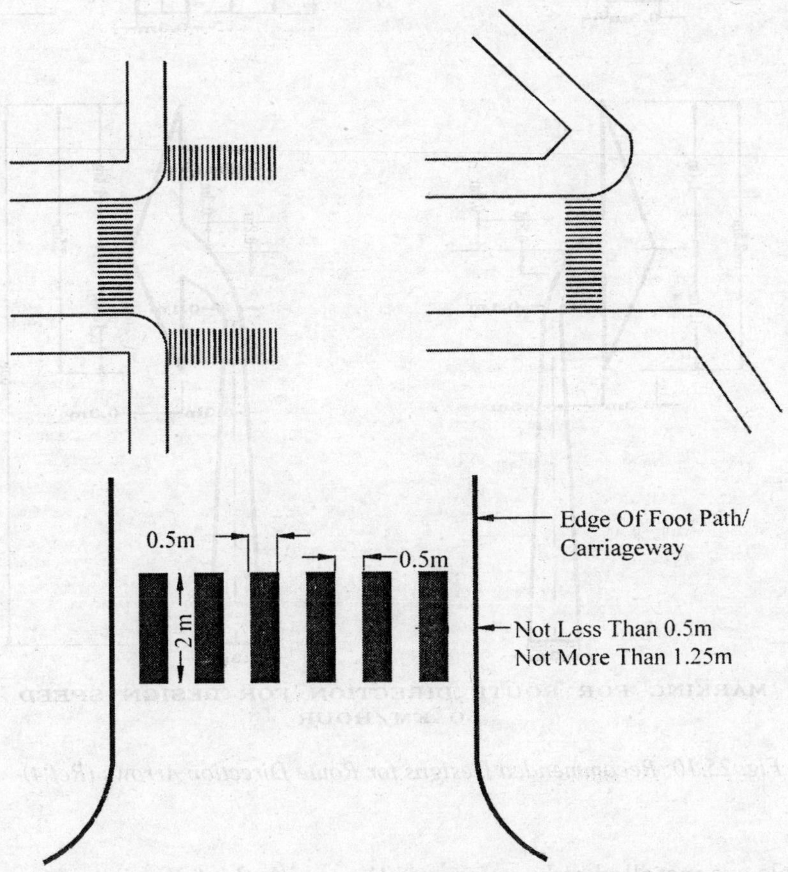

Fig. 25.9: Indian Practice for Pedestrian Markings

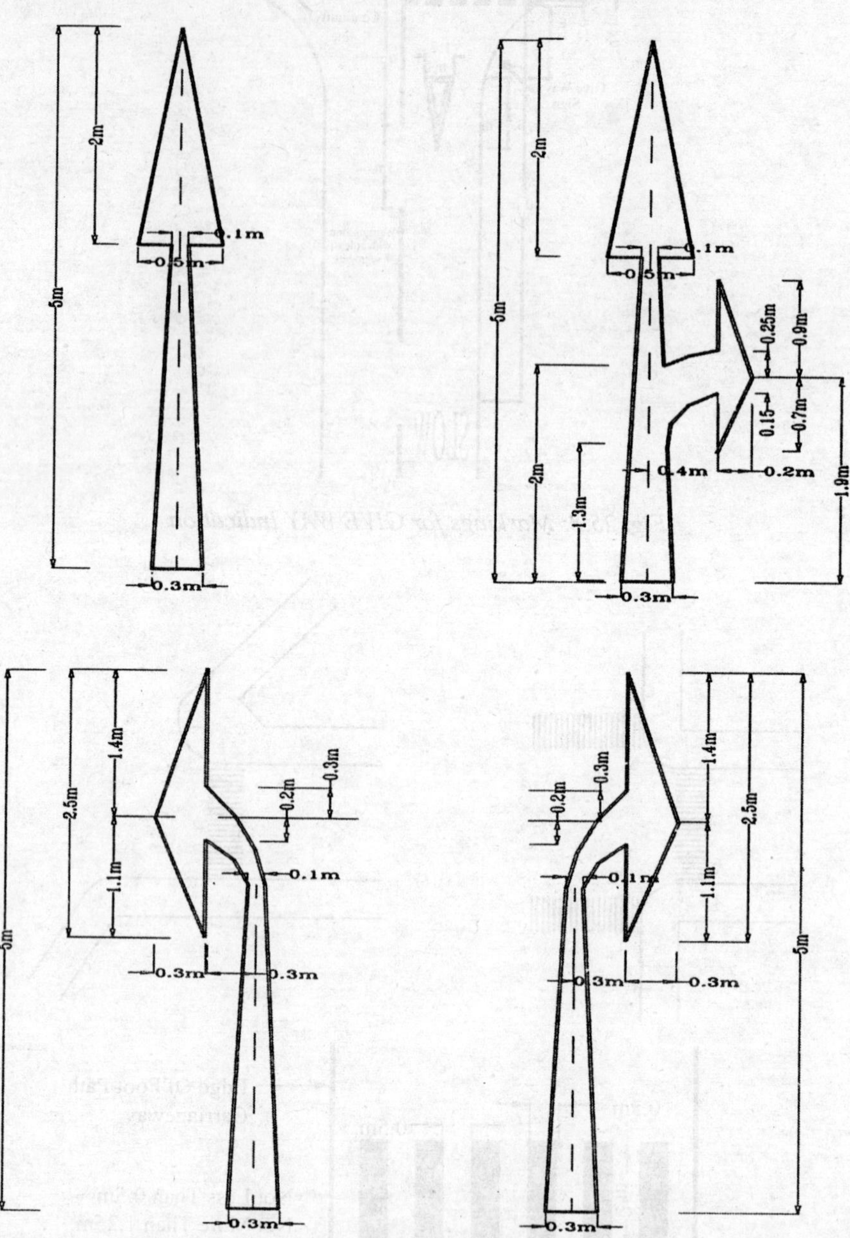

Fig. 25.10: Recommended Designs for Route Direction Arrows (Ref 4)

25.2.4. Signals

Road Traffic Signals are installed at busy intersections with the following warrants as per Indian practice (Ref 4).

Warrant 1–Minimum Vehicular Vol

No. of lanes for moving traffic on each approach		Motor Vehicles per hour on major street (total for both approaches)	Motor Vehicles per hour on higher volume minor street approach (one direction only)
Major Street	**Minor Street**		
1	1	650	200
2 or more	1	800	200
2 or more	2 or more	800	250
1	2 or more	650	250

Warrant 2–Interruption of Continuous Traffic

No. of lanes for moving traffic on each approach		Motor Vehicles per hour on major street (total of both approaches)	Motor Vehicles per hour on higher volume minor street approach (one direction only)
Major Street	**Minor Street**		
1	1	1000	100
2 or more	1	1200	100
2 or more	2 or more	1200	150
1	2 or more	1000	150

Warrant 3–Minimum Pedestrian Volume

(i) On the major street, 600 or more vehicles enter the intersection (total of both approaches), or where there is a raised median 1.5 m or more in width, 1000 or more vehicles per hour (total of both approaches) enter the intersection on major street, and

(ii) During the same 8 hours as in (i) above, there are 150 or more pedestrians per hour on the highest volume cross-walk crossing the major street.

Warrant 4–Accident Experience

Five or more reported accidents taking place in a year.

The Indian practice is to provide Red, Amber and Green circular lenses. The Amber interval is a transition interval between the termination of the green indication and the exhibition of the red indication. The amber period is generally 2 seconds.

Signals can be fixed-time signals or vehicle-actuated, the latter taking care of the volume of traffic entering the intersection.

The optimum cycle time is calculated from the following equation:

$$C_0 = \frac{1.5\,L + 5}{1 - Y} \text{ seconds} \qquad (25.1)$$

where C_0 = Optimum cycle time

L	=	Lost time per cycle in seconds
Y	=	$y_1 + y_2 + \ldots\ldots\ldots\ldots y_n$, where $y_1 + y_2 + \ldots\ldots\ldots\ldots y_n$ are the maximum ratios of flow to saturation flow

The saturation flow is the flow during which the flow is continuous. For new signal installations the saturation flow is obtained by the formula:

$$s = 525\ w\ \text{PCU/hr} \tag{25.2}$$

where s = saturation flow

w = width of approach road in metres

The traffic volume is converted into PCU as per the following factors

	PCU
Car	1.00
Heavy or medium goods vehicle	1.75
Light goods vehicle	1.00
Bus	2.25
Motor cycle, moped or scooter	0.33
Pedal cycle	0.20

The cycle time is normally restricted to 120 seconds.

Where signals are located close to each other on a road, it is desirable to design the system in such a manner that the platoons of vehicles can pass through all the signals without stoppage.

25.3. Parking

Parking of vehicles is a problem which has to be addressed by the traffic engineer. The space required for parking is as under:

Cars	:	2.5 m x 5 m, and preferably 3 m x 6 m
Two-wheelers	:	0.8 m x 2.5 m
Bicycle	:	1.4–1.8 sq m
Commercial Vehicles	:	3.75 m x 7.5 m

On-street parking can be parallel parking (where space is restricted) or angle parking which is preferred for making parking and unparking manoeuvres.

Surface off-street parking, multi-storey car parks and underground car parks are common in major cities. Modern multi-storey car parks can be totally mechanically operated by robots.

25.4. Street Lighting

Street lighting is needed for guiding road users to see the carriageway and the immediate surroundings clearly. The Indian standards for levels of illumination are given in Table 25.1.

Table 25.1: Levels of Illumination as Per Indian Practice

Classification of Lighting Installation	Type of Road	Average Level of Illumination on Road (lux)	Ratio Minimum -Average Level of Illumination on Road (lux)	Transverse Variation of Illumination	Type of Luminaire	
					Preferred	Permitted
Group A1	Important traffic routes carrying fast traffic	30	0.4	33	cut-off	semi-cut off
Group A2	Other main roads carrying mixed traffic like main city streets, arterial roads, throughways etc.	15	0.4	33	cut-off	semi-cut off
Group B1	Secondary roads with considerable traffic like principal traffic routes, shopping streets etc.	8	0.3	20	cut-off or semi-cut off	semi-cut off
Group B2	Secondary routes with light traffic	4	0.3	20	cut-off or semi-cut off	semi-cut off

In the cut-off system, the light is reduced as completely as possible above a 70^0 angle with the vertical, thus eliminating glare. In the non-cut-off system, there is considerable output near the horizontal plane, permitting glare. The semi-cut-off is in between the above two types.

The mounting height for heavily trafficked routes is 7.5–12 m, and 3–8 m for other streets.

The spacing of lanterns should not exceed 55 m and should preferably be 35-45 m on important traffic routes.

The lamps can be tungsten filament, tubular fluorescent, sodium vapour discharge or high pressure mercury discharge. High pressure sodium vapour discharge and mercury bulb fluorescent are the most suitable for streets with high traffic.

High mast lighting is provided at grade separators, car parks, important junctions and toll plazas.

25.5. Environmental Effect of Traffic

Road traffic causes several environmental effects such as:

(i) Air pollution

(ii) Noise

(iii) Vibration

(iv) Visual intrusion

(v) Severance

As regards air pollution, emissions caused by the exhaust of vehicles contain major pollutants like carbon dioxide, carbon monoxide, lead compounds, carbon particles (smoke) and unburnt petrol. Air pollution can be controlled by reducing the use of private vehicles, frequent checking of vehicle emissions, use of battery operated vehicles and by improving vehicle design and maintenance.

Noise levels are measured by decibels (A scale). The acceptable noise levels recommended by the Wilson Committee, U.K., are given in Table 25.2 (Ref 6).

Table 25.2: Acceptable Noise Levels (Wilson Committee)

Situation	Noise Levels (dB(A)	
	Day	Night
Country Areas	40	30
Suburban Areas	45	35
Busy Urban Areas	50	35

Noise can be controlled by changes in the design of vehicles, changes in tyres or road surfaces and modifications in traffic operations. Noise barriers can prevent noise reaching adjacent buildings.

Vibration is caused by the movement of vehicles on narrow streets flanked by buildings. High levels of vibration can cause structural damages and annoyance to the residents. Vibrations can be controlled by improving the surface finish of the road, improvements in vehicle design, reducing the number of commercial vehicles using the road and provision of wider streets.

Visual intrusion is caused by roadside advertisements, unregulated parking and ribbon development along highways. The aesthetics can be improved by planting avenue trees, control of access, restricting roadside advertisements and providing adequate off-street parking.

Severance is caused when a road is aligned through a village or habitation, dividing the life of the community, forcing them to cross the road and risking their safety. Great care is needed in selecting the alignment of the road to avoid severance.

25.6. Road Safety

Statistics about road accidents show a poor record of road safety in India. About 1,42,000 persons were killed in road accidents in India in the year 2011 (Ref 6). This means that every day abcut 400 people are killed on Indian roads, which is equivalent to two aeroplanes crashing in a day. The factors responsible for the road accidents are:

(i) The road

(ii) The vehicle

(iii) The drivers and other road users

(iv) Environmental Factors

(v) Lack of Enforcement

The design of road to make it safe for all road users can prevent accidents. Improvement of intersections, segregation of pedestrians and cyclists, provision of proper signages, road markings and signals, reducing pavement slipperiness and adequate illumination are some of the engineering measures that can prevent accidents.

The proper maintenance of vehicles in a roadworthy condition is the responsibility of the vehicle owner.

Drunken driving is the cause of several accidents. Periodic checking by Police using breathalysers can act as a deterrent.

Over speeding is another frequent cause of road accidents. Checking the speeds by Police personnel using radar guns can act as a deterrent. Enforcement measures by Police, empowered to impose heavy fines on the site for traffic violators can improve road safety.

Use of crash helmets, seat belts and inflatable airbags in vehicles can reduce the severity of accidents.

Education, particularly in schools, can prevent accidents. Training of professional drivers in driver training schools is another part of a safety education programme.

Propaganda by poster campaigns, use of TV and radio for creating awareness and use of catchy and imaginative slogans can go a long way in reducing road accidents.

Road accidents cause a serious economic loss to the society. Though latest data is not available, the cost as on 2009 are as under (Ref 7).

Table 25.3: Economic Cost of Accidents

S.No.	Type of Accident	Economic Cost of Each Accident (₹)
1.	Fatal	864,350
2.	Serious	391,800
3.	Major	172,650
4.	Minor	30,450
5.	Damages to Car	26,150
6.	Damages to Two-wheeler	6,650
7.	Damages to Three-wheeler	7,600
8.	Damages to Buses	76,050
9.	Damages to Heavy Commercial Vehicle	8,600
10.	Damages to Multi-Axled Vehicle	13,400

25.7. Intelligent Transport System (ITS)

Intelligent Transport System (ITS), also known as Transport Telematics, is a system that applies modern information technology to improve the operation of transport networks (Ref 8). ITS Can cover various activities such as:

(i) Monitoring incidents on the road and locating them on a Geographical Information System (GIS)

(ii) Monitoring traffic flow and giving Advanced Traveller Information (ATS) to travellers on weather, road closures, traffic congestion, accidents etc.

(iii) Electronic collection of toll using smart cards

(iv) Traffic responsive traffic control

Application of ITS has enhanced safety, comfort and lowered user costs considerably.

REFERENCES

1. Code of Practice for Road Signs, IRC:67-2010, Indian Roads Congress, New Delhi, 2010.
2. Guidelines for Variable Message Signs, IRC:SP:85-2010, Indian Roads Congress, New Delhi, 2010.
3. Code of Practice for Road Markings, IRC:35-1997, Indian Roads Congress, New Delhi, 1997.
4. Guidelines on Design and Installation of Road Traffic Signals, IRC:93-1985, Indian Roads Congress, New Delhi, 1985.
5. Code of Practice for lighting of public thoroughfares, IS:1944 (Part I and II), Bureau of Indian Standards, New Delhi, 1970.
6. Road Accidents in India, 2011, Transport Research Wing, Ministry of Road Transport and Highways, New Delhi, 2012.
7. Manual on Economic Evaluation of Highway Projects in India, IRC:SP:30-2009, Indian Roads Congress, New Delhi, 2009.
8. ITS Handbook, 2000 PIARC Committee on Intelligent Transport, Artech House, Boston, London, 1999.

SECTION—3

RAILWAY ENGINEERING

26

Chapter

Introduction to Railway Engineering

26.1. Importance of Rail Transport

Rail transport is one of the dominant transport modes in the world today, with a history of nearly 200 years. George Stephenson perfected the steam locomotive, and his "Rocket" locomotive was tried out on the Liverpool Manchester line in 1829. The first railway line was opened to traffic in 1825 between Stockton and Darlington in UK. The first railway line was opened to traffic in the USA in 1830 being a portion of the Baltimore-Ohio railroad. Between 1830 and 1850, several countries of Europe opened their first railway line. The growth of railway transport facilitated the industrial revolution in Europe, moving the raw materials and finished goods in a short time at an economical cost. The railways helped the westward movement and occupation of the early settlers in the USA.

26.2. Railway Engineering

The construction, maintenance and operation of a railway system requires the expertise of various engineering disciplines, such as Civil Engineering, Mechanical Engineering, Electrical Engineering, Telecommunication Engineering etc. Civil Engineers are concerned with the route selection, alignment, construction and maintenance of railway stations and goods yards. The civil engineering aspects of railway engineering will be dealt in this book.

27

Chapter

RAILWAY TECHNOLOGY

In a transport system, there are four components:

1. Containers
2. Propulsion unit
3. Pathway
4. Terminal

In Railways, the containers are the passenger coaches and goods wagons. The propulsion unit is the locomotive, which may be a steam engine, diesel locomotive or an electric locomotive. The pathway is the railway track on which the trains move. The terminal is the railway station and goods yard where passengers and goods are serviced.

The track in a railway system is exclusive, and consists of two steel rails separated at a distance, known as "gauge". The wheels of the coaches, wagons and locomotives are also of steel, thus providing movement of steel on steel with little friction. The track can be laid on the ground, on elevated structures or underground in tunnels. Modern innovations like monorail and pneumatic wheels on concrete rails are part of Light Rail Transit and are discussed in a separate Chapter.

The following are some of the areas where the Indian Railways should concentrate to upgrade the railway system:

1. Introduce fast freight trains of speed 100 Km per hour, if necessary, by constructing dedicated freight corridors.
2. Expedite gauge conversion
3. Hasten the pace of electrification
4. Attract more of container traffic
5. Introduce Piggy-back system, wherein trucks can be loaded on the flat rail cars and can travel long distance. This way, energy conservation, reducing pollution and road accidents can be achieved. The Konkan Railway Corporation has already started this operation.
6. Increase the speed of fast passenger trains from 150 to 250 Km per hour by upgrading the track and traffic control system, and by introducing "bullet trains".

28 Chapter

Role of Railways

28.1. Various Roles of Railways

Railway transport has contributed over the past many years in several ways in the development of the economy of many countries. The various roles of railways can be discussed under the following heads:

1. Facilitated the industrialisation
2. Serves as bulk carriers
3. Suitable for long haulage
4. Suitable for high density traffic
5. Serves as mass rapid system in metropolitan areas
6. Serves as a cheap mode for long distance passenger travel
7. Helps administer large areas
8. Supports military movements
9. Provides public utility services for movement of low-rated goods

28.2. Railways as Facilitators of Industrialisation

European countries, particularly Britain, were the pioneers in the industrial revolution in the eighteenth and nineteenth centuries. These countries colonized vast areas of Asia, Africa and the American continent. Huge natural resources were found in these colonised countries, and the colonial powers built railway lines to transport these to the ports for eventual shipments to Europe, where large factories were founded to manufacture steel, textiles and machinery.

In the USA, major industries got located along the rail corridors that were built to reach the western region. In India, the laying of the vast railway network helped the rise of the textile mills in Ahmadabad and Bombay, jute mills in Calcutta, the steel mill at Jamshedpur and cement units in Karnataka and Madhya Pradesh.

28.3. Railways as Bulk Carriers

Rail transport is particularly suited for the carriage of bulk commodies like mineral ores, coal, food grains, fertilisers, steel and cement. In 2004-05, Indian Railways moved 640 bil tonne-Km of freight.

28.4. Railways are ideally suited for long haul traffic

Railways are ideally suited for full-rake traffic over long hauls. For example, in India, the average lead of revenue-earning freight traffic is around 700 Km (2004-05) (Ref 1).

Full rake movement from one point to another over a long distance eliminates stoppages at intermediate stations and time lost in decoupling and coupling wagons.

28.5. Railways are suitable for high density traffic

Taking India's case, the four metro cities of Delhi, Bombay, Madras and Calcutta constituted the corners of quadrilateral. The sides and diagonals of this quadrilateral carry the bulk of the rail-based freight movement.

It is precisely for this reason that the Indian Railways are now planning to develop these corridors with a capability of running freight trains at a speed of 100 Km per hour.

28.6. Railways as Mass Rapid Transit System in Metro Cities

Most of the metro cities in the world today have rail-based Mass Rapid Transit System for facilitating passenger movement. The underground tube systems in London and New York are classic examples. The surface-based suburban railway system in Mumbai carrying nearly 5 million passengers daily is another example. The Calcutta and Delhi metros have benefitted the city dwellers immensely.

28.7. Railways as a cheap mode for long distance passenger travel

In developed countries, for example USA, with a very good system of highways, persons choose personalised road transport even for long distance of 1000-2000 Kms. But in a less developed country like India, where roads are in a poor condition and air travel is beyond the reach of many, rail transport is the most preferred mode for long-distance travel. With the introduction of high speed trains, the time taken for travel has come down. Overnight journeys by rail is a competitor to air travel.

28.8. Railways help administer large areas

Countries of a large size have to administer remote areas. Till air travel became popular, rail transport was the only mode to reach out to these areas. The Trans-Siberian Railway of the USSR cuts across vast areas. In India, the advent of the railways in the latter half of the nineteenth century facilitated the British in administering the vast country. Even today, the Assam rail link is vital for connecting the remote areas of the North-East with the rest of the country. China's recently opened Qinghai-Lhasa rail link is facilitator for establishing close contact between China and its recently annexed province of Tibet.

28.9. Railways Support Large Military Movements

In a large country like India, the military establishments are spread throughout the country. In the case of war with its neighbours (Pakistan and China), large movement of troops, supplies, and military hardware (armoured vehicles, tanks, guns etc) take place through the railway network.

28.10. Railways as Public Utility Providers

Where Railways are government-owned, as in India, they are obliged to provide reasonable facilities for receiving and forwarding traffic offered to them without prejudice or discrimination. Such obligation is not shared by other modes of transport. In addition, the central government in India directs the railways to give preference to transport of low rated goods such as coal, mineral ores, food grains, fertilisers, often at subsidised rate so that the common man is benefitted.

28.11. Advantages of Rail Transport

The advantages of rail transport over other modes can be listed as under:

1. Rail transport is suitable for carriage of bulk commodities over long haul (for example: coal, mineral ore, steel products, cement, petroleum products).
2. Rail transport enjoys an advantage when a full rake load is to be transported from one point to other.
3. Rail transport consumes less energy than road transport, because of the low friction between steel wheels and rail.
4. Rail transport can be operated using energy other than hydrocarbons which are in short supply globally. Rail transport can be operated on electricity.
5. Rail transport is pollution-free.
6. Rail transport is safer than road transport, as accidents are rare.
7. Rail transport is ideally suited for Mass Rapid Transit in cities.
8. Rail transport is economical for long distance passenger traffic.

28.12. Disadvantages of Rail Transport

1. Railway network cannot penetrate into every town and village. It is only the road network that can do so.
2. Providing a new rail link is economically feasible only if there is adequate freight and passenger traffic. Thus, rail transport is viable only on high traffic density corridors and not on low and medium traffic density corridors.
3. Rail transport has to depend upon road transport for collection and distribution of traffic. Thus, it is not a complete mode by itself. Door-to-door delivery is not possible, without the help of road transport.
4. A train can move only when a full load in all wagons is assembled. Movement of part loads is not financially viable. On the other hand, a truck can move when a load of 10 T is offered.
5. Rail transport is not economical for short haul of goods or passengers.
6. Careful packaging of goods is needed since handling of goods is not efficient and can cause damage.
7. Rail transport is not speedy. This disadvantage can be overcome only when dedicated freight corridors are built.

28.13. A comparison of Indian railways operations with some of the countries where railway transport occupies an important role will enable one to understand where India stands now and in what directions it should move forward. Table 28.1 gives a comparison (Ref 2).

Table 28.1: International Comparison of Railway Operations (Ref 2)

Country	Length of Railway Track (Km)	Passenger Km (mil)	Freight Tonne-Km (mil)
1. USA	222,000	22,500	2,075,000
2. Russia	151,000	144,700	1,205
3. Germany	87,207	66,500	71,356
4. Canada	65,403	1,458	299,508

(Contd...)

Country	Length of Railway Track (Km)	Passenger Km (mil)	Freight Tonne-Km (mil)
5. India	62,809	420,449	305,513
6. China	57,584	369,598	1,236,200
7. UK	37,849	38,300	18,400
8. Australia	35,780	11,510	127,400
9. Brazil	29,706	12,667	141,239

[Note: India's railways network length is the route length and not track length]

A glaring conclusion that can be drawn from the above statistics is that the Indian Railways are concentrating more on passenger movement than on freight movement. China, USA, Australia and Brazil, on the other hand use their railway system more for freight movement than for passenger movement. Low passenger fare, subsidised suburban rail travel, the political compulsions of introducing new express passenger trains in every railway budget and the high freight rates are responsible for this. As a result, goods move largely by roads, consuming more fuel. India needs to reverse the trend and improve its freight operations. The proposed dedicated freight corridors with train speeds of 100 Km/hr are a step in the right direction.

REFERENCES

1. Twelfth Five Year Plan, Planning Commission, Govt. of India, New Delhi.
2. Britannica Year Book 2007, Encyclopaedia Britannica, Chicago, 2007.

29

Chapter

History of Growth of Railways in India

29.1. Introduction of Railways In India

When the Britishers established their rule in India, transport was mainly by animal drawn vehicles travelling over roads in a poor condition. The British administration realized the importance of a good system of communication, particularly after the Sepoy Mutiny in 1857. This was the very period when railways were introduced in Britain, several European countries and the U.S.A.

The newly set up industries in UK needed cheap raw materials like cotton and jute, which were grown in India. These could be transported cheaply with a good railway network to the Indian ports for eventual shipment to Britain. A reliable and speedy transport system was also needed for consolidating their hold over the country, and this could only be railways.

The first railway line was laid in Bombay between Bori Bunder (later known as V.T., Victoria Terminus) and Thane, a distance of 31 Kms. It was extended to Kalyan in 1854 and later Igatpuri, traversing the difficult Thalghat hills. In the East, the first passenger train chugged out of Howrah station for Hoogly, a distance of 39 Kms in 1864. In the North, the Amritsar-Attari section was opened in 1862. In the South, the first railway line was opened in 1856.

In Assam, the Dibrugarh-Dinjan line was opened in 1882. Thus, the Britishers built the vast railway network in small bits and pieces, in difficult terrain, bridging great rivers and overcoming flooding problems. The task accomplished by the Britishers deserves great acknowledgement.

29.2. The Guarantee System (1853-1862)

The first phase of railway construction was mainly due to the efforts of a few companies formed in Britain. The Great Indian Peninsular (GIP) and East Indian (EI) were the first of these. The government not only offered free land to these companies, but also agreed to guarantee interest at 5 per cent per annum on the capital raised by the companies. The guarantee scheme gave rise to criticism of waste, extravagance and inefficient management, and the Indians felt that they were being made to pay taxes to make up for the losses incurred by the British capitalists. All the same, the Railways had a positive impact on India's economy.

The coal industry grew, a steel plant was established by the Tatas in Jamshedpur, several textile mills were started in Bombay and Ahmedabad and jute mills came up in Calcutta.

29.3. Subsidy Scheme (1862-1869)

The criticism of the guarantee scheme made the government to give it up and adopt the Subsidy Scheme, under which the government offered to give free land and an annual subsidy for 20 years. The scheme had to be abandoned since it failed to attract sufficient capital.

29.4. Direct State Construction (1869-1881)

With the abandonment of the Subsidy Scheme, the government took up direct construction of certain new lines. Government borrowings, surplus revenues and famine relief measures funded the projects.

29.5. The Gauge Decision

Lord Dalhousie, who was the governor General of India, was responsible for selecting a wide gauge (5'-6", known as the Broad Gauge, BG) in preference to the standard gauge of 4'-8½" widely prevalent in UK and Europe. The selection was based on the reasoning that a wide train would be able to withstand the high winds in the plains of India. It was soon found that the broad gauge was costly for construction and in many lines, the Metre Gauge (MG), 1.00m (3'-33/8") was adopted.

For hill stations and unremunerative line, the Narrow Gauge (0.610 m or 0.762 m) was also adopted. The adoption of several gauges was an unwise decision, leading to several problems of goods transhipment and passengers having to change trains. The government has taken a bold decision now to have only broad gauge under the unigauge policy.

29.6. Direct State Management

The government appointed Ackworth Committee recommended (1921) that the government should take over the railways and manage it. The government took over the principal railways one by one after the contracts expired.

The Railway Board was constituted. A number of development schemes were taken up. Competition from road transport was growing and steps to secure coordination between the two modes were initiated.

29.7. Railway Development Since Independence

The Indian Railways suffered serious dislocation when India was partitioned. One of the immediate tasks of independent India was to construct the Assam Rail Link, which was a formidable engineering challenge. The Chittaranjan locomotive factory, the Integral Coach Factory at Perambur and the Diesel Locomotive Factory at Varanasi were set up. Many schemes of doubling of track, electrification of selected routes, modernisation of workshops and construction of new lines were taken up. Some of the important new lines taken up are the Pathankot-Jammu line giving access to the Jammu & Kashmir State, Kandla rail link giving access to the new port of Kandla and the Konkan Railway between Panvel and Mangalore. The gradual conversion of several gauges to the broad gauge is yet another achievement

29.8. Growth of Railways In India

Table 29.1 gives the growth of railways in India.

Table 29.1: Growth of Indian Railways (Ref 1, 2, 3, 4)

Year	Route Length (Km)	Passengers Carried (mil)	Passenger Traffic Bil-Passenger-Km	Revenue Earning Freight (Mil T)	Revenue Earning Freight Traffic (Bil T-Km)
1950-51	53,596	1284	66.52	73.2	37.57
1956-61	56,247	1594	77.67	119.2	72.33
1970-71	59,790	2431	118.12	167.9	110.70

(Contd...)

Year	Route Length (Km)	Passengers Carried (mil)	Passenger Traffic Bil-Passenger-Km	Revenue Earning Freight (Mil T)	Revenue Earning Freight Traffic (Bil T-Km)
1980-81	61,240	3613	208.56	195.9	147.65
1991-92	62,500	4049	315.0	338.0	250.0
2001-02	63,100	5093	490.9	492.3	333.2
2004-05	63,500	5416	570.0	601.9	411.4
2010-11	64,460	7651	979.0	921.5	606.0

29.9. Railway Administration

With the single exception of USA, where private ownership and operation is practised, the railways in most countries are state-owned. In India, the railways are fully state-owned. The railways are administered by the Railway Board, consisting of:

1. Chairman
2. Member, Engineering
3. Member, Traffic
4. Member, Electrical
5. Member, Mechanical
6. Member, Staff
7. Financial Commissioner

The Railway Board is under the Ministry of Railways. Railway finances are separated from the general finances of the country. The Railways have their own budget.

For administering the vast network, the Railways are divided into Zones, each having a route length of about 3,000–10,000 Km.

Some of the important organisations connected with the Indian Railways are listed below:

1. Indian Railway Construction Corporation (IRCON)
   - which takes up railways and other civil works on a contract basis
2. Rail India Technical and Economic Services (RITES)
   - which takes up consultancy work in transportation, railways, roads and airports
3. Research, Design, and Standardisation Organisation (RDSO)
   - based in Lucknow, takes up research and standardisation work
4. Container Corporation of India (CONCOR)
   - provides container service
5. Indian Railway Finance Corporation (IRFC)
   - which raises funds through bonds, and provides finance to selected railway projects
6. Rail Vikas Nigan Ltd. (RVNL)
   - an organisation providing non-budgetary initiative for creating rail infrastructure on a commercial format

7. Railway staff College, Vadodara
   - providing training to railway Officers
8. Indian Railways Institute of Mechanical and Electrical Engineering, Jamalpur
9. Konkan Railway Corporation
   - Construction, maintenance and operation of the 760 Km long Konkan Railway
10. Indian Railway Catering and Tourism Corporation
    - providing catering services and facilitating tourism

29.10. Road vs Rail Transport in India

In the initial years of the railways operations (1850-1920), the railways enjoyed a near monopoly since the people realized the advantages of the "iron horse", the railway engine. With the advent of motor vehicle in India during and after the First World War, road transport began to make inroads into the monopoly enjoyed by railways. The Government, being the owners of the railways, were interested in protecting the rail transport. But the clamour for better roads could not be suppressed, the road transport, particularly goods transport, began to grow rapidly. The loss of rail share in the total road plus rail transport progressively is depicted in Table 29.2.

Table 29.2: Road : Rail Share in Transport (Ref 1, 2, 3)

Year	Passenger Km		Freight Tonne-Km	
	Share of Road (%)	Share of Rail (%)	Share of Road (%)	Share of Rail (%)
1950-51	26	74	11	89
1960-61	42	58	13	87
1970-71	59	41	34	66
1980-81	69	31	50	50
1999-2000	83	17	51	49

The following conclusions can be drawn:

1. The Railways continue to lose share of passenger traffic. Soon, the share may be just 10 per cent. Road transport and air transport are eroding the rail share.
2. In the freight movement, the share of rail seems to have stabilized at around 50%. The railways should take all measures to see that the share does not go down further. They enjoy several advantages and shouid consolidate their position, and in fact try to regain their share lost to the roads.

REFERENCES

1. Hundred years of Indian Railways, Railway Board, New Delhi.
2. Issues of Railway Year Book, Railway Board, New Delhi.
3. Issues of Economic Survey, Govt. of India, New Delhi.
4. Twelfth Five Year Plan, Planning Commission, Government of India, New Delhi.

5. Report of the National Transport Policy Committee, (Pandey Committee), Planning Commission, New Delhi, 1980.
6. Estimation of Total Road Transport Freight and Passenger Movement in India for the year 2000 AD, Engineering Consultants Pvt. Ltd., New Delhi, 1987.
7. Tenth Five Year Plan, 2002-07, Planning Commission, New Delhi, 2002.
8. Railways Year Book 2005, Impressions India, New Delhi, 2005.

30
Chapter

Surveys and Project Preparation

30.1. Stages in Preparing a Railway Project

The following flow-chart (Fig. 30.1) gives the stages involved in preparing a railway project.

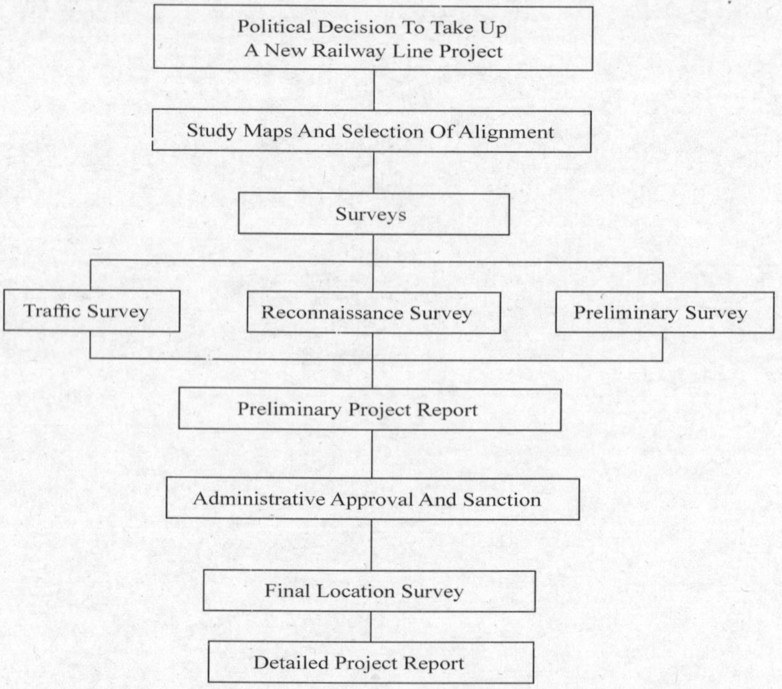

Fig. 30.1: Stages in Preparing a Railway Project

30.2. Selection of Alignment

A political decision have been taken to provide a rail connection between two points, the next step is to select the best possible alignment. Some of the factors that need to be considered in the selection of the alignment are:

1. The alignment should be as direct as possible, without unnecessary detours. A direct alignment is the shortest, and possibly the cheapest to build, maintain and operate.
2. The alignment should serve important towns and cities en route. If not directly touch the towns and cities, it should be as near to these as possible.
3. The alignment should avoid difficult areas like landslides, marshy ground, deep gorges and hills requiring expensive tunnelling.
4. The cost of construction and subsequent maintenance should be minimum.
5. The cost of operation of trains should be minimum. This can be achieved with an alignment having easy gradients.
6. Obligatory point such as sites for major bridges and tunnels often dictate the alignment.
7. Environmentally sensitive areas should be avoided.
8. Strategic and defence considerations must be kept in view.

30.3. Study of Maps

It is necessary to study all available maps (Survey of India Topo Sheets, Remote Sensing data), and make several trials of the alignment in the office. The advantages and disadvantages of each of the alternative alignment must be discussed.

30.4. Traffic Survey

The financial viability of a railway project depends upon the traffic that will use the facility. Estimation of traffic that will use the facility immediately on completion and at least over the next 20 years should be accurately assessed. Both passenger and freight movement should be assessed. The data to be collected include:

1. Traffic using the existing alternative facility like road and Inland Water Transport and the extent of its likely diversion to the railways. The diversion can be assessed based on the cost of movement of goods and passengers and time taken.
2. Important commodities that can be moved, including their tonnage and destination.
3. Economy of the region, its past trends and prospects for growth in the future.
4. Traffic that will be generated and induced by the facility.

Various scenarios shall be investigated and a realistic assessment of traffic shall be made.

30.5. Reconnaissance Survey

The several emerging alignments should be reconnoitred by walking, using simple instruments like the Abney Level, Ghat Tracer, Aneroid Barometer, Compass etc.

The Reconnaissance Survey helps in reducing the alternatives to a few, say two or three, which appear to be promising.

30.6. Preliminary Survey

The two or three alternatives emerging as promising from the Reconnaissance Survey should be taken up for a Preliminary Survey. This includes instrumented surveys like traversing and levelling (with a Theodolite or Total Station, aided by GPS data), collection of hydraulic particulars for the fixation of waterway and design of bridges, subsoil borings for the design of foundations of bridges and tunnels, material survey (for borrow area soil, stone aggregates, sand etc.).

30.7. Preliminary Project Report

The Preliminary Project Report should contain:

1. Selection of Alignment, including various alternatives, obligatory points, towns and cities served
2. Traffic Survey and Forecast
3. Design Standards adopted
4. Engineering Surveys
5. Preliminary Cost Estimate
6. Phasing of Expenditure
7. Financial Viability

Based on this Report, the authorities can accord administrative approval and sanction.

30.8. Final Location Survey

In this stage, the alignment is finally laid on the ground accurately using Theodolite/ Total Station and precision levels. Permanent reference marks are left for taking up the work. Detailed design and cost estimate are prepared. Land Acquisition proceedings are initiated and tender documents are prepared. The Detailed project Report should contain all the aspects.

31

Chapter

Geometric Design of Railway Track

31.1. Geometric Design Elements

Geometric design of a railway track is the subject of dimensioning of the visible elements of the track. It covers the following elements:

1. Cross-sectional elements
   - Width of land
   - Gauge
   - Formation
   - Embankment slopes
   - Slopes of cuttings
   - Distance between adjacent tracks
   - Horizontal clearance
   - Vertical clearance
2. Horizontal Curvature and its associated elements
3. Gradient
4. Vertical curves

31.2. Cross-Sectional Elements

31.2.1. Width of Land

The land width to be acquired for laying a new railway line depends upon the gauge selected and the number of tracks to be accommodated (initially or at a later date). It is customary to acquire a land width of 50-60 m for arterial broad gauge lines.

31.2.2. Gauge

The subject of gauge is discussed separately in Chapter 32 which deals with the Permanent Way.

31.2.3. Formation

In Railway Engineering, the embankment constructed on the natural ground to accommodate the railway track is known as the formation. The formation width is the top width of the embankment. In case of a track in cutting, the formation width is the base width of the cutting excluding side drains. Figs 31.1 and 31.2, give typical cross-sections of a broad gauge track in embankment and cutting.

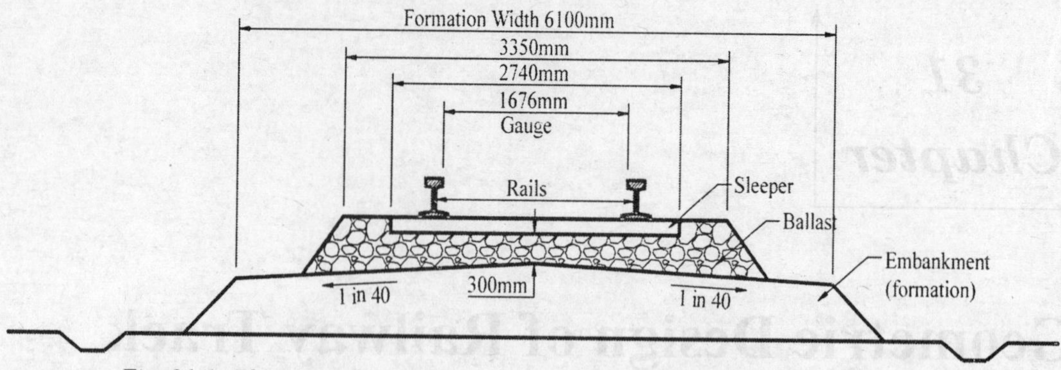

Fig. 31.1: Elements of Cross-section of a BG Track in Embankment

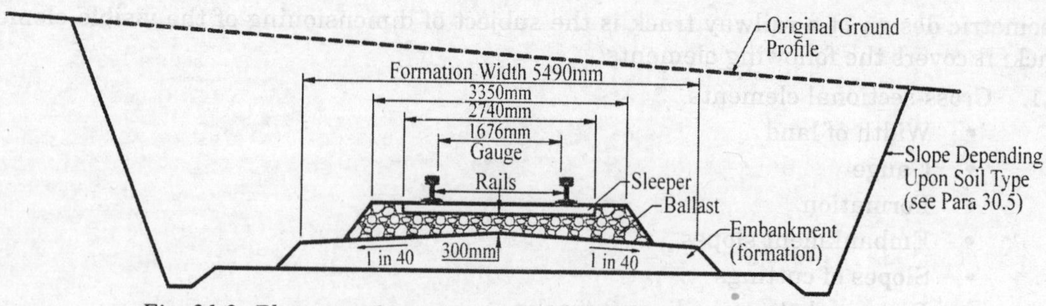

Fig. 31.2: Elements of Cross-section of a BG Track in Cutting

It may be noted that the top of the formation has an outward slope of 1 in 40 on either side of the centre line to facilitate drainage.

31.2.4. Embankment Slopes

The slope of the embankment depends upon the soil type and height of the embankment. For normal height, a slope of 2:1 (2 Horizontal : 1 Vertical) is adopted. The embankment slopes are turfed to prevent rain cuts and erosion.

31.2.5. Slope of Cuttings

The slope of cuttings depends upon the soil type. The slopes generally adopted are:-

Table 31.1: Slopes of Cuttings

Soil Type	Slope (1 in H, H being the horizontal dimension)
Wet clay	1 in 2.5 to 1 in 4
Well drained day	1 in 2 to 1 in 1.5
Silt	1 in 1 to 1 in 1.5
Sandy soil	1 in 1.3
Gravel (Murram)	1 in 1

(Contd...)

Soil Type	Slope (1 in H, H being the horizontal dimension)
Disintegrated Rock	1 in 1/2
Soft Rock and Shale	1 in 1/4 to 1 in 1/8
Medium Rock	1 in 1/12 to 1 in 1/16
Hard Rock	Near Vertical

31.2.6. Distance between Tracks

The minimum centre to centre distance between tracks is 4265 mm for BG and 3660 mm for MG (Fig. 31.3).

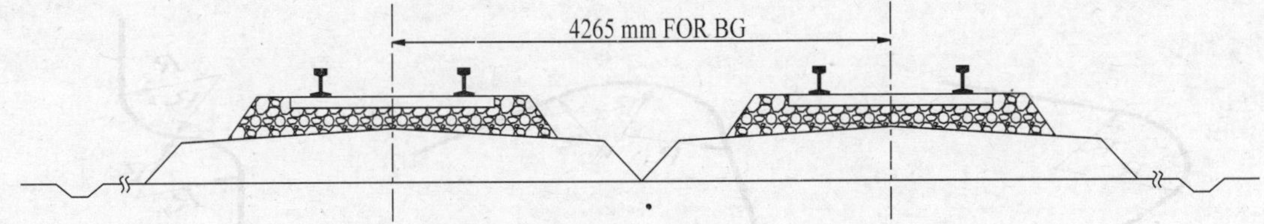

Fig. 31.3: Distance between Tracks

31.2.7. Horizontal Clearance to Structures

The minimum horizontal clearance to buildings and structure is as below:

	BG
• Minimum distance from centre of track to any structure from Rail Level to 305 mm above Rail Level	1675
• Minimum distance from centre of track to any structure (except a platform) from 305 mm above Rail Level to 4420 mm above Rail Level for BG and 3320 mm above Rail Level for MG	2135

31.2.8. Vertical Clearance

For BG, the vertical clearance for overhead structures is 4875 mm. Most of the routes in India are likely to be electrified. For DC traction, the clearance required is 5410 mm, and for AC traction it is 5870–6200 mm.

31.3. Horizontal Curvature and its Associated Elements

31.3.1. Need for Curves

It is obvious that alignment cannot be straight and there are bound to be changes in alignment depending upon the topography. Horizontal curves are inserted at every change in the alignment to obtain a gradual change. Railway curves have to be as gentle as possible, with large radius. Care should be taken to avoid curves at:

(i) bridges
(ii) approaches to bridges
(iii) tunnels
(iv) viaducts

(v) station and their approaches

(vi) level crossings

(vii) deep cuttings

31.3.2. Types of Horizontal Curves

Three types of curves are commonly met with:

(i) single circular curve

(ii) compound curve

(iii) reverse curve

Fig. 31.4 gives the three types.

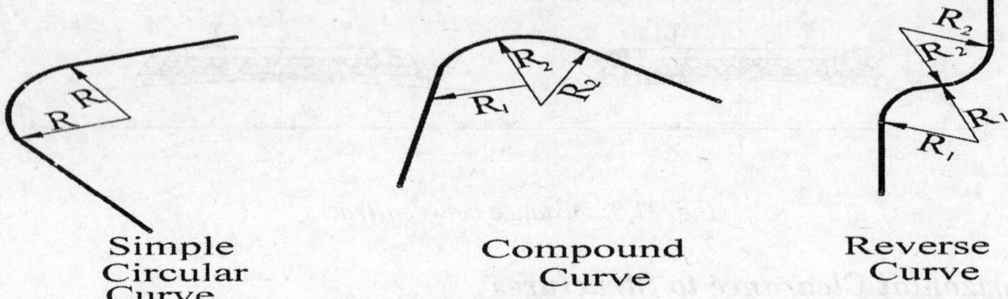

Fig. 31.4: Types of Curves

31.3.3. Degree of Curve

There are two ways of defining a horizontal curve, viz,

(i) by its radius, as is common in highway engineering

(ii) by its degree, as is common in railway engineering

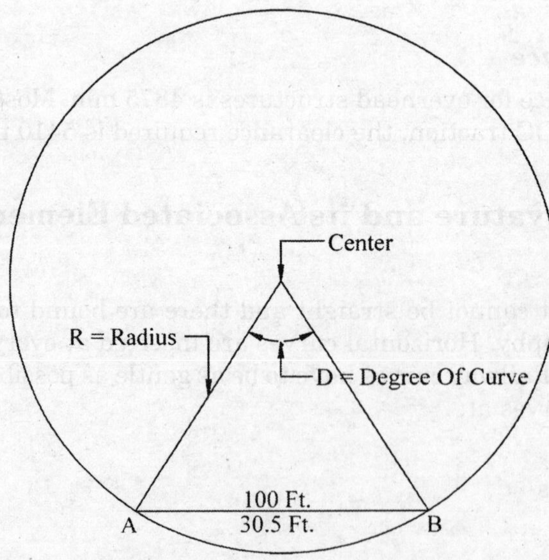

Fig. 31.5: Degree of a Railway Curve

The degree of a curve is the angle subtended at the centre by a chord 100 ft (30.5 m) (Fig. 31.5). For a large circle and small degree of angle, the chord AB can be considered to be equal to the arc AB.

Thus, angle subtended at centre by chord/ arc AB, Degree of curve

$$= \frac{260}{2 \pi R} \times 30.5 = \frac{1748.4}{R}$$

$$= \text{say } \frac{1750}{R} \qquad (31.1)$$

Thus, a one degree curve has a radius of 1750 m, and a three degree curve has a radius of

$$\frac{1750}{3} = 583.3 \text{ m}$$

Thus, the larger the value of the degree of a curve, the smaller is the radius.

The maximum degree of curvature permitted in the Indian Railway is:

Broad Gauge

Plain line	10^0 or 175 m radius
Turnout	8^0 or 218 m radius

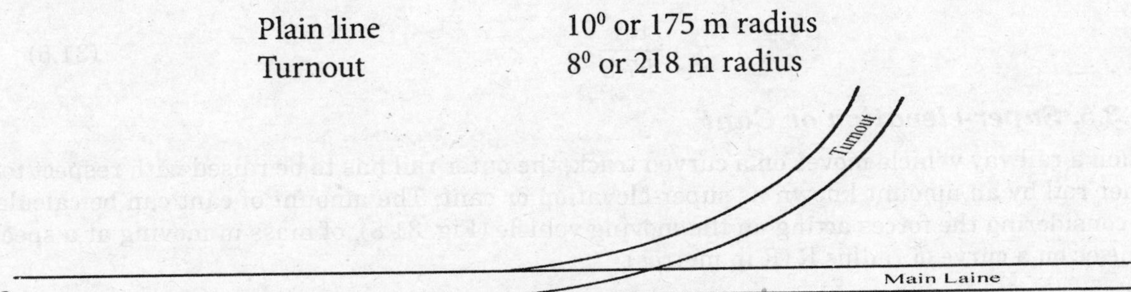

Fig. 31.6: Turnout from a Main Line

31.3.4. Some Simple Relationships in a Circular Curve

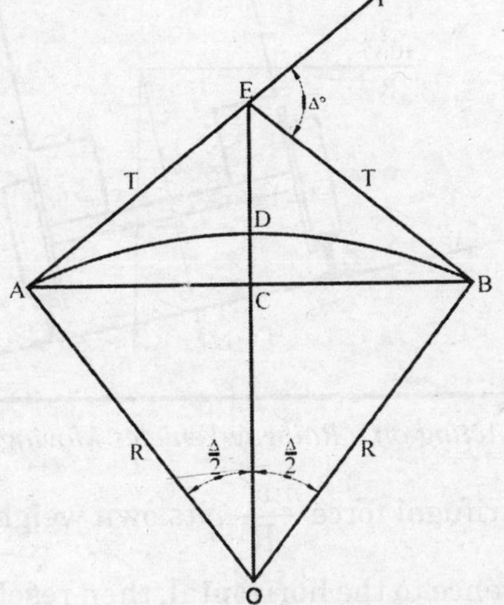

Fig. 31.7: Circular Curve Relationships

Let the radius be R

Let Δ be the deflection angle in degree (∠ FEB)

It is also equal to the subtended angle (∠ AOB)

$$T = \text{Tangent length} = R \tan \frac{\Delta}{2} \tag{31.2}$$

$$\text{Apex Distance ED} = OE - OD$$

$$= R \sec \frac{\Delta}{2} R$$

$$= R\left(\sec \frac{\Delta}{2} - 1\right) \tag{31.3}$$

$$\text{Chord length AB} = 2 \times AC$$

$$= 2R \sin \frac{\Delta}{2} \tag{31.4}$$

$$\text{Curve length AB} = R\,\phi, \text{ where } \phi \text{ is in radians}$$

$$= R\,\frac{\Delta \times 2\pi}{360} \text{ where } \Delta \text{ is in degrees}$$

$$= \frac{R\Delta}{57.3} \tag{31.5}$$

31.3.5. Super-elevation or Cant

When a railway vehicle moves on a curved track, the outer rail has to be raised with respect to the inner rail by an amount known as super-elevation or cant. The amount of cant can be calculated by considering the forces acting on the moving vehicle (Fig. 31.8), of mass m moving at a speed of v m/sec on a curve of radius R (R in metres).

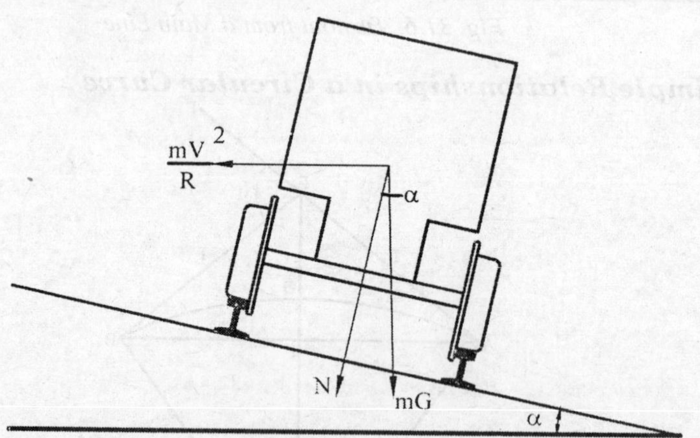

Fig. 31.8: Forces Acting on a Railway Vehicles Moving on a Curved Path

The vehicle is subjected to a centrifugal force $\frac{mv^2}{R}$, its own weight mg acting vertically. If α is the inclination of the track with reference to the horizontal, then resolving the forces along the inclined plane and neglecting lateral friction,

Example

$$\frac{mv^2}{R}\cos\alpha = mg\sin\alpha$$

$$\therefore \frac{v^2}{gR} = \tan\alpha$$

$$\therefore \tan\alpha = \text{superelevation, e}$$

$$= \frac{v^2}{gR}$$

If the speed is expressed in Km/ hr

$$e = \left(\frac{V}{3.6}\right)^2 \times \frac{1}{9.81} \times \frac{1}{R}$$

$$= \frac{V^2}{127\ R} \qquad (31.6)$$

The value of e thus calculated is in decimals. Since one is interested in the value of superelevated in mm, e has to be multiplied by the gauge (G).

$$\text{Thus, superelevation in mm} = \frac{GV^2}{127\ R} \qquad (31.7)$$

All railway vehicles (for example goods wagons and passenger coaches) do not travel at the same speed. The question then arises for what speed should the cant be calculated and provided. Too much of cant can cause overloading of the lower rail, whereas too little can cause overloadings of the upper rail. A compromise is generally considered and the speed so selected is called the "equilibrium speed".

The maximum cant permitted on Indian Railways is:

Broad Gauge

165 mm for group A, B and C

140 mm for group E and E routes

A value of 185 mm is recommended for planning on all group A routes.

When a railway vehicle travels at a speed higher than the equilibrium speed, cant deficiency occurs. The maximum cant deficiency permitted is:

Broad Gauge

75 mm

100 mm for speeds in excess of 100 Km/hr with special permission.

Example: Calculate the cant on a railway curve of 2 degree in a BG track for a maximum speed of 80 Km/hr.

Solution

$$\text{Cant in (mm)} = \frac{GV^2}{127\ R} \quad \text{(Formula 31.7)}$$

$$G = 1676 \text{ mm}$$

$$R = \frac{1750}{\text{Degree}} \text{ (Formula 31.1)}$$

$$= \frac{1750}{2} = 875 \text{ m}$$

$$V = 80 \text{ kmph}$$

$$\therefore \text{Cant} = \frac{GV^2}{127\,R} = \frac{1676 \times 80 \times 80}{127 \times 875}$$

$$= 96.5 \text{ m}$$

31.3.6. Transition Curve

31.3.6.1. Need

When a railway vehicle on a straight portion of the track (*i.e.,* of infinite radius) enters a curve of finite radius, it is suddenly subjected to the centrifugal force which causes shock and sway. In order to avoid this, it is customary to provide a transition curve at the beginning of the circular curve, having a radius equal to infinity at the end of the straight and gradually reducing the radius to the radius of the circular curve where the curve begins. Incidentally, the transition curve also facilitates the provision of the cant, from a value of zero at the beginning of the transition to the calculated full value at the beginning of the circular curve. Transition curves are provided at either end of the circular curve, *i.e.,* at the beginning and at the end.

31.3.6.2. Elements of a circular curve provided with transition

Fig. 31.9 gives the elements of a circular curve with transitional ends. It may be noted that the circular curve has to be shifted by an amount known as "shift" at half the transition length to accommodate the transition length.

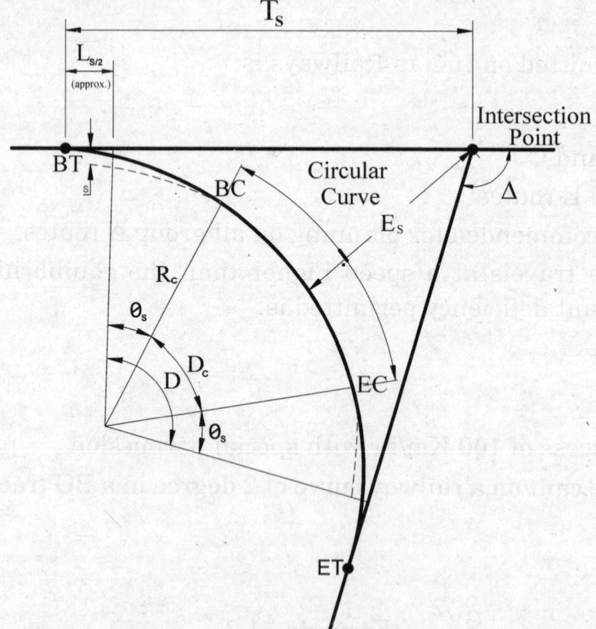

Fig. 31.9: Main Elements of a Circular Curve provided with Transitions

BC = Beginning of circular curve

BT = Beginning of Transition

R_c = Radius of circular curve

s = Shift

Δ = Deflection Angle

θs = Angle subtended by half the Transition Length

EC = End of circular Curve

ET = End of Transition

L_s = Length of Transition

T_s = Tangent Length

Δc = Angle subtended by Circular Curve

E_s = Apex distance

The cubic parabola is the type of transition curve generally adopted in the Indian Railways. The following relationships may be noted, with reference to Fig. 31.9.

$$\text{Shift} = \frac{L_s^{\,2}}{24R_C} \tag{31.8}$$

$$T_s = \frac{L_s}{2} + (R_C + s)\tan\frac{\Delta}{2} \tag{31.9}$$

$$E_s = (R_C + s)\sec\frac{\Delta}{2} - R_C \tag{31.10}$$

31.3.6.3. Length of Transition Curve

The desirable length of transition is calculated from considerations of permissible rate of cant, cant deficiency and radial acceleration. The length is calculated from one of the following equations:

1. $L = 0.008\ C_a \times V_m$ (based on rate of gain of cant and of cant deficiency of 35 mm per second) (31.11)
2. $L = 0.008\ C_d \times V_m$ (31.12)
3. $L = 0.72\ C_a$ (based on maximum cant gradient of 1.4 mm per metre or 1 in 720 (31.13)

Where

L = Transition length in metres

C_a = Actual cant in mm

W_w = Cant deficiency in mm

V_m = Maximum permissible speed in Km/hr

The maximum value is adopted.

Problem

Calculate the length of transition curve for a BG track on Group A for a one degree curve and a maximum speed of 160 Km/hr.

Solution

$$R = \frac{1750}{1} = 1750$$

V = 160 kmph

Cant (in mm) = $\frac{GV^2}{127\,R}$

= $\frac{1676 \times 160 \times 160}{127 \times 1750}$

= 193 m

This may be restricted to 185 mm as the maximum permitted value (see para 31.3.5).

Using formula 31.11,

L = 0.008 x 185 x 160

= 236.8 m

Using formula 31.13,

L = 0.72 x 185

= 133.2 m

Provide 236.8 m or 240 m, the higher of the two values.

31.3.6.4. Setting out Cubic Parabola

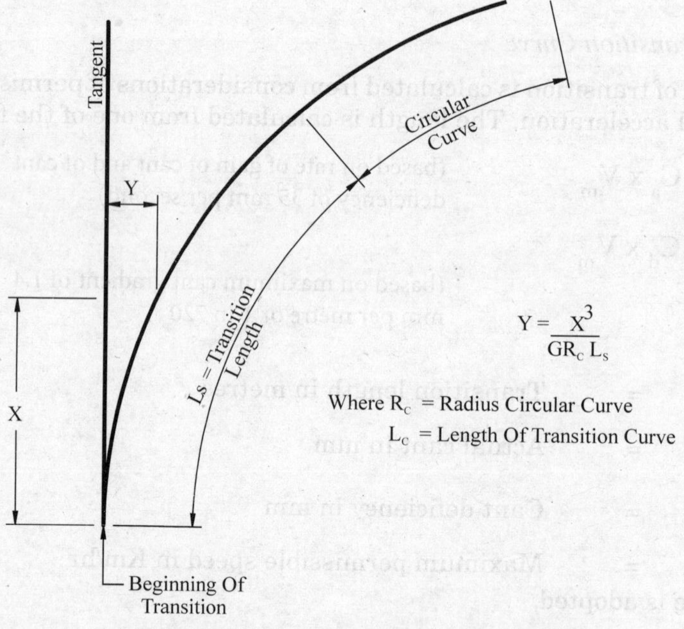

Fig. 31.10: Equation of Cubic Parabola

The equation of a cubic parabola is:

$$y = \frac{X^3}{6 R_c L_s} \qquad (31.14)$$

Thus, knowing y, the distance of any point from the beginning of the transition, the offset y from the Tangent to the curve can be calculated.

31.3.6.5. Method of Attainment of Cant

The cant can be attained by any one of the following methods:

1. Maintain the level of the inner rail constant, and raise the outer rail gradually over the transition length to attain the desired cant.
2. Maintain the centre line level of the track constant, raise the outer rail gradually to attain half the desired cant and simultaneously depress the lower rail by an equal amount.
3. Maintain the outer rail level constant and depress the inner rail gradually to attain the desired level.
4. Maintain the level of the inner rail constant for half the transition length, the outer rail getting raised by half the cant. In the remaining half, maintain the outer rail at the level obtained, and gradually depress the inner rail to half the cant at the end of the transition.

Method 1 is popular among the maintenance staff, especially when manual maintenance is done.

31.3.7. Safe Speed on Curves

The basic formula governing speed, radius and cant is:

$$\text{Cant (in mm)} = \frac{GV^2}{127\ R} \quad \text{(Formula 31.7)}$$

The cant is composed of C_a (actual cant) and C_d (cant deficiency).

Thus
$$C_a + C_d = \frac{GV^2}{127\ R}$$

$$\therefore V = \sqrt{\frac{(C_a + C_d)127\ R}{G}}$$

For Broad Gauge,

$$V = 0.27\sqrt{(C_a + C_d)\ 127\ R} \quad (31.15)$$

[Note: G in the above derivation is 1750 mm for BG, which is the centre to centre distance between the rail heads, and not the gauge].

31.3.8. Gauge Widening on Curves

When a railway vehicle travels on a curved track, the front axle tends to push the outer rail, and the flange gets worn out, and so also the outer rail head. The vehicle will have a smooth passage if the gauge is slightly widened, and the wear of the flange and the rail head is reduced. But excess of gauge widening increases the play of the wheel, which could result in increasing the angle of attack of the wheel flange on the rail, again leading to extra wear of both the flange and the rail. Too much of slack in the gauge can also lead to derailment. Thus, a compromise has to be achieved in gauge widening. From experience, the Indian Railway have arrived at the following optimum gauge widening policy for BG.

1.	Straight, including curves of radius 400 m and more	3 mm tight, *i.e.*, gauge is 1673 mm

2. Curves of radius less than 400 m — upto 5 mm slack, *i.e.*, gauge can be upto 1681 mm

31.3.9. Check Rail of Curves

When a railway vehicle travels on a curve, it has been mentioned earlier that the flange of the wheel rubs against the outer rail, leading to excessive wear of the rail head and the flange. Besides, there is a risk of the wheel climbing over the rail, which may lead to derailment. To prevent this, a "check rail" is provided parallel to the inner rail on the inside (between the two rails) at a certain distance from it. On Indian Railways, such check rails are provided on curves of more than 8^0 on BG. The minimum clearance prescribed is 44 mm for BG and is shown as x in Fig. 31.11.

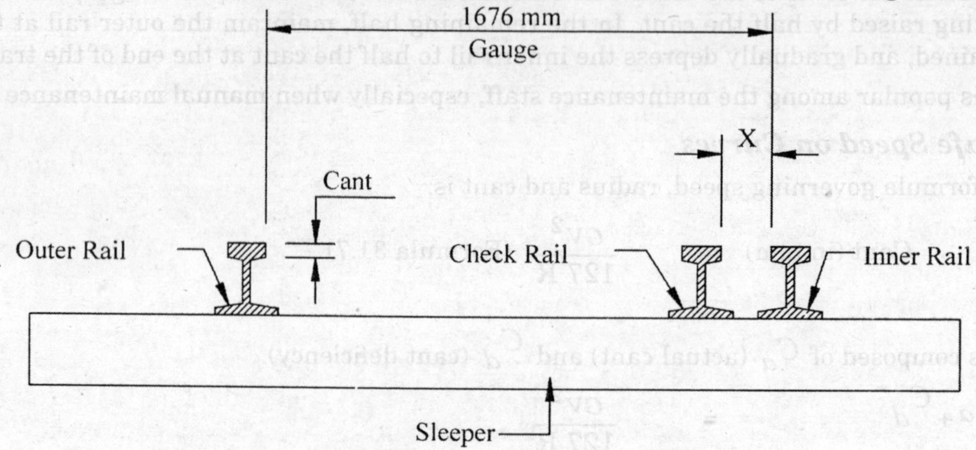

Fig. 31.11: Check Rail on Curves

31.4. Gradients in Railway Track

31.4.1. Desirable Gradient for Drainage

The topography of a region requires that the railway track rises and falls in elevation in keeping with the general ground level. Ideally, the gradient should be flat so that the tractive force developed by the locomotive is not spent on overcoming the effect of gravity. It may also be noted that perfectly flat gradient may not be conducive to drainage. Hence a minimum gradient of 0.5 per cent (1 in 200) is generally desirable.

31.4.2. Ruling Gradient

In railway engineering, the steepest grade which exists in a particular section is called the ruling gradient. In plain areas, generally the ruling gradient adopted is 1 in 200; but this can be made steeper for short stretches without having significant effect on the load traction capacity. In hills, gradients as steep as 1 in 50 are unavoidable.

31.4.3. Momentum Gradient

When a train goes down a falling gradient, it gains momentum and if the falling gradient is followed by a rising gradient, the momentum so gained can overcome the rising grade effect to some extent. The rising gradient can be made slightly steeper than the ruling gradient and is then called the "Momentum Gradient". In order to take advantage of the momentum, it is necessary that the section be a "no signal zone" so as to prevent the trains from slowing down or stopping.

31.4.4. Pusher Gradient

In some ghat section (steep hilly terrain), it is often unavoidable that steep gradients are introduced. For instance, the Bhor Ghat section of Bombay-Pune line has steep gradients (about 1 in 40). It then becomes necessary to have an additional engine (which is known as the banking engine) pushing the train from behind. Such a stretch is said to be having a pusher gradient.

31.4.5. Gradients in Station Yards

Trains stop and are stationary in Station Yards. In such sections, a strong wind combined with the effect of gravity can move a train if the gradients are steep. Moreover, when trains start, steep gradients require additional tractive effort, which may not be possible to attain with the engine. Thus, in station yards, it is customary to limit the gradient to 1 in 400, and a flatter gradient of 1 in 1000 is recommended.

31.4.6. Grade Compensation on Curves

When a horizontal curve and gradient are combined, the tractive effort has to overcome the grade resistance and curve resistance. To keep the resistance within allowable limits, the gradient is somewhat reduced to compensate for the curve effect. The reduction in the grade, known as grade compensation, is calculated on the basis of 0.04 per cent per degree of the curve for BG track. Thus, if the ruling gradient is 1 in 200 (*i.e.,* 0.5 per cent), and the curve is of 3 degrees, the grade compensation to be provided is:

3 x 0.04 = 0.12

Thus, the gradient to be provided is:

0.5–0.12 = 0.38 per cent or 1 in 263

31.5. Vertical Curves

When two vertical grades meet, it is necessary to avoid the abrupt kink by providing a vertical curve. Vertical curves can be summit curves (convexity upwards) or sag curves (convexity downwards) (Fig. 31.12).

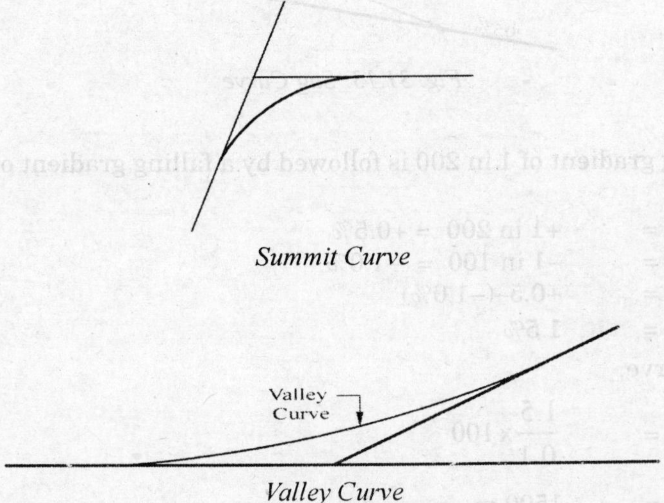

Fig. 31.12: Summit and Valley Curve

The length of vertical curve is governed by the maximum permissible rate of change of grade. In Indian Railways, the rate of change of grade is 0.1 per cent for summit curves and 0.05 per cent for valley curves. The length is given by the formula:

$$L = \frac{N}{r} \times 100 \qquad (31.16)$$

Where L = Length of vertical curve in m

N = Deviation angle expressed in radian/ or Tangent of angle in decimals

r = permissible rate of change of grade

Problem

A railway line has a rising gradient of 1 in 200 intersecting another rising gradient of 1 in 100. Calculate the length of the vertical curve.

First gradient = +1 in 200, +0.5%

Second gradient = +1 in 100, +1.0%

∴ Algebraic difference in successive grades

= +1.0–0.5 = 0.5%

Since it is valley curve, the permissible rate of change of gradient is 0.05%.

$$\therefore L = \frac{0.5}{0.05} \times 100$$

$$= 1000 \text{ m}$$

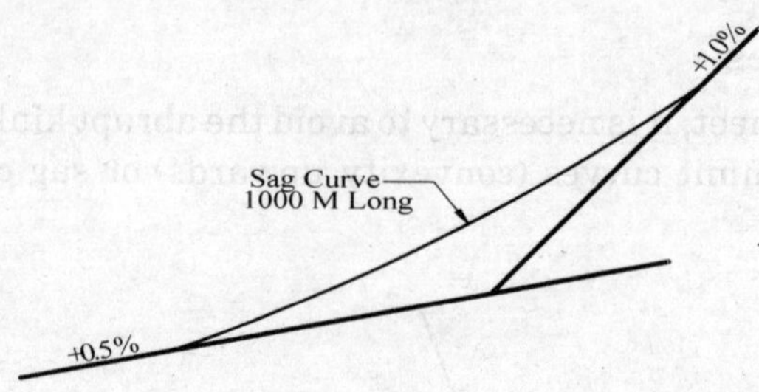

Fig. 31.13: Sag Curve

Problem a

On a BG track, a rising gradient of 1 in 200 is followed by a falling gradient of 1 in 100. Design the vertical curve.

Rising gradient = +1 in 200 = +0.5%

Falling gradient = –1 in 100 = –1.0%

Algebraic difference = +0.5–(–1.0%)

= 1.5%

Since it is a summit curve,

$$L = \frac{1.5}{0.1} \times 100$$

$$= 1500 \text{ m}$$

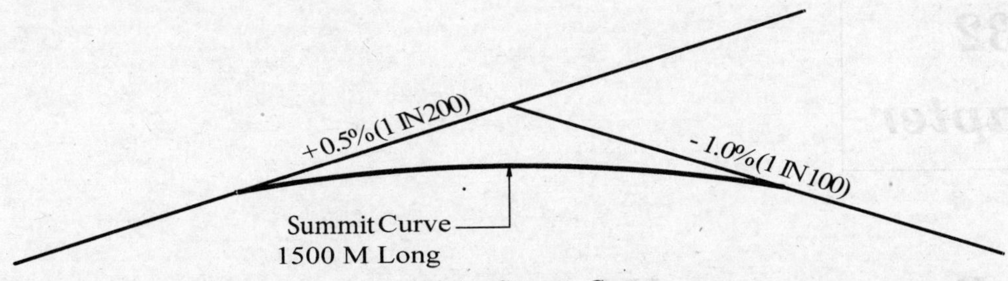

Fig. 31.14: Summit Curve

Vertical curves are generally in the shape of a parabola, which can be easily laid out. The vertical distance of any point from the tangent line, y, at a distance of x from the beginning of the curve is given by:

$$y = \frac{Nx^2}{2L} \quad (31.17)$$

Where N = Algebraic difference in grades

L = Length of vertical curve

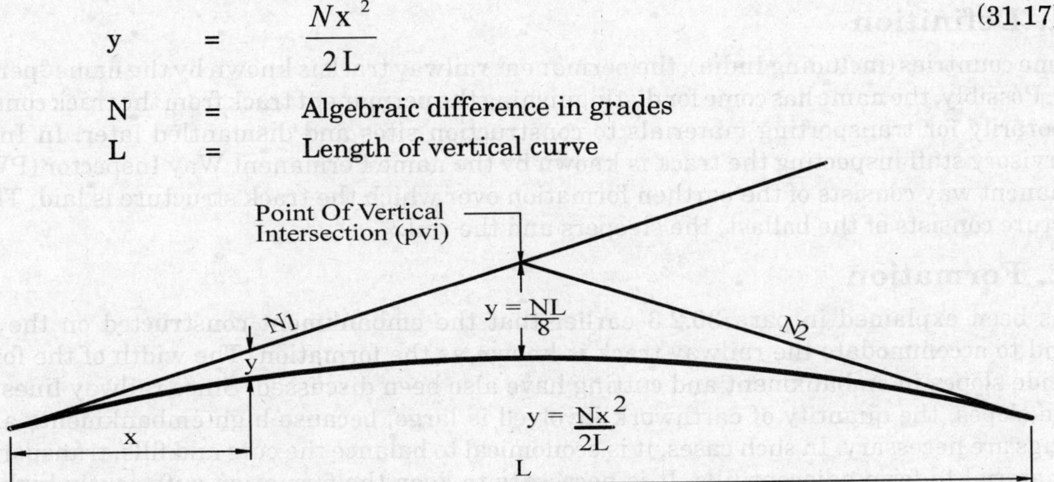

Fig. 31.15: Parabolic Vertical Curve

Thus at $\frac{L}{2}$, y, the maximum distance of the curve point from the point of vertical intersection is given by

$$y = \frac{N.L}{8} \quad (31.18)$$

32

Chapter

The Permanent Way

32.1. Definition

In some countries (including India), the permanent railway track is known by the name "permanent way". Possibly, the name has come for distinguishing the permanent track from the track constructed temporarily for transporting materials to construction sites and dismantled later. In India, the supervisory staff inspecting the track is known by the name Permanent Way Inspector (PWI). The permanent way consists of the earthen formation over which the track structure is laid. The track structure consists of the ballast, the sleepers and the rail.

32.2. Formation

It has been explained in para 30.2.3 earlier that the embankment constructed on the natural ground to accommodate the railway track is known as the formation. The width of the formation and side slopes in embankment and cutting have also been discussed. Since railway lines require gentle slopes, the quantity of earthwork involved is large, because high embankments and deep cuttings are necessary. In such cases, it is economical to balance the cuts and fills, transporting the cut material to form adjacent fills. It is necessary to keep the formation sufficiently high so that it does not get submerged under flooding conditions. This requirement is particularly important in flat terrain liable to get inundated from water from nearby major rivers. In some cases, the construction of a railway embankment blocks the natural drainage and may cause ponding. The formation level in such cases has to be above the affluxed flood level. Construction issues connected with the formation are dealt with in a separate Chapter.

32.3. Drainage

Water can cause damage and harm to a railway track in several ways. These are:

(i) Rain water falling on the track enters the formation through the ballast. The formation gets softened and its bearing capacity is jeopardised.

It is for this purpose that the top of the formation is given an outward slope of 1 in 40 on either side of the centre line (Fig. 32.1).

(ii) Rain water falling on the embankment slopes causes rain cuts and erosion. This can be prevented by turfing the side slopes (Fig. 32.2).

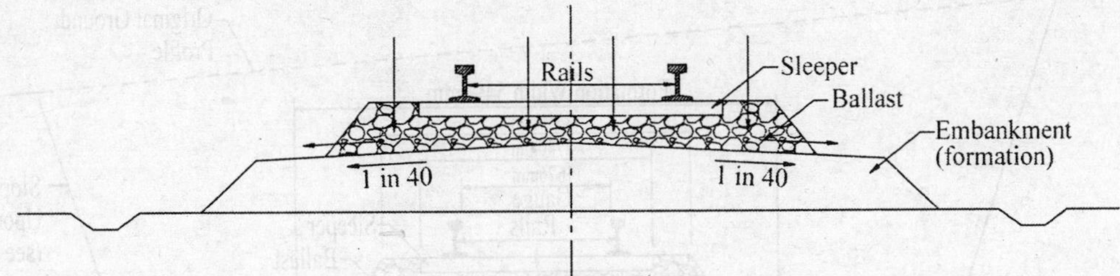

Fig. 32.1: Drainage of rain water

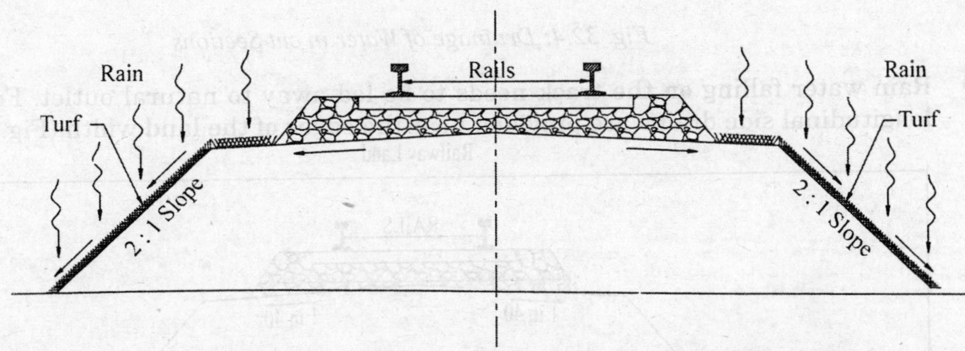

Fig. 32.2: Turfing Embankment Slopes to Prevent Rain-cuts and Slope Erosion

(iii) By the capillary rise of water from the high water table in the natural ground, or from the water standing against the slopes, the moisture enters the top of the formation and softens it, causing reduction in its bearing capacity. This action is particularly observed in highly expansive clay, such as Black Cotton soil found in certain parts of central and peninsular India. The capillary rise can be stopped by providing a sand blanket (Fig. 32.3).

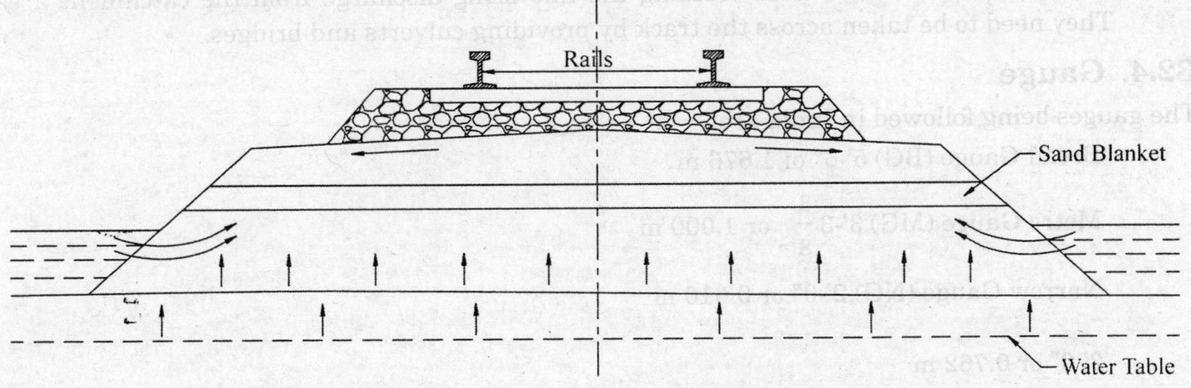

Fig. 32.3: Use of Sand Blanket to Prevent Capillary Rise

In modern times, synthetic geo-fabrics are available and they perform the same function as a sand blanket.

(iv) In one-sided or two-sided cuts, water from the hillside flows down the hill slopes and needs to be drained away. This can be done by providing side drains of adequate capacity and additionally by providing catch-water drains uphill (Fig. 32.4).

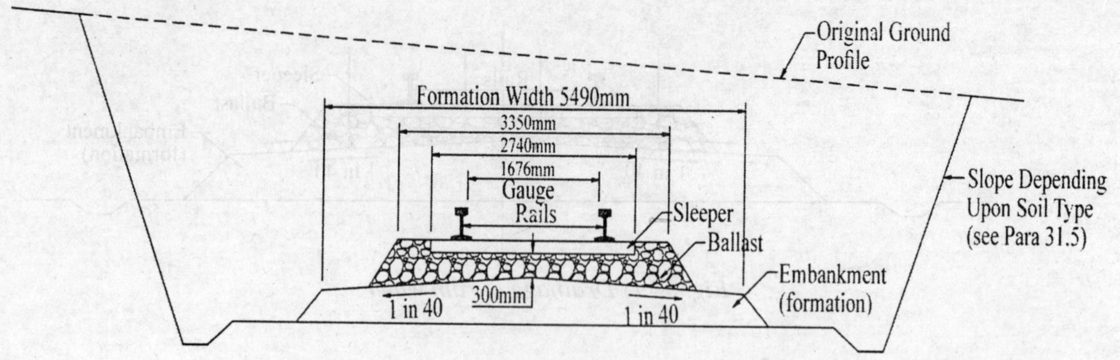

Fig. 32.4: Drainage of Water in cut Sections

(v) **Rain water falling on the track needs to be led away to natural outlet. For this purpose, longitudinal side drains are provided near the edges of the land width (Fig. 32.5).**

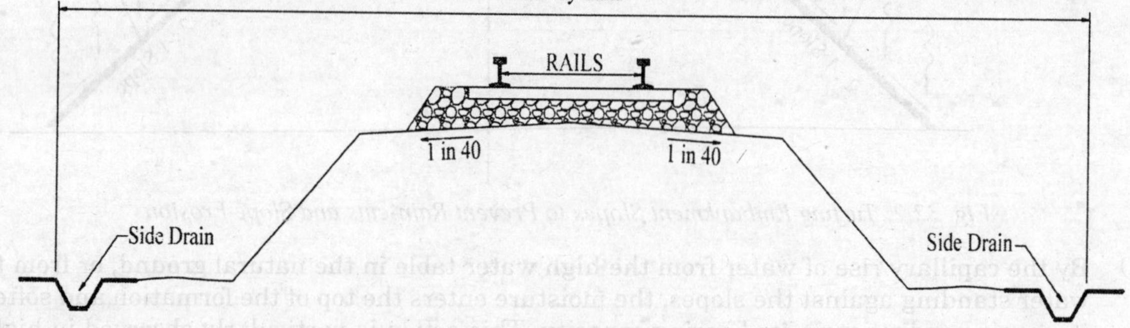

Fig. 32.5: Longitudinal Side-Drains

(vi) **The several natural streams crossing the line bring discharge from the catchment area. They need to be taken across the track by providing culverts and bridges.**

32.4. Gauge

The gauges being followed in India are:

Broad Gauge (BG) 5'-6" of 1.676 m

Metre Gauge (MG) 3'-3 $\frac{3}{8}$" or 1.000 m

Narrow Gauge (NG) 2'-0" or 0.610 m

2'-6" or 0.762 m

The standard gauge used in UK and USA etc is 4'-8½" or 1.435 m

As per the Unigauge policy of Indian Railways, the plan is to convert all MG sections to BG gradually. The NG sections are generally in the hilly terrain and will have to be retained as such. NG sections in the plains will have to be either converted to BG or abandoned since the traffic is very low and alternative road connections exist.

32.5. Rails

32.5.1. Function of Rails

Railway technology is based on the movement of steel wheels of locomotives, coaches and wagons on steel rails. The two rails are separated by the gauge distance. The rails act as the continuous support, countering the vertical and horizontal loads of the moving axles, transmitting these loads to the sleepers, ballast, formation and the natural ground and enabling the train to move smoothly.

32.5.2. Requirement of Rails

For performing the above functions, rails must meet with the following requirements:

(1) **High Strength :** The axle loads in railways are very high (as high as 23T), and in addition to the static effect, dynamic effect is also predominant. To resist the high stresses developed, the quality of steel must be very good. The tensile strength specified is in the range of 700-800 MPa.

(2) **Adequate Flexural Stiffness :** The rail acts as a beam supported on the sleepers which are spaced at a certain distance. Thus, it must have a high moment of inertia and flexural stiffness. Apart from vertical loads, the rails are also subjected to lateral forces, particularly on curves. They should have adequate lateral stiffness to resist these lateral forces.

(3) **Resistance to Wear :** The wheels of the trains constantly rub against the rails, causing considerable wear. Also, at the joints the wheels constantly pound the ends of the rail. The steel, of which rails are made, must be of such a metallurgical composition that it resists the wear. The rail head should have sufficient thickness that even if some wear does take place, the remaining section is strong enough.

(4) **Sufficient Width of Rail Head to Provide Enough Contact Area with the wheel Flange** : The wheel flange has a certain width. The width of the rail head should be adequate to accommodate the wheel flange.

(5) **Adequate width at the base to spread the load to the sleeper :** The rail section should have an adequate base so that the wheel load is spread on the sleeper and the sleeper is able to withstand it.

(6) **Adequate depth of head :** The rail head should have adequate depth to take care of wear in the vertical direction.

(7) **Fillets to be rounded :** The rail section has three components, viz, the head, the web and the base. At the junctions of these, heavy concentration of stresses takes place, and this can be countered by rounding off the fillets.

(8) **Centre of Gravity to be Approximately at Mid-height :** The centre of gravity of the rail section should be approximately at mid-height of the section, such that the maximum tensile and compressive stresses developed are equal.

(9) **Sufficient Height of Section and Rib Area :** In order to resist the shear stresses, the rib should have adequate height and width.

32.5.3. Historical Evolution of Shape of Rail

When the rail era began, the shape of the rail section was what is known as the "double-headed", (Fig. 32.6), with the rail head and rail bottom of identical section. It was argued that if the top section got worn out, the rail could be inverted with the bottom serving as the top, thus increasing the life. The rail was seated on chairs, and the base got dented with the pounding by the wheels, and could not be used as the rail head when inverting.

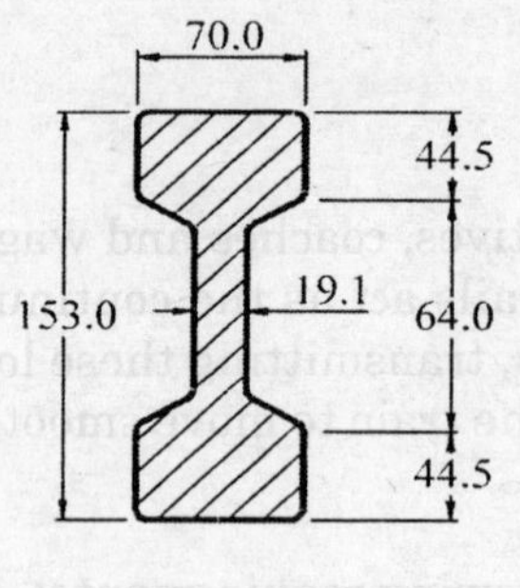

(a) Double-headed Rail

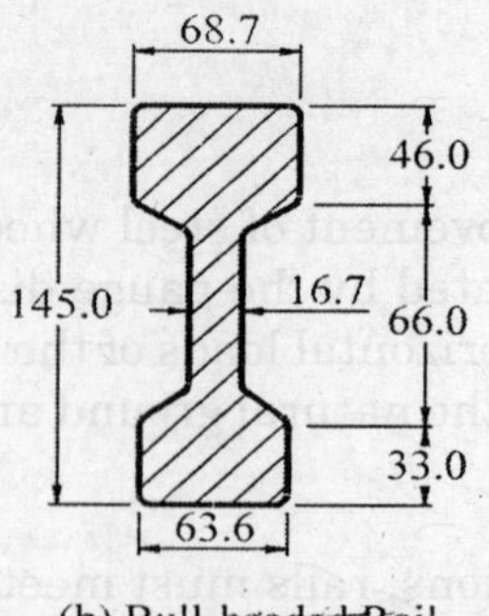

(b) Bull-headed Rail

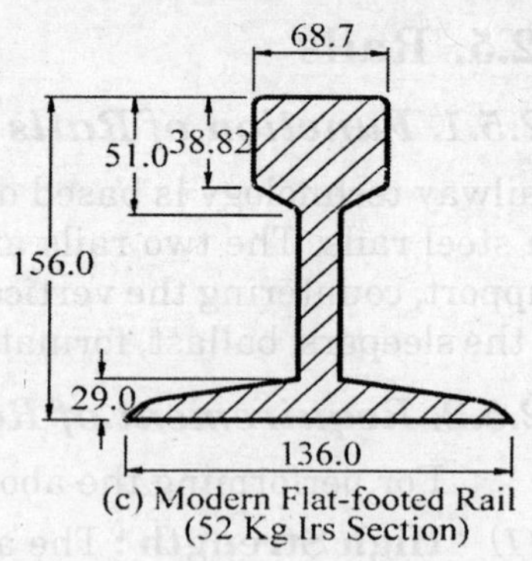

(c) Modern Flat-footed Rail (52 Kg Irs Section)

Fig. 32.6: Different Shapes of Rail

An improvement was then made to get over the above deficiency by making the head larger than the foot. The rail rests on a cast iron chair, which rests on the sleepers. The cast iron chair is of large area and spreads the load to the sleeper, thus increasing the longevity of the sleeper. This rail is known by the name "bull-headed" rail, Fig. 32.6(b).

The latest development is the flat-footed rail, Fig. 32.6(c), having the following advantages:

(1) Far better rigidity, strength and stiffness to resist vertical and lateral roads
(2) No chairs are required to fix the rail to the sleepers, resulting in economy
(3) The arrangement at points and crossings is simpler
(4) The flat-foted rail maintains a regular top surface and develops less kinks
(5) Inspection is simple
(6) Cost of maintenance is low

However, in the opinion of some, a bull-headed rail gives a better alignment of and a more solid track, and a smoother ride than a flat-footed rail.

In India, all the rails are flat-footed. The same is the case in many other countries.

32.5.4. Section of Rail

The rail section is denoted by the weight. In the metric system, it is in Kg/metre. Thus a rail 55 R signifies 55 Kg/m in India. The weight of the rails is governed by several factors, such as:

1. Gauge
2. Speed
3. Sleeper spacing
4. Axle Load likely to use
5. Depth of ballast cushion
6. Permissible wear, whose upper limit is 5 per cent

An empirical rule for selecting the rail section is:

$$\text{Rail section (Kg/m)} = \frac{\text{Axle Load (in Kg)}}{510}$$

Thus, for an axle load of 22 T, the rail section needed is $\frac{22,000}{510}$ = 45 Kg/m

In India, the rail section commonly used on BG tracks is the 52 R, having a weight of 52 Kg/m.

32.5.5. Steel for Rail

In view of the need for high strength and resistance to wear, a high carbon steel is used. The composition typically used is:

Carbon	0.55 - 0.68%
Manganese	0.05 - 0.9%
Silicon	0.05 - 0.3%
Sulphur	0.05 or less
Phosphorous	0.05 or less

32.5.6. Rail Length

Theorctically it is advantageous to have rails of long length since the number of joints is reduced. But, other factors like facilities for transport, ease of handling and cost of production have to be reckoned.

In India, a length of 13 m for BG is adopted.

Very short lengths are undesirable, and the rule followed is that the minimum length should be the axle spacing (3.6 m) in India.

32.5.7. Coning of wheels and Canting of rails

The flat portion of the wheel is known as the "rim" or "tread" which is flanked by the "flange". The flanges enable the wheels to move without derailment.

The distance between the inside edge of the flanges of the two wheels forming the axle is always kept less than the gauge to prevent the flanges from rubbing against the inside face of the rail (Fig. 32.7).

If the rim of the wheels is kept flat, slight irregularities in the track would cause lateral movement of the wheels over the rails till the flange comes in contact with the face of the rail. This would, in turn, cause lurching and wear of the rail face.

To avoid the lateral movement of the wheels and rubbing of the flange against the rail face, the rim is sloped at a slope of 1 in 20 (Fig. 32.7), a feature known as "coning".

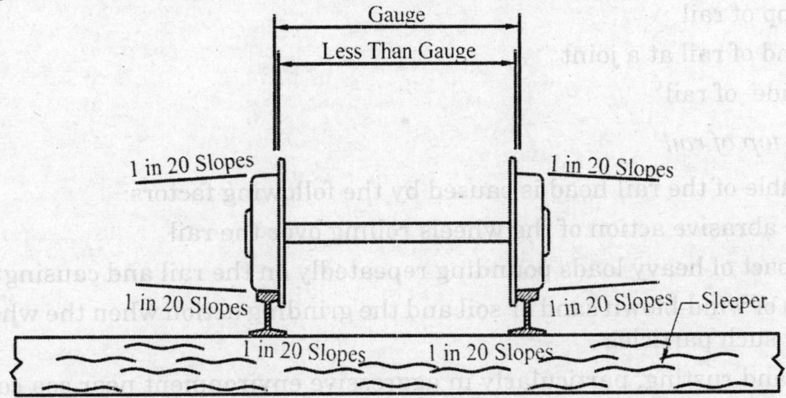

Fig. 32.7: Coning of Wheel and Canting of Rail

Since the wheels are coned, the rail is also canted inwardly at the same slope. In timber sleepers, the sleeper is suitably cut, a process known as "adzing". In metal sleepers, the base plate is suitably inclined.

The following are the advantages of coning of wheels and canting of rails:

1. Lateral movement of the wheels caused by slight irregularities in the track is avoided. Consequently, the wheels do not rub the inner face of the rail, and this prevents wear of the rail head.
2. Wheels are always maintained in a central position with respect to the track.
3. On curved tracks, the outer wheel has to cover a distance slightly greater than the inner wheel. As the wheel has a tendency on curves to move sideways towards the outer rail, the flange almost touches outer rail and the circumference of the outer wheel is thus greater than that of the inner wheel, which travels a lesser distance.
4. The wear of the head of the rail is uniform
5. A smooth ride results

Some of the disadvantages of coning of wheels and canting of rail are:

1. There is a tendency of developing a horizontal thrust, resulting in increasing the gauge.
2. Lateral bending stresses are developed in the rails.
3. Non-uniformly distributed load tends to crush the wooden sleepers.

32.5.8 Distresses in Rails

Various types of distresses in rails manifest themselves over the service life. These are:

1. Wear
2. Corrugations
3. Buckling
4. Cracks
5. Hogging
6. Corrosion and rusting

32.5.9 Wear of Rails

32.5.9.1. Three Types of Wear of Rails are Commonly Noticed:

1. Wear on top of rail
2. Wear at end of rail at a joint
3. Wear on side of rail

32.5.9.2. Wear on top of rail

Wear on top or table of the rail head is caused by the following factors:

1. Due to the abrasive action of the wheels rolling over the rail
2. Due to impact of heavy loads pounding repeatedly on the rail and causing the metal to flow
3. Deposition of wind-blown sand or soil and the grinding action when the wheels travel on rail containing such particles
4. Corrosion and rusting, particularly in aggressive environment near sea coast

When the top of the rail gets hammered, the metal tends to flow to the sides. As a result, burrs are created (Fig. 32.8).

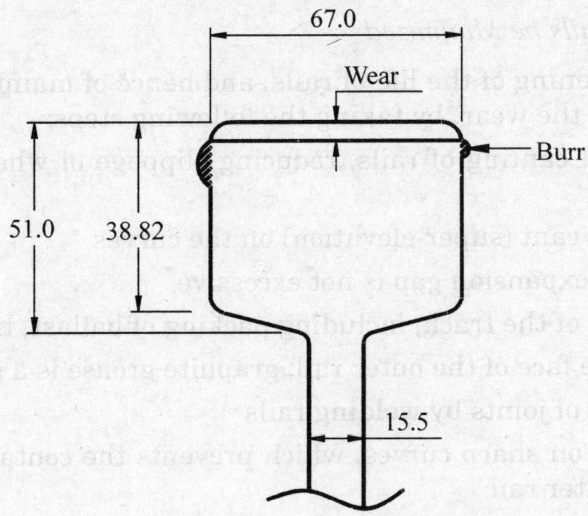

Fig. 32.8: Wear of Rail on Top and Burr

32.5.9.3. Wear on Sides of the Rail

The rail head suffers wear on the gauge side on curved track due to the following factors:

1. **The flanges of the wheel rub against the side when negotiating a curve because the axle is rigid**
2. **The centrifugal force causes the vehicle to move towards the outer rail, and consequently the flanges rub against the rail face**
3. **The slippage that occurs when the outer wheel has to travel a longer distance than the inner wheel, resulting in the wear on the inner rail face.**

32.5.9.4 Wear at end of rail near a joint

This is a common type of wear and is caused by the battering action of the wheel as it negotiates the expansion gap (Fig. 32.9).

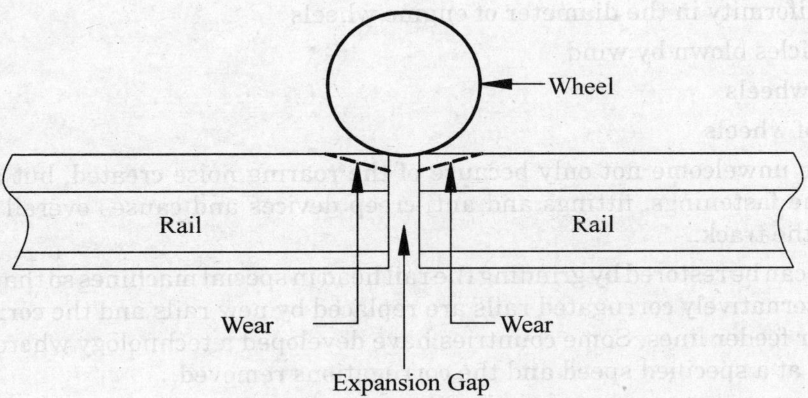

Fig. 32.9: Wear at End of Rail

32.5.9.5. How can Wear of Rails be Minimized?

Wear of rails causes a shortening of the life of rails, and hence of maintenance cost. Thus, efforts should be made to minimize the wear, by taking the following steps:

1. Coning of wheels and canting of rails, reducing slippage of wheels on the inner rail in a curved track
2. Providing the desired cant (super-elevation) on the curves
3. By ensuring that the expansion gap is not excessive
4. Regular maintenance of the track, including packing of ballast, tightening of the fish bolts
5. Lubricating the gauge face of the outer rail; graphite grease is a good lubricant
6. Reducing the number of joints by welding rails
7. Providing check rails on sharp curves, which prevents the contact between the wheel and the rail face on the outer rail
8. Use of special wear resistant quality of steel, particularly on curves
9. Metallurgical treatment of ends of rails and the rail head in the steel rolling mills

32.5.10. Corrugations

A wavy surface developing on the head of the rail is known as corrugations. The depressions that are formed are irregular in depth and at varying intervals. When a train passes over a corrugated rail surface, an unpleasant roaring noise is produced and as such the defect is also known as "roaring rails". Certain locations are prone to the development of corrugations. These are:

1. where the formation is soft and yielding
2. where trains stop and start, due to the braking and accelerating action
3. on long tunnels, perhaps due to humid conditions
4. on steep gradients
5. on electrified tracks, due to high speeds

Some other causes that have been identified for the formation of corrugations are:

1. high nitrogen content in rails
2. lack of uniformity in the diameter of engine wheels
3. sand particles blown by wind
4. coning of wheels
5. slippage of wheels

Corrugations are unwelcome not only because of the roaring noise created, but also because the defect loosens the fastenings, fittings and anti-creep devices and causes overall deterioration in the condition of the track.

Corrugated rails can be restored by grinding the rail head in special machines so that the depressions are removed. Alternatively corrugated rails are replaced by new rails and the corrugated rails are used in branch or feeder lines. Some countries have developed a technology whereby a special rail grinding train is at a specified speed and the corrugations removed.

32.5.11. Corrosion

Corrosion is caused by moist and humid conditions, particularly in aggressive environment near the sea, in industrial areas, and near ash-pits. Corrosion can diminish the area of the rail section and render it weak and liable to cracks and failure. Corrosion can be prevented by coating the rail with an anti-corrosive coat of red oxide followed by a few coats of bituminous emulsion.

32.5.12. Kinks in rails

Kinks in rails are formed, particularly at their ends near joints and they produce jerks in the trains. They are formed due to:

1. Loose ballast packing at joints
2. Defective alignment
3. Defective gauge

They should be rectified as soon as they are noticed since they cause jerks in trains and, if followed to deteriorate, can result in overturning of trains.

32.5.13. Buckling

Expansion joints are provided at the ends of rails to allow the rails to expand and contract due to temperature changes. If the gap is sufficient, the rail cannot expand and it buckles. Buckling can be avoided by:

1. leaving sufficient expansion gap
2. lubricating the contact surfaces of the fish plates and rails at regular intervals
3. not tightening the joints too much as to prevent expansion/ contraction of the rails

32.5.14. Rail failures

A rail is deemed to have failed if it is found necessary to replace it immediately, for reasons other than normal wear. Failure manifests itself in the form of cracks fractures, split head, crushed head etc.

Failures are caused by:

1. inherent defect in the rail at the manufacturing stage
2. derailments
3. fatigue on account of repetition of loads
4. corrosion
5. improper maintenance, particularly at the joints
6. defects in welding of rails
7. Some of the typical rail failures are illustrated in Fig. 32.10.

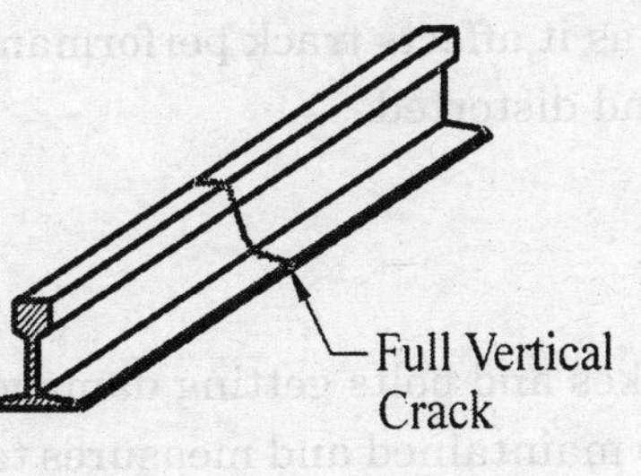

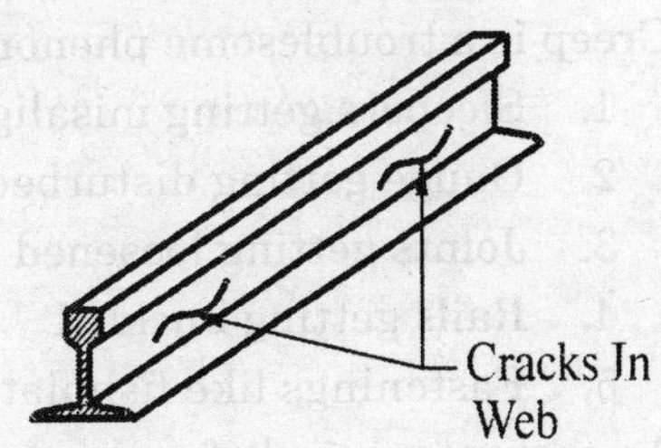

Horizontal
Crack

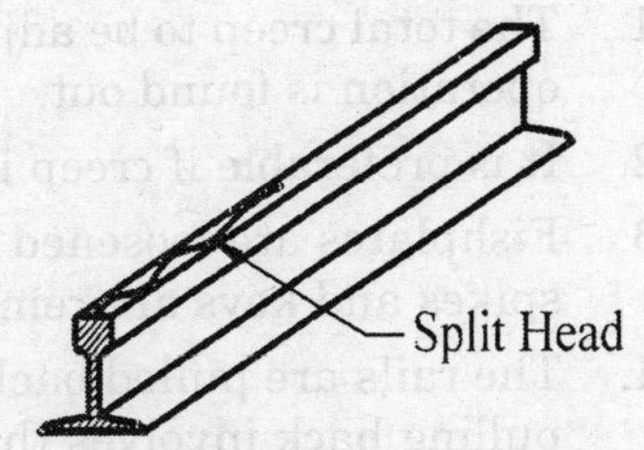

Fig. 32.10: Some Typical Rail Failures

32.5.15. Creep of Rails

It is observed that rails have a tendency to move gradually in the direction of predominant traffic. This forward movement of rails is known as creep. The extent of creep varies from location to location. At some places it may be negligible, but at places it can be as high as 15 cm in a month. Creep may take two forms:

1. Movement of rails alone with sleepers retaining their position
2. Movement of rails and sleepers together

Causes that lead to creep are:

1. Rails being too light to carry the loads
2. Inadequate and improper expansion joints
3. Rails not fastened adequately to the sleepers
4. Imperfect and badly maintained rail joints
5. Ballast offering inadequate resistance to forward movement of track due to inadequate ballast, poor packing
6. Wear of rail seat in metal sleepers
7. Uneven spacing of sleepers
8. In case of wooden sleepers, decay of sleepers
9. Inadequate drainage
10. Yielding formation

Creep is a troublesome phenomenon as it affects track performance in several ways such as:

1. Sleepers getting misaligned and distorted
2. Gauge getting disturbed
3. Joints getting loosened
4. Rails getting buckled
5. Fastenings like fishplates, spikes and bolts getting damaged

An accurate record of creep should be maintained and measures taken to determine its causes and to overcome the causes.

There are several operations in which creep can be adjusted, and these are:

1. The total creep to be adjusted is determined and the length of track to be pulled back in one operation is found out.
2. It is preferable if creep is adjusted before the onset of summer.
3. Fishplates are loosened from one end and removed from the other. Sleeper fittings such as spikes and keys are removed.
4. The rails are pulled back with the help of rail after inserting expansion gap liners. Manual pulling back involves the deployment of a large labour force. Modern practice is to employ mechanical and hydraulic devices.

There are anchors available for arresting creep. These are fixed at the rate of at least six per rail.

32.5.16. Testing of rails

The standard test for acceptance of a rail is the falling weight test or tup test. The rail is supported at two points at a designated distance and a weight is dropped at the centre of the two supports from a designated height. The rail should not fail or get fractured. Other tests like a tensile test on a test piece are also conducted.

32.5.17. Check Rails and Guard Rails

Check rails and guard rails are additional rails not intended for movement of wheels directly on them, but as checks and guards to the train movement at selected locations.

Check rails are used at the following locations:

1.2 Points and crossings with the purpose of guiding the wheels at the narrow nose crossing and preventing them hitting the nose.

2. On sharp curves (fitted to the inside rail of the curve) to prevent excessive wear and on the outer rail.

Fig. 32.11 shows the check rail on a curve.

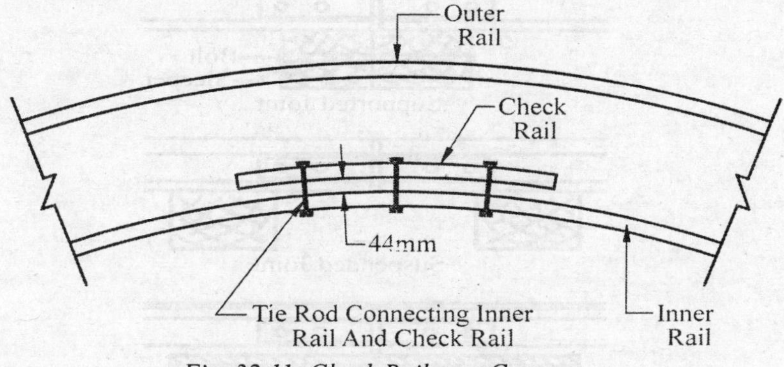

Fig. 32.11: Check Rail on a Curve

Guard rails are provided at the following locations:

1. On rail road level crossings
2. On large girder bridges

Fig. 32.12 shows the guard rail.

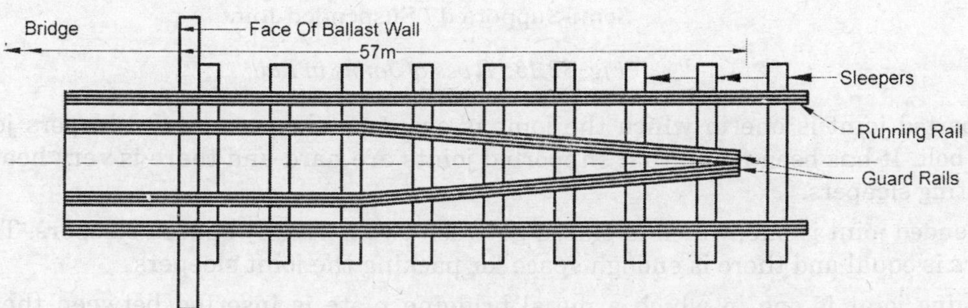

Fig. 32.12: Guard Rail at End of a Bridge on BG Track

32.5.18. Rail Joints

As the rails are produced in standard lengths, joints between two rails become necessary. Being vulnerable parts of the track, they need very careful attention and should meet the following requirements:

(i) a good joint should hold the two rail ends at the same level and in a straight line

(ii) it should permit free expansion and contraction of the rails caused by temperature variations

(iii) It should provide the same stiffness and strength as the rails
(iv) it should permit easy replacement of a rail
(v) it should involve the least maintenance
(vi) it should have a low initial cost and low maintenance cost

The most common type of joint is a fish plated joint. The types of fish plates are discussed later.

A joint can be (i) supported joints, (ii) suspended joints, (iii) bridging joints or (iv) semi-supported/ semi-suspended joints. These are illustrated in Fig. 32.13.

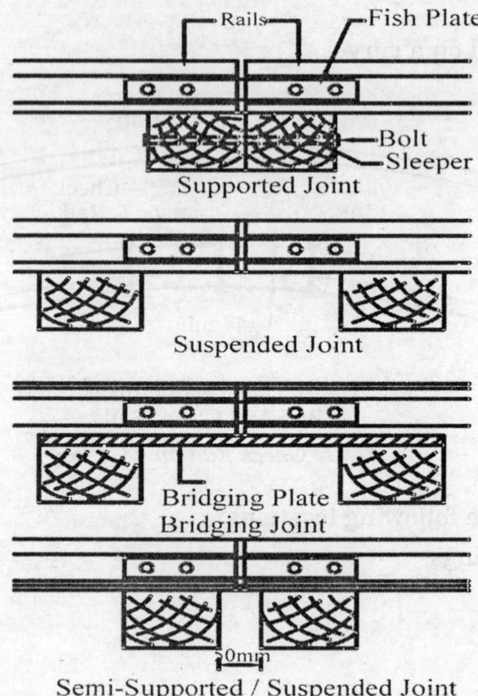

Fig. 32.13: Types of Joints in Rail

A supported joint is one in which the joint is over one sleeper or two sleepers joined together with a bolt. It has been found that supported joints are hard and there is very heavy wear of the supporting sleepers.

A suspended joint is one, which is placed between two normally spaced sleepers. The load on the sleepers is equal and there is enough space for packing the joint sleepers.

A bridging joint is one in which a metal bridging plate is inserted between the rails and the sleepers.

A semi-supported/ semi-suspended joint is one which the sleepers are brought closer, the rail ends remaining suspended. A bridging plate may or may not be provided. This type of joint is adopted as the standard in India. On straight tracks, joints in the two rails are opposite each other (square joints), but on curves the joints are often staggered.

32.5.19. Expansion Gap

Expansion gap is provided to permit expansion and contraction of the rail. The gap is governed by the simple equation:

$$e = L\alpha t \times 10^3$$

where e = expansion gap in mm

L = Rail length in m

α = Coefficient of expansion of rail steel, which is 0.00001152 per ^{0}C

t = Temperature variation

The temperature variation is calculated by the difference in prevailing rail temperature at the time of laying the track and an expected maximum temperature of 70^{0}C.

32.5.20. Welding of Rails

Joints described earlier have the disadvantage of resulting in jerky ride and wear of rails. In order to overcome these, the rails are welded. A welded rail of 200 m is common. Welding is done by (i) flash butt welding in which a strong electric current is passed raising the temperature of the rail ends to about 1300^{0}C and causing the rail ends to melt and get welded and (ii) thermit process in which a mixture of iron oxide, aluminium and ferroalloy is ignited in a crucible and poured into the rail ends.

32.5.21. Sleepers

Sleepers hold the rails to the correct gauge and distribute the load to the ballast.

The following are the types of sleepers commonly used by the Indian Railways:

1. Wooden
2. Cast-iron
3. Steel
4. Pre-stressed Concrete

Wooden sleepers are not favoured now since they are obtained from trees which involves depletion of forests, and are susceptible to decay and attack by vermin. They have the advantage that they permit track circuiting. They have a life of 10-15 years. Cast Iron and Steel Trough sleepers have a long life (30 to 50 years), have scrap value and maintain the gauge firmly. Indian Railways now favour prestressed concrete sleepers. They are manufactured in a factory under strict control of quality.

The sleeper density is the number of sleepers provided in the rail. Thus, if M is the length of the rail in metres, M+1 means there are M+1 sleepers. For high speed routes on BG, the density specified is M+9 with a 13 m long rail.

32.5.22. Ballast

Ballast consisting of stone aggregates is provided below the sleeper and above the earthen formation. Its functions are:

(i) to distribute the load and transfer it to the earthen formation.
(ii) to provide a firm, even and elastic bed for the sleeper
(iii) to absorb the impact forces caused by the wheels
(iv) to provide a convenient medium for packing the track to the required level and grade
(v) to prevent capillary rise of moisture
(vi) to inhibit growth of vegetation under the track

The stone aggregates are of a cubical nonflaky size 50 mm and down. Under points and crossings, a smaller size of 40 mm is used. The thickness of ballast for high density BG track is 300 mm, and for medium density track it is 250 mm. The standard profile of a ballast is given in Fig. 32.14.

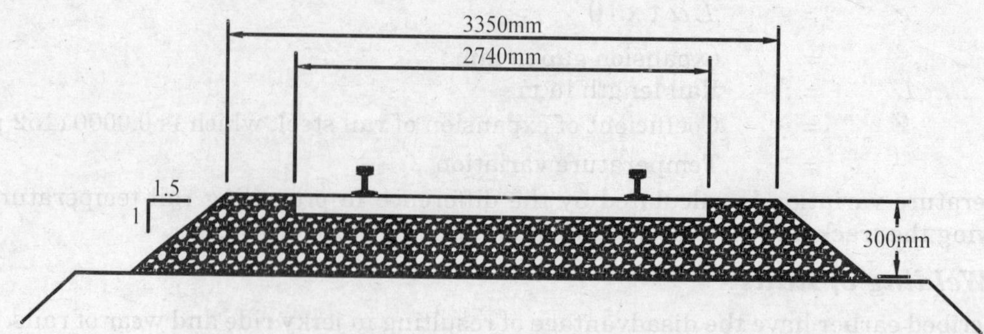

Fig. 32.14: Standard Profile of a Ballast

32.5.23. Sleeper Fastenings

Fasteners are required to bind the rail to the sleeper firmly. With wooden sleepers, the fasteners are dog spikes and screw spikes. For steel, cast iron and prestressed concrete sleepers, spring jaws with keys and Pandrol clips are used (Figs. 32.15 and 32.16).

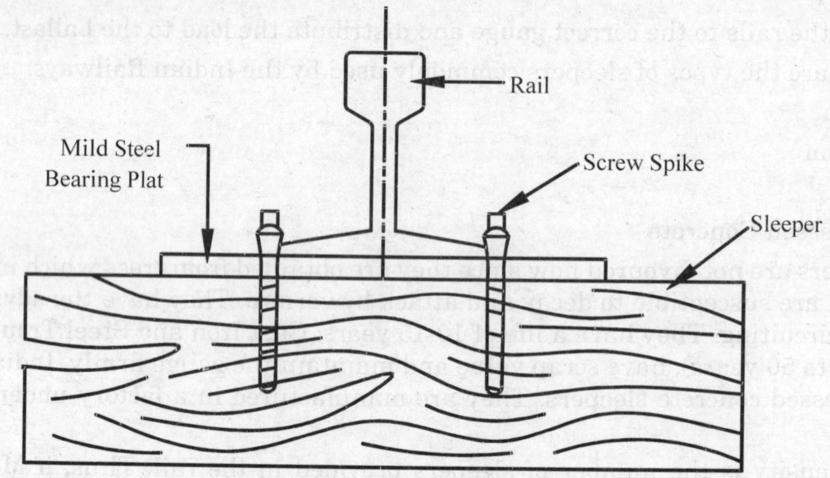

Fig. 32.15: Screw Spike Fastener

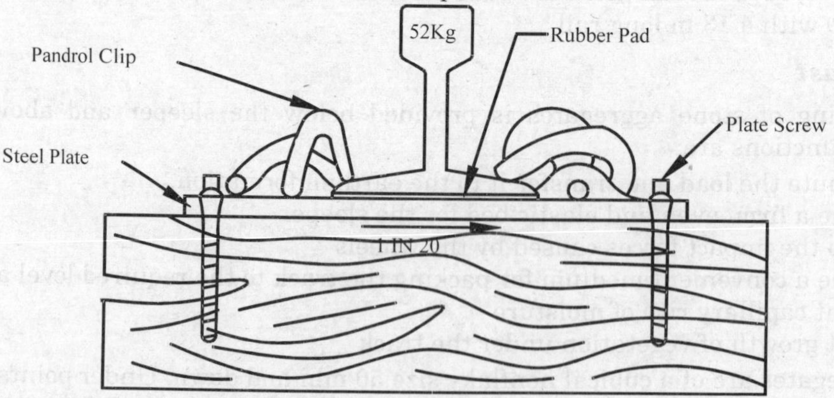

Fig. 32.16: Pandrol Clip Fastener

33

Chapter

Points and Crossings

33.1. Function

In railway technology, trains move on exclusive track. If the trains have to move from one track to another, some special arrangements are needed. Points and crossings provide the necessary arrangement. These locations are the weakest and most vulnerable part of a railway track from the point of view of safety, and hence they deserve special attention in design, fabrication, inspection and maintenance.

33.2. Turnouts

33.2.1. Turnout

A turnout is an assembly of several components enabling a train running on a track is diverted to another. A turnout may be "left-handed" or "right-handed" depending upon whether the train travelling on main track is diverted to the left or the right.

33.2.2. Crossing Number

In Indian Railways, turnouts are designed with respect to the "Crossing Number". The Crossing Number is an indication of the angle of crossing. The smaller the angle, the smoother the passage of the train from one track to the other. The Crossing Number is defined as the ratio of the spread of the crossing to the length of the crossing measured from its theoretical nose, as seen from Fig. 33.1.

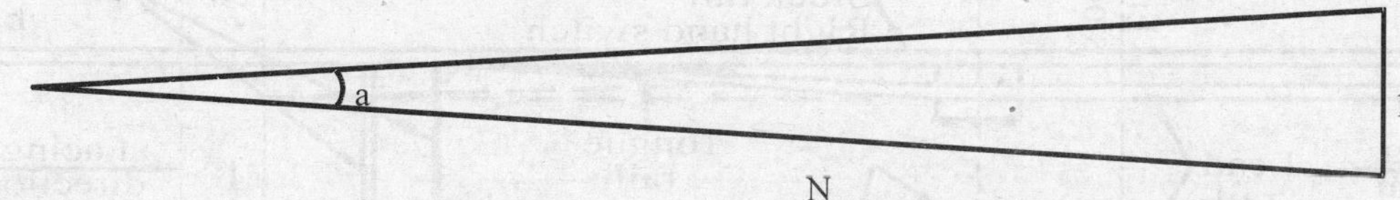

$\cot \alpha = N$ Crossing Number is N, or 1 in N Fig. 33.1: Crossing Number

The following are the Crossing Numbers and speeds permitted in the Indian Railways:

1 in 8½	Used for goods trains, station yards and locations where space is restricted and speeds are low. Speeds permitted are in the range of 15-20 Kmph
1 in 12	Used for passenger trains and station yards of main lines. Speeds permitted are in the range of 25-30 Kmph
1 in 16	Used for high speed trains. Speeds permitted are in the range of 40-50 Kmph.

33.2.3. Components of a Turnout

A sketch of a left-handed turnout is given in Fig. 33.2.

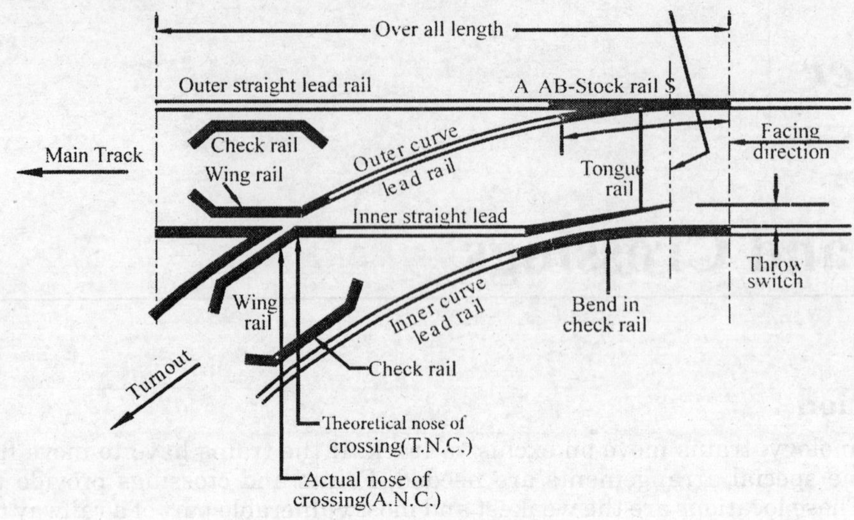

Fig. 33.2: A Left-handed Turnout

The following are the components of the turnout:

1. A switch assembly
2. Vee crossing
3. A pair of check rails
4. Four lead rails

32.2.4. Switch Assembly

A switch assembly, also known as "points", consists of four rails (Fig. 33.3). The two outer rails are known as the stock rails and the two inner rails are known as the tongue rails. The tongue rails pivot about a point known as the "heel" of the switch. The tongue rails end at a point known as the "toe" of the switch. The two tongue rails are held firmly by stretcher bars. The tongue rails are machined out of standard rail sections.

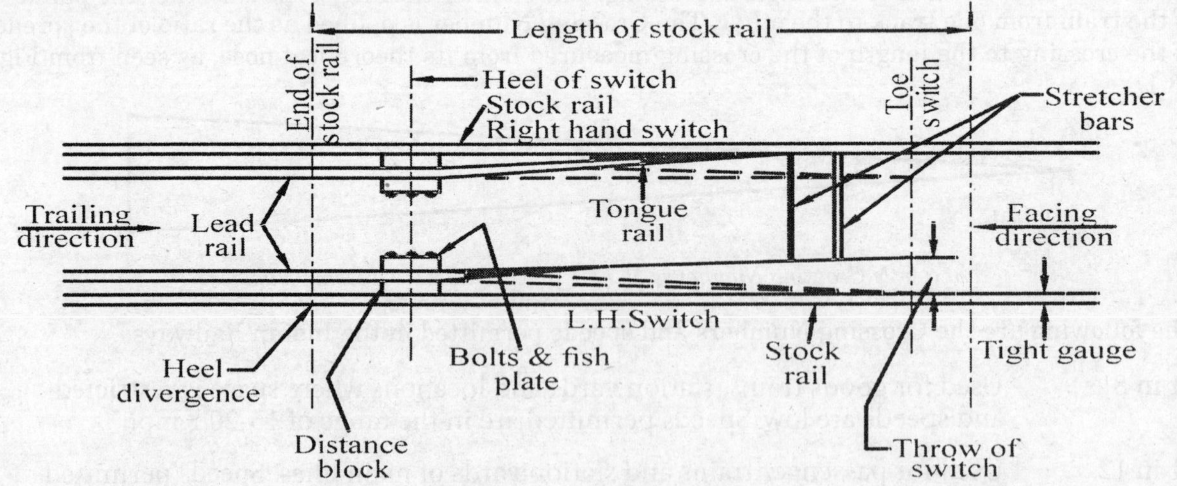

Fig. 33.3: Switch Assembly

32.2.5. Geometry of a Turnout

The geometric layout and calculations

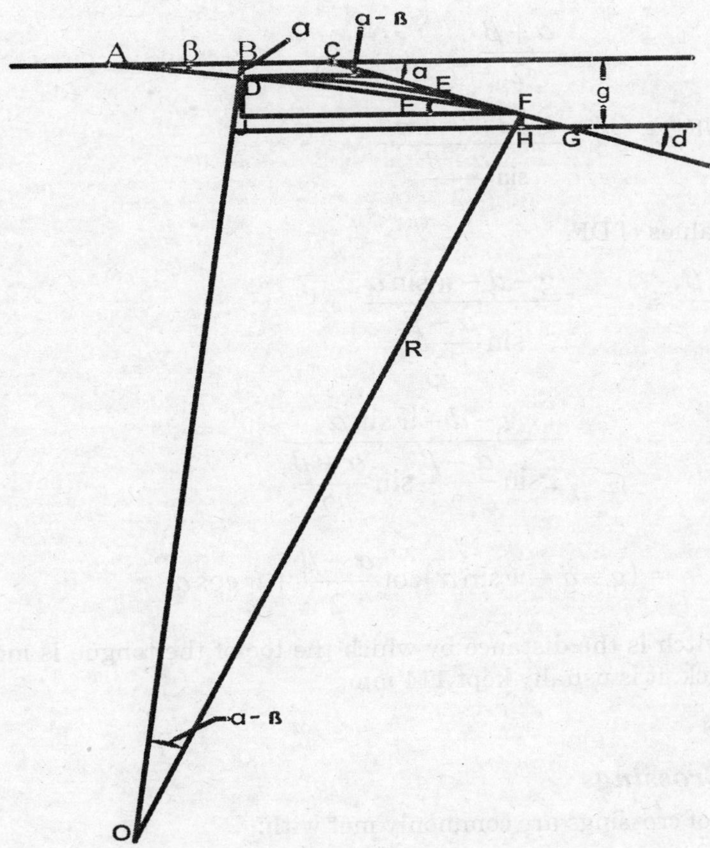

Fig. 33.4: Layout of Turnout

The layout assumes that the turnout curve is tangential to the tongue rail and the front leg of the Vee crossing, both of which are assumed to be straight. In the above figure,

AB	=	Tongue rail
FG	=	Front leg of the Vee crossing
DF	=	Curved turnout rail
BD	=	Heel divergence = d
g	=	gauge
α	=	Turnout angle
β	=	Switch angle
JG	=	Lead distance
FG	=	w
∠CED	=	$\alpha - \beta$
∠EDF	=	$\dfrac{\alpha - \beta}{2}$

$$DF = 2R \sin\frac{\alpha-\beta}{2}$$

$$\angle DFJ = \alpha - \frac{\alpha-\beta}{2}$$

$$= \frac{\alpha+\beta}{2}$$

$$\therefore DF = \frac{g-d-w\sin\alpha}{\sin\frac{\alpha+\beta}{2}} \quad (33.1)$$

Equating the two values of DF,

$$2R\sin\frac{\alpha-\beta}{2} = \frac{g-d-w\sin\alpha}{\sin\frac{\alpha+\beta}{2}} \quad (33.2)$$

$$\therefore R = \frac{g-d-w\sin\alpha}{2\sin\frac{\alpha-\beta}{2}\sin\frac{\alpha+\beta}{2}} \quad (33.3)$$

$$JG = \text{Lead} = (g-d-w\sin\alpha)\cot\frac{\alpha+\beta}{2} + w\cos\alpha \quad (33.4)$$

The throw of the switch is the distance by which the toe of the tongue is moved from its contact position. For BG track, it is usually kept 114 mm.

33.3. Crossings

33.3.1. Types of Crossings

The following types of crossings are commonly met with:

1. Diamond crossing
2. Scissors crossing
3. Triangles

33.3.2. Diamond Crossing

When two tracks cross each other, a diamond crossing results. Fig. 33.5 gives the sketch of a diamond crossing.

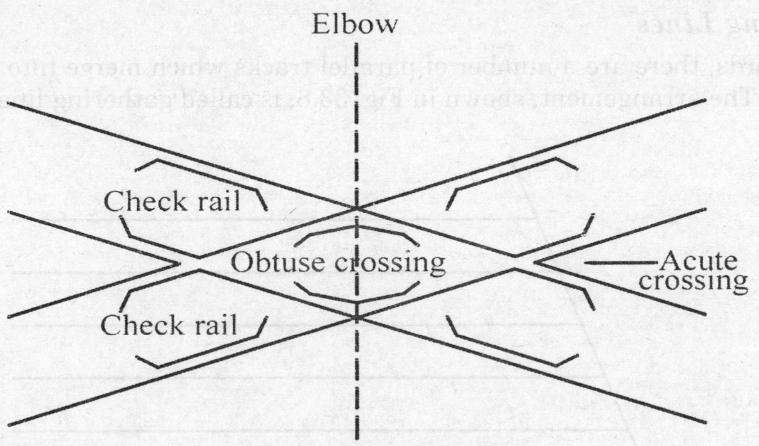

Fig. 33.5: Diamond Crossing

The details of the acute crossing in Fig. 33.5 given in Fig. 33.6.

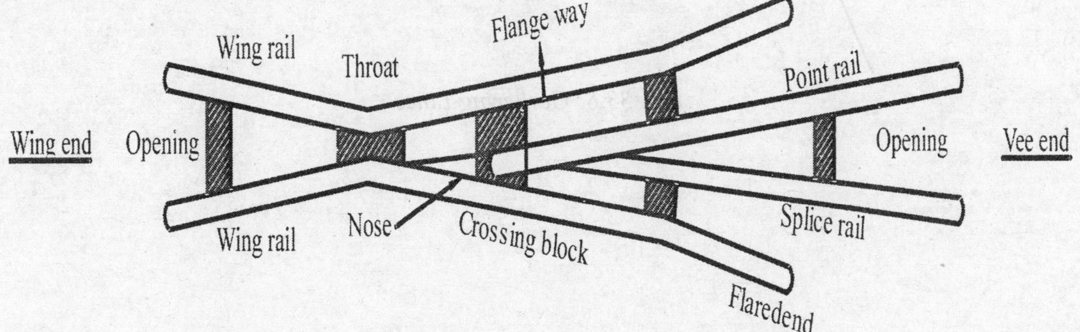

Fig. 33.6: Acute Crossing

33.3.3. Scissors Crossing/ Cross-over

When there are two tracks running close to each other (and they need not necessarily be parallel to each other), the arrangement for crossing from one track to the other is known as a cross-over. A scissors crossing/ cross-over results where the two cross-overs overlap each other. Fig. 33.7 gives the details.

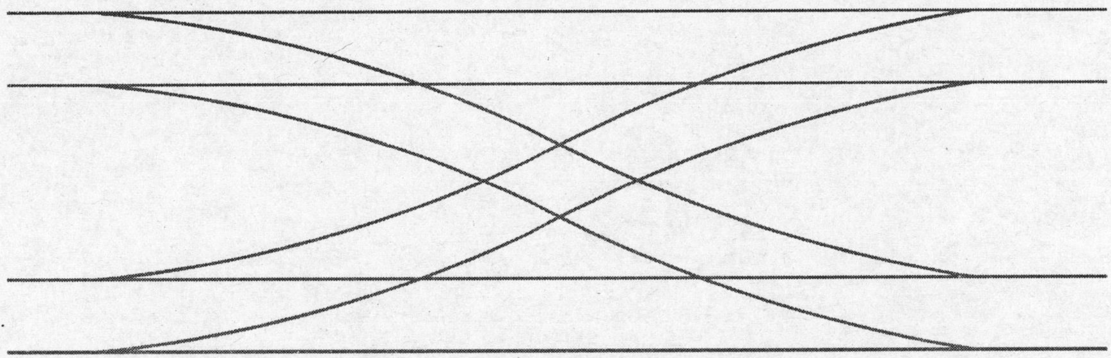

Fig. 33.7: A Scissors Crossing/ Cross-over

33.3.4. Gathering Lines

In marshalling yards, there are a number of parallel tracks which merge into a single track in a ladder formation. The arrangement, shown in Fig. 33.8, is called gathering lines.

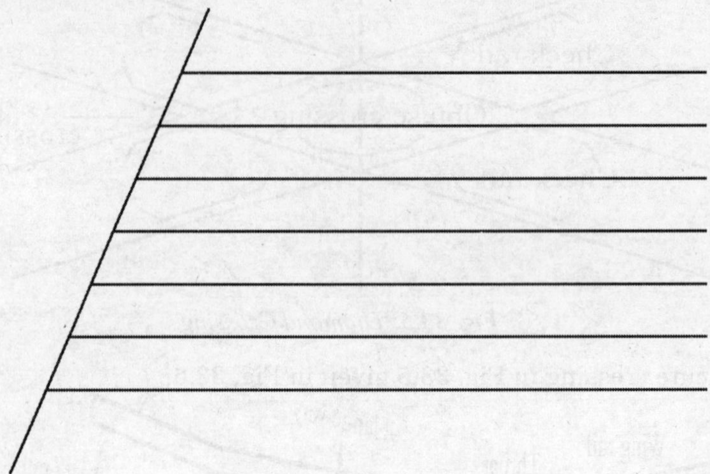

Fig. 33.8: Gathering Lines

34 Chapter

Signalling and Traffic Control

34.1. Need

Railway technology involves the movement of trains and exclusive tracks. Thus, when a train moves a track, there cannot be a train moving on the same track in the opposite direction or in the same direction at a faster speed. Crossings and overtakings can take place only at stations where more than one track exists. Signalling and traffic control systems are needed for the following reasons:

1. To warn the driver whether the track ahead is clear and it is safe for him to proceed.
2. For regulating the movement of trains so that the maximum use is made of the existing track, where trains of different speeds (slow passenger trains, express passenger trains, goods trains) travel and safety and efficiency are given prime importance.
3. To facilitate shunting operations in a station yard.

34.2. Signalling

34.2.1. Types of Signals

In the present day, two types of signals are in use in Indian railways:

1. Semaphore signals
2. Colour light signals

34.2.2 Semaphore Signals

Semaphore signals are those with an arm at the top of the signal pole. There are two systems in vogue:

1. Lower Quadrant System, adopted in India
2. Upper Quadrant System, adopted in some countries abroad

In the lower quadrant system, the arm can take up two positions:

1. Horizontal, which indicates "stop" or danger
2. Lowered at an angle of 45^0, which indicates "proceed"

In the upper quadrant system, the arm can take up three positions:

1. Horizontal, indicating "stop"
2. Raised at an angle of 45^0, indicating "proceed with caution"
3. Raised to a vertical position, indicating "proceed"

The semaphore arm is of two types,

1. Free end is square, which is used as a stop signal (Fig. 34.1)
2. Free end is V-shaped, which is used as a warner signal (Fig. 34.2)

The semaphore signal, Fig. 34.1, has the following parts:

1. Horizontal arm
2. Spectacle frame, which consists of red coloured glass at top and green coloured glass at the bottom. A lamp is provided at night to give the colour light indication.
3. Crank rod
4. Weight and balance lever
5. Wire and pulley leading to the signal cabin, from where the signal man operates a lever to manipulate the position of the arm.

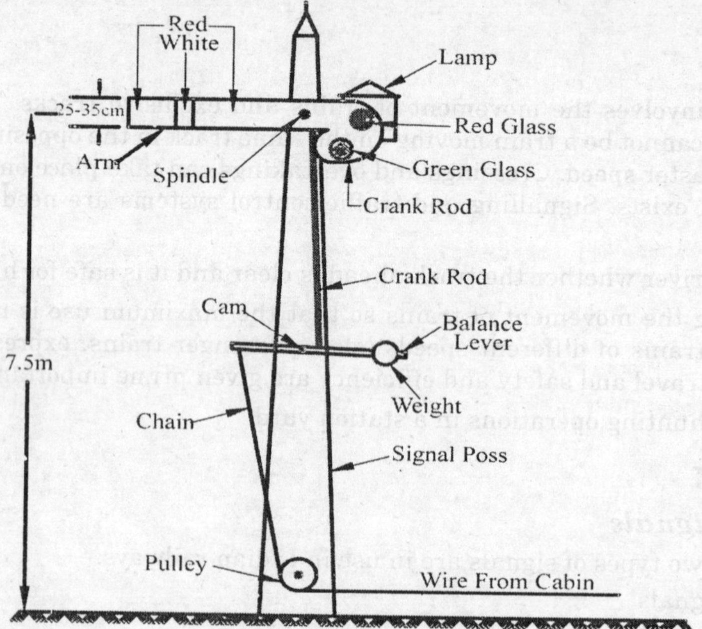

Fig. 34.1: Semaphore Signal

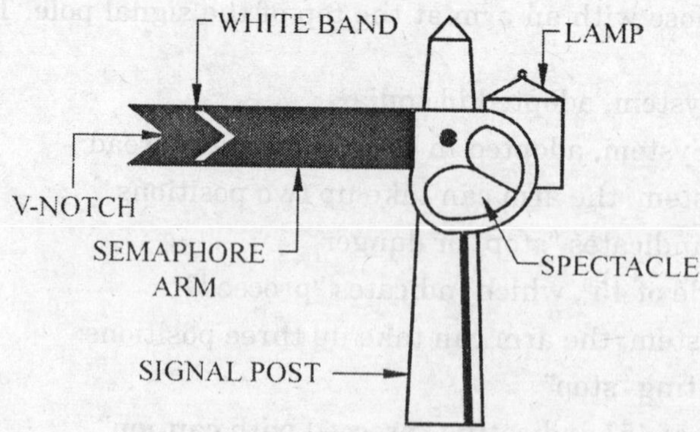

Fig. 34.2: Warner Semaphore Signal

When the stop and warner signals are on the same post, the various possible positions give different messages to the driver, as given in the sketches at Fig. 34.3

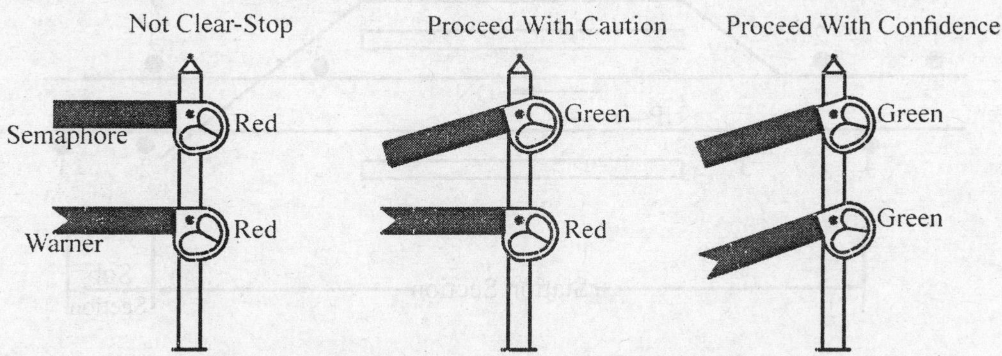

Fig. 34.3: Warner and Stop Semaphore Signals on the Same Post

34.2.3. Colour Light Signals

In the colour light signals, electric lights are used (both in day and night), using special lenses and hoods to give high intensity beam light. They are used in suburban train operations and metro rail systems in India. They are also replacing the conventional semaphore signals in important stations.

Fig. 34.4 gives a sketch of the coloured light signals, having three colours, viz, green, yellow (amber) and red.

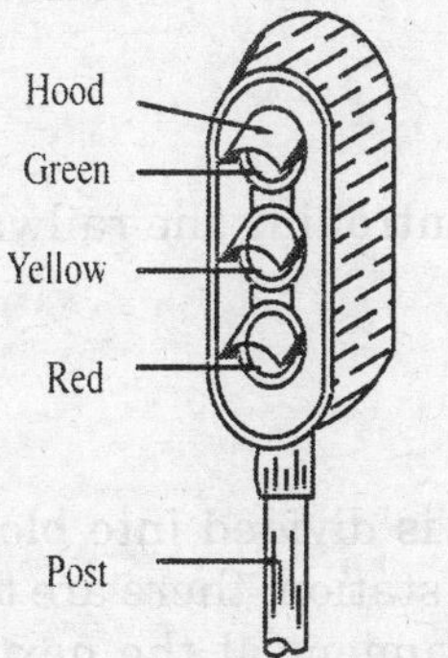

Fig. 34.4: Coloured Light Signal

34.2.4 Location of Signals

When a train approaches a station, the first signal a driver sees is the "outer" signal. This signal enables the driver to bring the train to a halt ahead of the next signal, which is called the "home" signal. The home signal is placed at the outer end of the station limit. For a BG track, the distance between the outer signal and home signal is 540 m. The home signal is placed at least 180 m ahead of the point which governs the track on which the train is to be received. When a train starts to proceed from a station, the first signal the driver sees is the starter signal, which is located just ahead of the station. If "green", the driver has permission to leave the platform on way to the next station. The next signal, "advance starter", is located beyond the outer-most set of points. A diagram showing the location of the signals is given in Fig. 34.5.

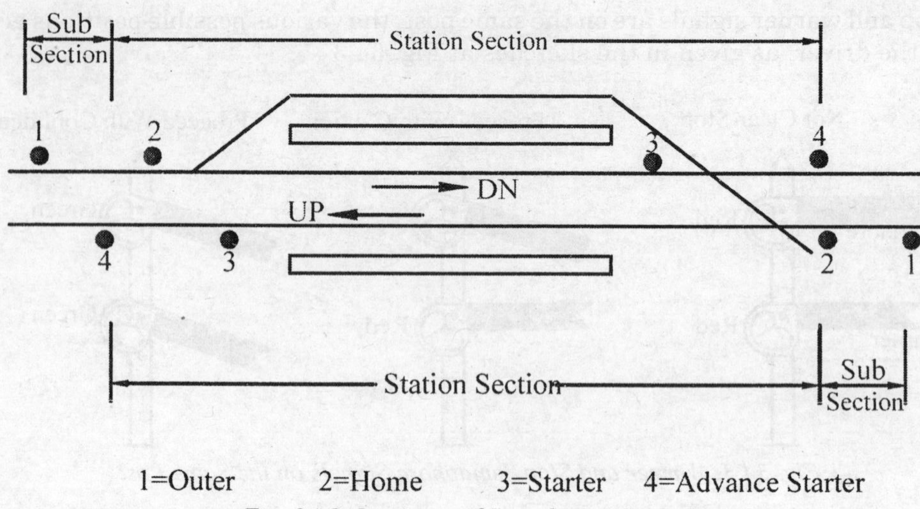

Fig. 34.5: Location of Signals at a Station

34.3. Interlocking

Interlocking is a system by which the setting of points and signals through operating levers in the signal cabin is rendered fail-safe against conflicting train movements. For example, when the points are set for receiving a train on a particular track, it should not be possible for the cabin man to lower the signal for any other track. Similarly, when the signal is lowered for receiving a train on a particular track, the points can be set only for that track. The interlocking device renders the operation of wrong levers impossible. Various mechanical devices are available for achieving interlocking.

34.4. Traffic Control System

The following systems are in use for controlling the railway traffic:

1. Block System
2. Centralised Traffic Control
3. Automatic Train Control

In the Block System, the railway line is divided into blocks, a block section generally being the distance between two stations. At each station, there are two block instruments. An electric circuit connects the corresponding block instrument at the next station. For single line operations, the manipulation of the instruments under dual control results in the release of a tablet. The tablet is enclosed in a token, which is handed over to the driver, giving him authorisation to proceed to the next block. It may be noted that station master of the next station was already involved in the release of the tablet under the dual control of the block instruments. On double lines, the instruments do not release tablets, but support appropriate signal indications.

Accordingly, the station master of the station where the train is standing gives signals for the start of the train.

In the Centralised Traffic Control (CTC) system, the entire movement of trains in a designated area is controlled by an operator in a centralised control room cabin. There is an illuminated diagrammatic sketch of the entire track network controlled by the CTC. The operator manipulates the switches to direct the train movements and his operations interlock the signals and points. There are no levers as in the case of manually operated semaphore signals and points.

In India, several sections of the railway lines operate under CTC. It has been found that train delays are reduced and the line capacity gets enhanced.

In the Automatic Train Control (ATC) system, the track has electric circuit. Electrical fittings at the wheel brakes also exist. Under the system, the engine driver observes and obeys the visual signals, but in case of failure to do so, the train automatically reduces speed and comes to a halt. ATC has been introduced on certain sections of Indian railways.

35

Chapter

Traction and Traction Resistance

35.1. Types of Locomotives In India

The following types of locomotives exist in India:

1. Steam Locos
2. Diesel Locos
3. Electric Locos

Steam locos, which used to be the only traction vehicle some decades ago, have been gradually phased out since they have a low thermal efficiency (about 7 per cent), are heavy and are, therefore, costly to operate. Their only advantage was that they run on coal, which is abundantly available in India. Diesel locomotives have a thermal efficiency of 25 per cent and is one of the most popular propulsion system in Indian railways now. But the global shortages of petroleum crude is a serious issue in the coming years and the Railways have to look for alternative source of traction energy. In this respect, electric traction offers the best alternative. To add, the thermal efficiency of electric traction is as high as 90 per cent. Electric locos are lighter in weight when compared to diesel locos and very high speeds are possible. Thus, there is no doubt that electric traction will be the preferred mode in the years to come. The Indian Railways are aware of this and a programme of electrification is being given high priority. The only drawback is that high initial investments are needed to generate electricity and install overhead power lines all along the track. But once a track is electrified, the operating and maintenance costs are low.

35.2. Traction Resistance

A train has to overcome the following types of resistance when in motion:

(i) Train resistance, composed of friction between wheels and rails (rolling friction), friction due to wave action of rails and resistance caused by the internal parts of the locomotive, wagons and coaches. The sum total of the above is 0.0016 W where W is the weight of the train.

(ii) Resistance dependent on speed, which is given by 0.00008 WV, where W is the weight of train in tonnes and V is the speed in Km/hr.

(iii) Atmospheric Resistance, which is given by 0.0000001 WV^2, where W and V have the same meaning as in (ii).

(iv) Grade Resistance, which is given by

$\frac{W \times g}{100}$ where W is the weight of the train and g is the gradient in per cent.

(v) Curve Resistance, given by 0.0004 x Degree of Curve x W (for BG)

(vi) Resistance due to starting which is given by:

$0.01\ W_1 + 0.005\ W_2$

where W_1 is the weight of locomotive and W_2 is the weight of wagons and coaches

(vii) Resistance due to acceleration given by:

0.028 Wa, where W is the weight of the train and a is the acceleration in Km/sec^2

(viii) Wind Resistance, due to wind velocity, and given by $0.000017\ AV^2$, where A is the exposed area in sq. m and V is the speed in Km/hr.

35.3. Hauling Capacity of a Locomotive

The hauling capacity of a locomotive is the product of coefficient of friction between the wheels and rail (varies from 0.03 to 0.25 depending upon wetness and weather condition) and the weight on the driving wheels.

36

Chapter

Stations, Yards and Station Equipment

36.1. Stations

36.1.1 Types of Stations

Stations are the places where trains stop for picking up and disembarking passengers and loading and unloading of goods. Stations are of three types:

(i) Wayside stations where trains move in two directions only, as for example Faridabad between Delhi and Mathura.

(ii) Junctions where railway lines exist for more than two directions, as for example Delhi.

(iii) Terminal stations where railway lines end, as for example Howrah in Kolkata, Madras Central in Madras and V.T. in Mumbai.

36.1.2. Selection of Site for a Station

The following criteria should be kept in view where a new station is to be built:

(i) Since a station is intended to pick up and disembark passengers, the site selected should be close to the town it intends to serve.

(ii) Where considerable goods traffic is expected, the station should be near to the place where goods are produced. For example, if a station is expected to pick up coal, it should be near to the coal mine.

(iii) The ground should be flat and well drained.

(iv) Possibility of future development and expansion should be kept in view in acquiring land.

(v) The railway track at the station should be straight as a curved track affects visibility.

(vi) Facility for obtaining good water should be available not only for the passengers in the station but also for watering the trains.

36.1.3. Station Platform

The trains stop at the platform where passengers embark and disembark. The essential features of a platform are:-

(i) The platform should be long enough to accommodate the longest train. On BG lines, a minimum length of 305 m is desirable.

(ii) The height of the platform above the rail should be such that the passengers can easily embark and disembark. For BG lines, a height of 840 mm is desirable (Fig. 36.1).

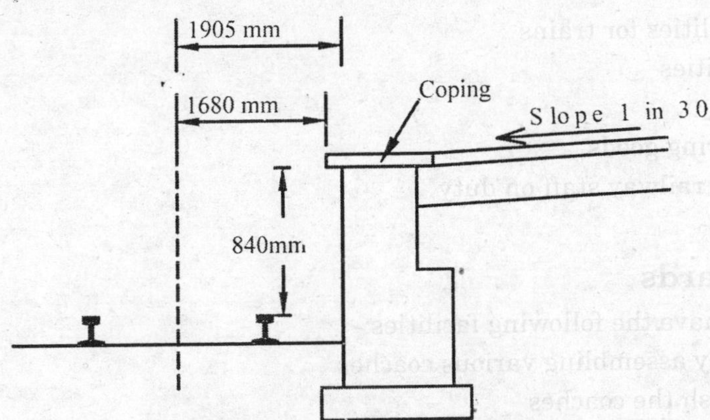

Fig. 36.1: Cross-section of a Platform

(iii) The horizontal distance from the centre of the BG track to the face of the front wall of the platform should be 1680 mm (Fig. 36.1).

(iv) The platform should have a good and strong pavement, preferably of cement concrete, so that maintenance is minimal and heavy parcels can be efficiently handled.

(v) A gentle slope of 1 in 30 towards the railway track will facilitate drainage.

(vi) After the full length of the platform, a slope of 1 in 6 (also paved) should bring down the level to the ground level.

(vii) The platform should have a covered roof preferably of AC Sheet to protect the passengers and goods from rain and sun. The roof should be at least 60 m long.

(viii) The width of the platform should be sufficient to accommodate facilities like vending shops, seats for passengers and place for storing goods. A width of 3.67 m is the barest minimum.

(ix) If there are more than one platform in a station, a foot over bridge to cross the railway tracks should be provided. A ramped approach is desirable so that baggage trolleys and disabled persons on wheel chairs can use the facility.

36.1.4. Facilities to be provided in a Station

The following facilities shall be provided at the station:

(i) Booking office for passenger tickets with ticket vending machines
(ii) Booking office for goods, with weighing facilities
(iii) Platform for receiving trains
(iv) Space for offices of the station master, ticket collectors and other railway staff
(v) Enquiry office
(vi) Facilities of waiting rooms
(vii) Toilet facilities
(viii) Cloak room for storing baggages of passengers
(ix) Public address system
(x) Electronic display of train number, destination, coach number
(xi) Clock
(xii) Public telephone
(xiii) Parking facilities for cars, taxies, two-wheelers, autorickshaws, and cycles

(xiv) Watering facilities for trains
(xv) Fuelling facilities
(xvi) Signal cabins
(xvii) Sheds for storing goods
(xviii) Rest room for railway staff on duty
(xix) Lighting

36.2. Station Yards

Station yards shall have the following facilities:-

(i) Form trains by assembling various coaches
(ii) Clean and wash the coaches
(iii) Goods yards shall have facilities for receiving wagons, weighing, goods sheds, forming goods trains by assembling wagons

The station yards should be away from the main line so that normal train operations are not disturbed. A typical station marshalling yard should have facilities for fuelling, repair and maintenance of locos, sorting sidings, turntable for reversing the engine.

A typical layout of a marshalling yard is given in Fig. 36.2.

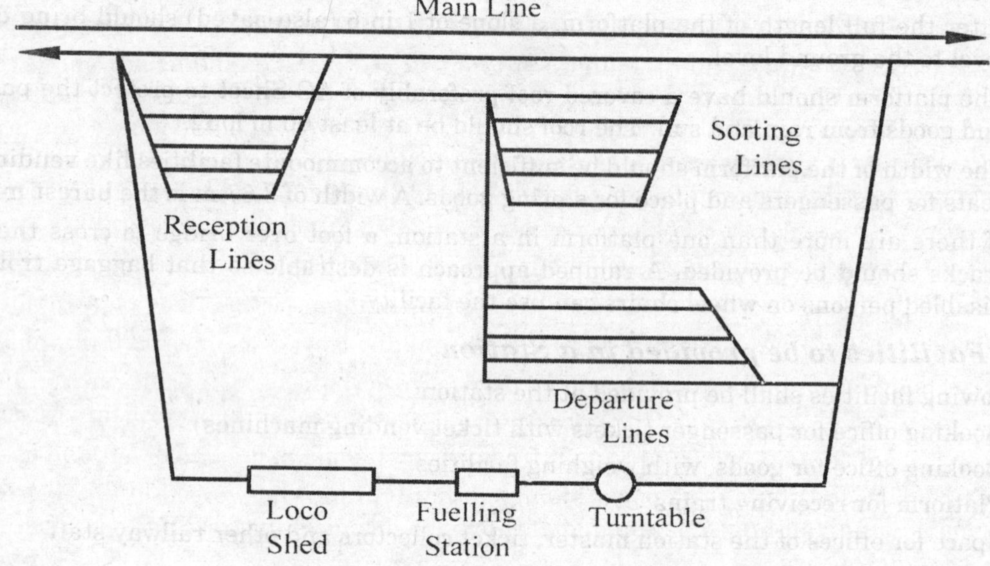

Fig. 36.2: Layout of Typical Marshalling Yard

36.3. Station Equipment

The equipment needed at stations are:

(i) Engine sheds
(ii) Watering facilities
(iii) Turntable (Fig. 36.3) or triangle (Fig. 36.4) for reversing engine
(iv) Facilities for cleaning coaches
(v) Examination pit for inspecting locos, coaches, and wagons from below

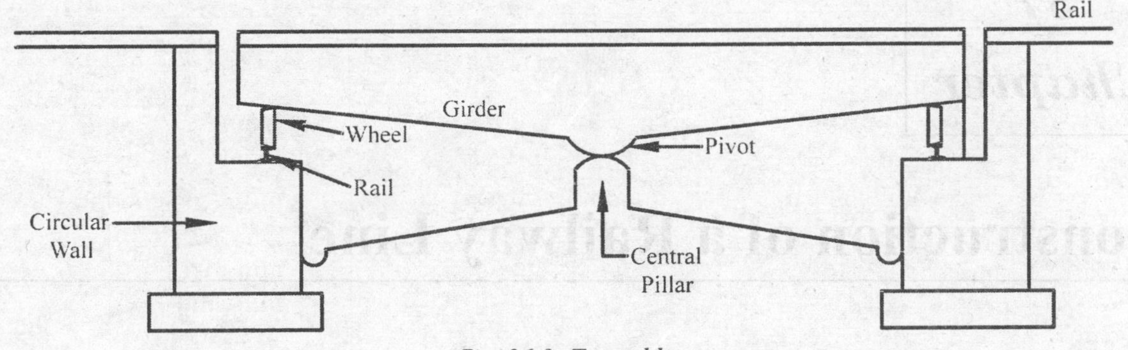

Fig. 36.3: Turntable

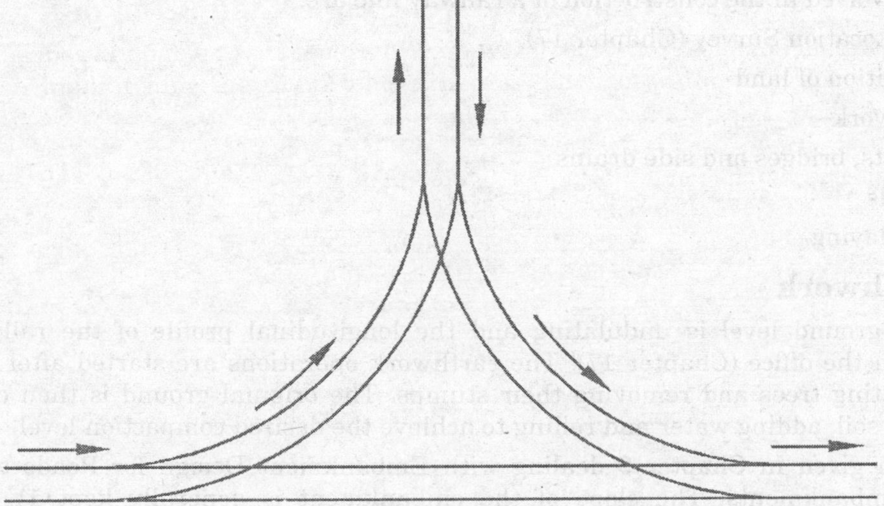

Fig. 36.4: Triangle

37

Chapter

Construction of a Railway Line

37.1. Stages Involved

The stages involved in the construction of a railway line are:

(i) Final Location Survey (Chapter 17)

(ii) Acquisition of land

(iii) Earthwork

(iv) Culverts, bridges and side drains

(v) Tunnels

(vi) Track laying

37.2. Earthwork

The natural ground level is undulating and the longitudinal profile of the railway track is determined in the office (Chapter 17). The earthwork operations are started after clearing and grubbing, cutting trees and removing their stumps. The original ground is then compacted by loosening the soil, adding water and rolling to achieve the desired compaction level.

Consideration given in Chapter 9 dealing with Embankment Design for Roads apply equally to railway embankments. The slope of the embankment is generally kept 1½ horizontal : 1 vertical or 2 horizontal : 1 vertical for normal heights. The slopes of cuttings depend upon the nature of the soil. In ordinary soil a slope of ½ horizontal : 1 vertical is usual, and in rocky structure a slope of $^1/_8$ horizontal to 1 vertical is common. In very hard rocks, a near vertical slope can be allowed.

The earthwork is laid in layers, each layer not exceeding 25 cm. The compaction is done by smooth wheel static rollers, smooth wheel vibratory rollers, sheepsfoot rollers. The compaction is done to achieve a density of about 98 per cent of the Maximum Dry Density at the Optimum Moisture Content (OMC). Tests are done to ensure that the desired density is achieved.

37.3. Railway Bridges

Railway bridges can be of the following types:

(i) Masonry arch

(ii) RCC slabs

(iii) RCC girders with slabs

(iv) Prestressed concrete girders with slabs

(v) Steel

The first four types require sleepers and ballast before laying the rail sections.

Steel bridges are very common for Railway bridges. They can be of the following types:

(i) Plate Girders

(ii) Trussed

(iii) Arched

The steel plate girders are fabricated outside and placed in position using cranes.

Trusses can be assembled and rolled into position.

The sleepers are laid on the girders and fastened to them dispensing with the ballast.

37.4. Tunnels

37.4.1. Importance of Accurate Surveying

Tunnelling work has to be preceded by accurate surveying which can prevent mismatch in alignment and levels when the work is done from both ends. Modern survey equipment such as total stations, referencing by Geographical Position System can be extremely valuable. When the work advances, the alignment and levels should be periodically checked.

37.4.2. Tunnel Portals

The ends of the tunnels are called tunnel portals. The exact location where the tunnel should end and the normal railway line starts should be decided on the basis of the contours of the hill face and the nature of strata met with.

37.4.3. Tunnel Excavation

Tunnel Excavation can be done by three methods:

(i) No-Blast Technique

(ii) Drilling and Blasting Technique

(iii) By using Tunnel Boring Machines

No-blast technique is used where surface subsidence is not desirable and vibrations caused by blasting cannot be permitted. Machines like tunnelling shields and road headers are used.

In the drilling and blasting technique, holes are drilled at predetermined locations and controlled blasting is done. The excavated material is taken out.

In the Tunnel Boring Machine (TBM) which is very popular, a rotating circular head with several cutters is employed for boring. Only circular sections can be bored.

37.4.4. Tunnel Lining

If the rock is hard and there is no possibility of rock pieces falling down, no lining is needed. But where the exposed periphery is not capable of standing on its own, tunnel lining is done. Lining is generally of concrete and is cast-in-situ. The formwork is assembled and concrete is pumped. The concrete is cured. Precast segments can also be used for tunnel lining.

Where there are poor ground conditions like fissured rock and suspected flow of seepage water, the exposed rock is grouted before lining.

37.4.5. Ventilation and Lighting

Short railway tunnels do not need ventilation as the only emissions come from diesel locos which are not continuously emitting as in road tunnels. But on very long tunnels, artificial ventilation may be needed.

Lighting is needed for inspecting the tunnels.

37.4.6. Drainage

The tunnel sides are provided with drains to drain away the water. A suitable gradient depending upon the expected discharge and cross-sectional area is provided.

37.5. Track Laying

If a track already exists and doubling is to be done, the track materials (rails, sleepers and ballast) are transported in special material trains and dumped sideway on the new formation. The track is then assembled.

For a new line to be laid, the telescopic method is used, wherein the track already laid is used for transporting fresh material in wagons for extending the track forwards.

Another method involves assembling the sleepers and rails in units at a central workshop, transporting the assemblies to the site and linking.

38
Chapter

Maintenance of a Railway Line

38.1. Need for Maintenance

A railway track needs constant maintenance because of the following reasons:

(i) Maintenance of accurate gauge is the key to safe train movement
(ii) Comfortable ride is possible if the track is maintained in good condition
(iii) Energy consumed by traction can be minimised on a well-maintained track
(iv) Wear and tear of rail, fasteners, points and crossings and sleepers occurs and needs to be attended promptly.
(v) Environmental factors like rain, floods and temperature variations cause damages
(vi) Landslides and slips damage the track
(vii) The track rests on the soil embankment which settles due to the pounding effect of the train, and has to be restored to the designed level
(viii) The ballast gets pounded and crushed and needs to be replenished.

38.2. Maintenance Operations

The maintenance operations are looked after by the engineering staff. They are assisted by Permanent Way Inspectors (PWIs), Assistant Permanent Way Inspectors (APWIs) and Bridge Inspectors. The PWIs have a jurisdiction of around 40 to 90 Kms. Manual maintenance is carried out by gangs. A gang is generally in charge of 6–8 Km of track and may consist of 10–18 gangmen headed by a mate and a gang reside closely to the track assigned to it. The PWIs and APWIs use trolleys for inspection. The gangs inspect the track every day, checking the gauge, cant and packing the ballast. The APWIs carry out inspection by push trolleys twice a week and PWIs carry out inspections every week. The Assistant Engineer carries out the inspection once in a month, and the Divisional Engineer carries out the inspection once in three months. Besides, inspections in the last vehicle and the footplate of the engine are carried out.

In modern practice, an instrumented track recorder is used for recording the speed, alignment, level and super-elevation.

The maintenance operations include:

(i) Maintenance of earthwork, restoring rain cuts and erosion
(ii) Ballast packing and renewal
(iii) Rectification of alignment

(iv) Checking and correcting gauge

(v) Maintenance of rails, joints, fasteners

(vi) Maintenance of bridges and culverts

(vii) Maintenance of points and crossings

(viii) Maintenance of signals

(ix) Maintenance of electrification components

(x) Maintenance of stations, yards

Mechanised equipment is available for fast and efficient maintenance.

39

Chapter

Track Renewal

39.1. Need for Track Renewal

Normal maintenance can keep the railway track in a good condition to a limited extent. Due to increase in the number of trains, loads and over-aging, permanent damage is caused to the railway track and it needs to be renewed.

39.2. Types of Track Renewal

Track renewal can be of three types:

(i) Complete Track Renewal

(ii) Through Rail Renewal

(iii) Through Sleeper Renewal

Thorough inspection can assess the need for and type of renewal needed.

39.3. Methods of Track Renewal

Traditionally in India, track renewal used to be accomplished by manual methods, such as lifting the rails, renewing the ballast, resetting the sleepers and laying the rails. But in modern times, track renewal by mechanical means is resorted to. The new track is assembled outside, brought to the site in wagons and lifted by cranes or track-laying machine and lowered into position, a process known as "pre-assembled renewal".

39.4. Programme of Track Renewal

During the Twelfth Five Year Plan, it is planned to undertake 19,000 Km of track renewal and 1,500 Km renewal for replacement of 52 Kg rails with 60 Kg rails.

40 Chapter

Bridges and Tunnels

40.1. Bridges

Bridges on railways are generally of steel. Plate girders or trussed girders are commonly employed. The sleepers are laid on girders and fastened to them dispensing with ballast. RCC and prestressed concrete or masonry arches are also used. In such designs, ballast is needed.

40.2. Tunnels

Tunnels are unavoidable in railway lines laid in hilly terrain. Examples are the Mumbai-Pune line, Konkan Railway line and the line linking Jammu and Srinagar.

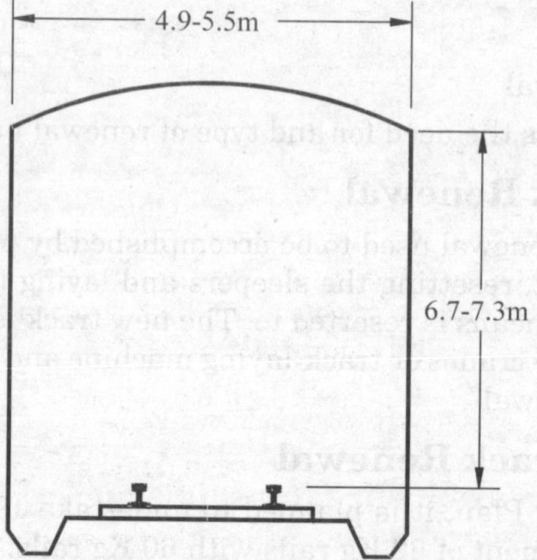

Fig. 40.1: Rail Tunnel for BG Track

Tunnels in hard rock do not need any lining, but those in soft rock need concrete lining. No elaborate ventilation system is needed as the only polluting vehicle is the solitary diesel or electric locomotive. But in long tunnels, artificial ventilator may be needed. A typical cross-section of a BG tunnel is given in Fig. 40.1. Drainage of tunnel should be provided for since water can seep through the unlined surfaces. Side drains as shown in Fig. 40.1 should be provided, and the water led away from the tunnel. The gradient must be gentle and facilitate the drainage and the alignment should preferably be straight though gentle curves can be provided. The construction is done by controlled blasting of the rock and advancing forward or by a Tunnel Boring Machine (TBM) (Chapter 37).

41 Chapter

Rail Accidents

41.1. Causes Leading to Accidents

Railway accidents are caused by:

(i) Error of judgement of driver

(ii) Unsatisfactory track, relating to gauge, cant, condition of rails, joints, fasteners, points and crossings

(iii) At unmanned road level crossings

(iv) Cattle crossing the track

(v) Floods causing erosion of earthwork

(vi) Sabotage

(vii) Failure of rolling stock or engine

(viii) Failure of signal system

41.2. Steps To Reduce Rail Accidents

41.2.1. The conversion of road level crossings into Road Overbridges (ROBs) is an important step. The present policy of the Indian Railways is to replace level crossings by ROBs if the product of the number of road vehicles and the number of trains termed as TVUs exceeds 50,000.

41.2.2. Several initiatives in signalling or Telecommunications will reduce accidents. These include track circuiting, interlocking at level crossings with signals, provision of Train Protection and Warning System (TPWS), Anti Collision Drive (ACD) and Train Collision Advance System (TCAS), provision of Mobile Communication and Train Radio Communication (MCTRC), provision of GPS-based fog safe device and provision of intelligent fire surveillance and extinguishing system.

42

Chapter

Modern Technological Advances

42.1. Types of Technological Advances Planned in India

Two significant technological advances have been planned by the Indian Railways. These are:

(i) High Speed (Bullet) Trains *(ii)* Dedicated Freight Corridors

42.2. High Speed (Bullet) Trains

High Speed (Bullet) Trains are in operation in several countries. These are:

(i) Shinkan.e in Italy, with a commercial speed of 300 Km/hr.
(ii) TGV in France, with an average speed of 255 Km/hr.
(iii) Eurostar in UK, Belgium and France, with a top commercial speed of 300 Km/hr.
(iv) Pandoline in Italy, with a commercial speed of 300 Km/hr.

India has recently signed an agreement with Japan for the construction of a high speed train between Mumbai and Ahmedabad.

42.3. Dedicated Freight Corridor

The Dedicated Freight Corridors (DFC) on the Western and Eastern routes are being implemented in India. The western DFC connects the western region (Maharashtra and Gujarat) with the northern hinterland. The eastern DFC is intended to carry predominantly coal, steel and containers. The average speed of freight trains is expected to go up from 25 Km/hr to 70 Km/hr, thus reducing the transit time by half the present values. The capacity thus released by freight trains can be used for running more passenger trains at higher speeds.

42.4. Ballast Less Track

For Metro lines, ballast is omitted. The rails are fixed on a concrete slab directly. Pandrol clips are used for fastening. Rubber pads are provided between the rail and the concrete slab to dampen the vertical movement.

42.5. Light Rail Transit (LRT)

Light Rail Transit (LRT) involves light weight coaches, sharp curves (desirable radius: 120 m, minimum radius: 30 m) and steep gradients (maximum 4 per cent). Speed of operation is in the range of 25 to 60 Km/hr. Trains can operate with a headway of 60 to 90 seconds. LRT can handle about 3,000 to 20,000 passengers per hour per direction (PPHPD), which corresponds to 15,000 to 200,000 passengers per day.

SECTION—4

AIRPORT ENGINEERING

43
Chapter

Role of Air Transportation

43.1. History of Growth of Air Transportation

The first power driven plane was flown by the Wright brothers in 1903 at Kitty Hawk, North Carolina, U.S.A. The early planes were driven by propellers, which were later replaced by turbo-props and turbo-jets.

The First World War and the Second World War used aeroplanes for the war effort. In 1927, Lindburg made the first solo flight between New York and Paris. In India, in 1911, Henri Piquet, a Frenchman flew the first flight from Allahabad to Naini. J.R.D. Tata made a solo flight in 1932 from Karachi to Bombay in a single engine Puss Moth.

In the last century, air transport has made dramatic strides. Whereas the propellers driven planes travelled at speeds of 250-300 km per hour, modern jet aircraft can travel at a speed of 1000 km per hour. Supersonic aircraft, for example Concord, can fly at a speed of 2300 km per hour which is greater than the speed of sound. Modern jets can carry more than 300 passengers.

43.2. Advantages and Disadvantages of Air Transportation

43.2.1. Advantages

(i) ***Speed:*** The biggest advantage of air transport is its high speed. Speeds of 1000 km per hour are quite common

(ii) ***Encourages Tourism:*** International and domestic travel by air has encouraged the tourism industry.

(iii) ***Promotion of Business and Trade:*** High value goods and perishables are transported quickly by air, giving a boost to business.

(iv) ***Military Use:*** Air transport is invaluable for the defence of the country. It can transport military personnel and defence equipment quickly.

(v) ***Relief and Rescue Operations:*** In times of natural and man-made calamities, air transport can provide relief to effected people and rescue them to safety quickly. The recent floods in Uttarakhand and Chennai have amply proved the benefits of air transport in this regard.

(vi) ***Accessibility to Remote Areas:*** For example, Ladakh in India is difficult to access by road. The airport at Leh provides good connectivity to Ladakh.

(vii) ***Safety:*** Air transport is very safe compared to road transport and railways. Modern technology and rigid regulations for licencing of planes and pilots make air travel very safe.

43.2.2 Disadvantages

(i) ***High Consumption of Liquid Fuel:*** The only fuel for aircraft now is petroleum crude based aviation petrol. As availability of crude worldwide is shrinking, this is a serious disadvantage.

(ii) ***High Noise Pollution:*** Persons residing near airports and the path of landing and take-off are subjected to high noise levels.

(iii) ***High Cost:*** Compared to road transport and rail transport, air travel is very costly and only the rich can avail of it.

43.3. Air Transport in India

Till a few years ago, air transport was the monopoly of the government sector. Indian Airlines, Air India and Vayudoot were all owned by the government. But, now private airlines also operate, both on domestic routes and international routes. The private sector is also involved in developing airports. India now has the ninth largest civil aviation market in the world (Ref. 1). The passenger handling capacity has risen three-fold from 72 million in 2006 to 220 million in 2011. The cargo handling capacity has risen six fold from 0.5 million tonnes in 2006 to 3.3 million tonnes in 2011. During the Twelfth Five Year Plan, domestic air travel is expected to grow at an average annual growth rate of 12 per cent, and the international air travel is expected to grow at 8 per cent per annum. The cargo traffic is expected to grow annually at 12 per cent and 8 per cent in the domestic and international sectors respectively. Several new airports in Tier II and Tier III towns will have to be built soon. To sum up, air transportation is expected to grow at a healthy rate in the coming years.

43.4. Aircraft Characteristics

43.4.1. Speed

The airport infrastructure facilities must match the speed of aircraft. Modern wide-bodied jets can travel at speeds of 1000 km per hour.

43.4.2. Size and Dimensions

The size and dimensions of the aircraft expected to use the airport determine the layout of the airport. Fig. 43.1 gives a typical sketch of an aircraft showing the controlling dimensins

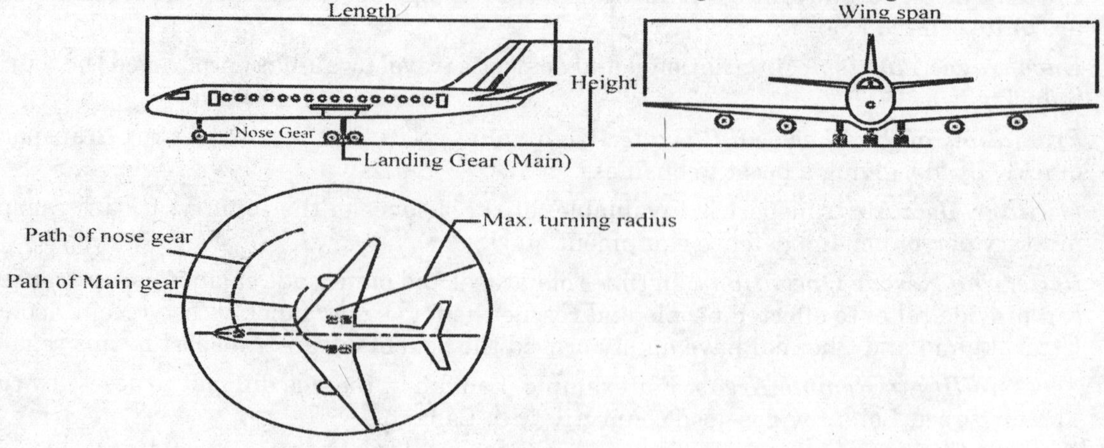

Fig. 43.1 : Aircraft Dimensions

The dimensions of a few common aircraft types are given in Table 43.1.

Table 43.1 : Principal Dimensions and Characteristics of Aircraft

Aircraft type	*Wing span (m)*	*Length (m)*	*Maximum weight at take off (kg)*	*Wheel track (m)*	*Wheel base (m)*	*Height (m)*	*Cruising speed (km/h)*
Boeing 747	59.6	70.5	351,500	11.0	25.6	19.3	940
Boeing 737	28.3	30.5	45,600	5.3	11.4	11.3	850
Airbus 300	44.8	53.8	152,000	9.6	18.6	16.6	900

43.4.3. Weight, Landing Gear and Tyre Pressure

The maximum weight of an aircraft is when it is loaded with fuel and is at take-off. Table 43.1 gives the maximum weight of some aircraft. The load of the aircraft is transmitted to the airport pavement by wheels conforming to various configurations. This is dealt with in para 43.7.

Tyre pressures are high in aircraft. Typical values are as under:

Table 43.2: Aircraft Tyre Pressure

Aircraft Type	*Tyre Pressure (M Pa)*
Boeing 747	1.43
Boeing 737	1.04
Airbus	1.18

The main tyres are in the vicinity of 50 × 16 size. The nose wheels are of the 40 x 15 size.

43.5. Airport Planning

43.5.1. Types of Airports

Airports are broadly classified into Civil Airport meant for use by civilian aircraft and Military Airports exclusively serving defence needs.

Civil Airports are of three types, viz.

(i) Air carrier airports which serve air carriers. Example : Mumbai Airport

(ii) General aviation airports, which cater to flying clubs, medical services, fire fighting etc. Example : Safdarjung Airport in Delhi

(iii) A mixture of the above operations. Example : Ludhiana Airport which allows air carrier operations and use by flying club.

Airports can also be (i) International, which have facilities for customs and immigration or (ii) Domestic, meant only for use by domestic services.

The types of landing that are possible are:

(i) ***CTOL:*** Conventional Take-off and Landing

(ii) ***STOL:*** Short Take-off and Landing, with runways of less than 600 m length.

(iii) ***VTOL:*** Vertical Take-off and Landing – Example : Heliport

(iv) ***RTOL:*** Reduced Take-off and Landing with a runway 1000 – 1200 m long.

43.5.2. Demand Forecast

The planning for an airport is dictated by the future demand it is expected to cater for. The forecast is generally done for a period of 20 years. For new airports it must be admitted that traffic forecast is based on trends in growth of national air travel, growth of the economy and the growth of business and industry in the region. If an airport already exists, and the planning is for its expansion and moderation, the historical growth of air travel provides a good basis, combined with forecast of the economy, industry and business in the region.

43.5.3. Site selection

The site selection is governed by the following factors:

Engineering and Physical Factors:

(i) The site should have fairly level area so that the earthwork cost is minimal.

(ii) The land required for a small airport is about 20 to 40 hectares, but for large airports it can be as large as 6,000 to 15,000 hectares. This quantity of land has to be acquired.

(iii) The site should permit future expansion.

(iv) Airport approaches must be clear of obstructions like tall buildings, towers, hills, and transmission lines.

(v) The site should be easily accessible to all users, such as passengers, industry and business.

(vi) The area should be well drained, free from floodability and have a low water table.

(vii) A good and firm soil can reduce the cost of pavement of runways, taxiways and apron.

(viii) The site should not create noise nuisance to residential areas and hospitals.

(ix) Services like electricity, water, gas, sewerage and telephone should be easily available.

(x) Bird hits cause serious damages to the aircraft and endanger the safety. The site should, therefore, be away from slaughter houses and refuse dumps that attract birds.

(xi) Good quality construction materials should be available nearby.

Meteorological Factors

(i) Wind velocity, direction, frequency, turbulence and gusts influence aircraft operation and these should be favourable. Wind direction and velocity influence the orientation of the runway.

(ii) Visibility conditions like fog, dust-storms influence safe landing and take-off. These should be favourable.

43.6. Airport Layout

43.6.1. Principal Components of Airport

The aircraft has three main components:

(i) Airfield

(ii) Terminal area

(iii) Flight support facilities

43.6.2. Airfield

The airfield consists of:

(i) landing strip, consisting of runways (one or more), shoulders and stopways.

(ii) Taxiways.

(iii) Apron

The runway is the place where the aircraft lands and takes off. It has a strong paved surface with shoulders on either side for serving as safety zones in case an aircraft goes off runway sideways. Stop-ways are provided at the ends of the runway to accommodate an aircraft that overshoots a runway during landing or has an aborted take-off. The taxi-way is a paved strip connecting the runway with the apron or connecting runways with one another. The apron is the hard-standing area where aircraft is parked for embarking/disembarking the passengers or where cargo is loaded/ unloaded. Fuelling and catering services are also carried out here. A holding apron (also called a run-up area) is that portion provided near the ends of the runway for aircraft waiting for take-off.

43.6.3. Terminal Area

The terminal area of an airport consists of:

(i) Terminal Building

(ii) Gates

(iii) Aircraft service facilities

The terminal buildings are where the passengers are serviced and cargo and mail are handled.

Gates are the openings in the terminal building where the embarking passengers are allowed to proceed to the aircraft. The transfer of passengers from the gate to the aircraft and vice versa is accomplished by buses or through aerobridges.

The aircraft servicing facilities include hangars which are covered area intended for repairing and servicing of aircraft.

43.6.4. Flight support facilities

These include structures for air traffic control, navigational aids and fuelling the aircraft.

43.6.5. Illustrative Layout

An illustrative layout of an airport is given in Fig. 43.2.

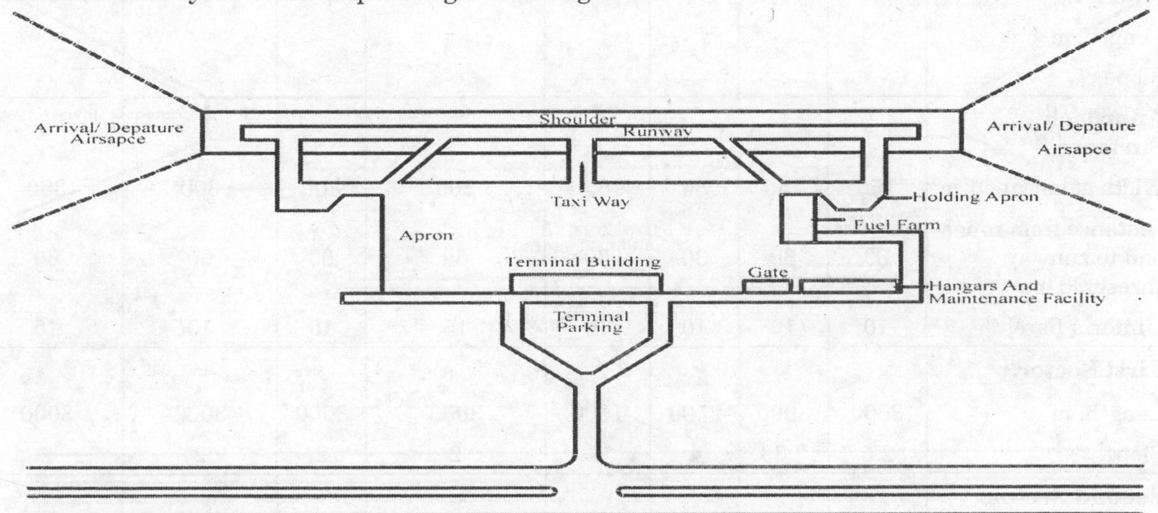

Fig. 43.2 : Illustrative Layout of Airport

43.6.6. Obstruction Criteria

Landing and take-off of aircraft requires that the zones at either end of the runway be clear of obstructions. In order to achieve this, the dimensions and shapes of obstacle surfaces have been standardised. The International Civil Aviation Organisation (ICAO) recommendations for approach runways and take-off runways are given in Tables 43.2 and 43.3 respectively. As per ICAO classification, airports may be categorised in A, B, C, D and E categories, the highest among them being category A. The primary surface is one which is longitudinally centred on a runway and which extends 60m beyond each end of the runway.

Table 43.3 : ICAO Standards for Approach Runways

Obstruction Clearance Surface	***Approach Conditions and Runway Code Designations***							
	Visual Runways				**Non-precision Instrumentation Runways**		**Precision Instrumentation Runways**	
							Category I	***Category II***
	A, B	***C***	***D***	***E***	***A, B***	***C***	***A, B, C***	***A, B, C***
Conical Surface:								
Slope, %	5	5	5	5	5	5	5	5
Height, m	100	75	55	35	100	75	100	100
Horizontal Surface:								
Height, m	45	45	45	45	45	45	45	45
Radius, m	4000	4000	2500	2000	4000	4000	4000	4000
Primary Surface:								
Width, m	--	--	--	--	--	--	--	--
Length, m	--	--	--	--	--	--	--	--
Slope, %	--	--	--	--	--	--	--	--
Approach Surface:								
Width at inner end, m	150	150	80	60	300	300	300	300
Distance from inner end to runway threshold, m	60	60	30	30	60	60	60	60
Uniform flare, %	10	10	10	10	15	15	15	15
First Sector:								
Length, m	3000	3000	2500	1600	3000	3000	3000	3000
Slope, %	2.5	3.33	4	5	2	2	2	2
Second Sector:								
Length, m	--	--	--	--	3600	3600	3600	3600
Slope, %	--	--	--	--	2.5	2.5	2.5	2.5

(Contd...)

Horizontal Sector:								
Length, m	--	--	--	--	8400	8400	8400	8400
Total length of sector, m	3000	3000	2500	1600	1500	1500	1500	1500
Transitional Surface:								
Slope, %	14.3	14.3	20	20	14.3	14.3	14.3	14.3
Internal Transition Surface:								
Slope, %	--	--	--	--	--	--	--	3.33

The horizontal surface is one which is a horizontal plane 45 m above the established airport elevation and is a circle whose centre is at the airport reference point (geographic centre of the airport) and whose radius varies from 2000 to 4000 m.

Table 43.4 : ICAO Standards for Take-off Runways

Obstruction Clearance Surface	***Main Runways***		***Other Runways***	
	A, B, C	**A, B, C**	**D**	**E**
Take-off Clime Surface:				
Width at inner end, m	180	180	80	60
Distance from inner end to runway threshold, m	60	60	30	30
Uniform flare, %	13	13	10	10
Width at outer end, m	1200	1200	580	580
Length, m	15000	12000	2500	1600
Slope, %	2	2.5	4	5

The conical surface extends outwards and upwards from the periphery of the horizontal surface at a slope of 20 to 1.

The approach surface is one longitudinally centred on extended runway centre line. The approach surface slopes at an angle to the horizontal, which is known as the approach angle. A slope of 20:1 is commonly adopted for visual runway.

Fig. 43.3 explains the above terms.

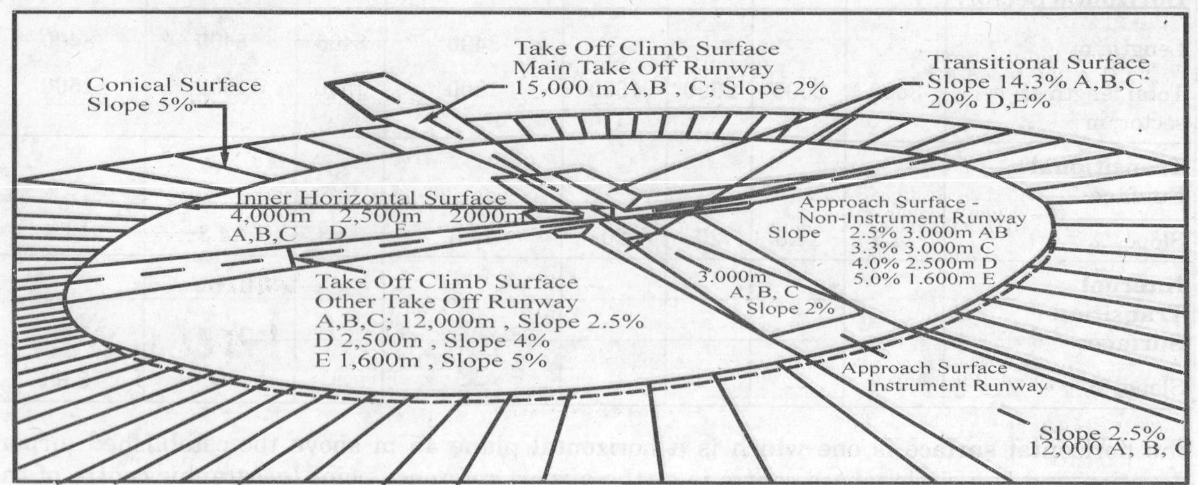

Fig. 43.3 : Perspective View of surfaces Connected with Airport

43.6.7. Configuration of Runways

Various configurations of runways is given in Fig. 43.4.

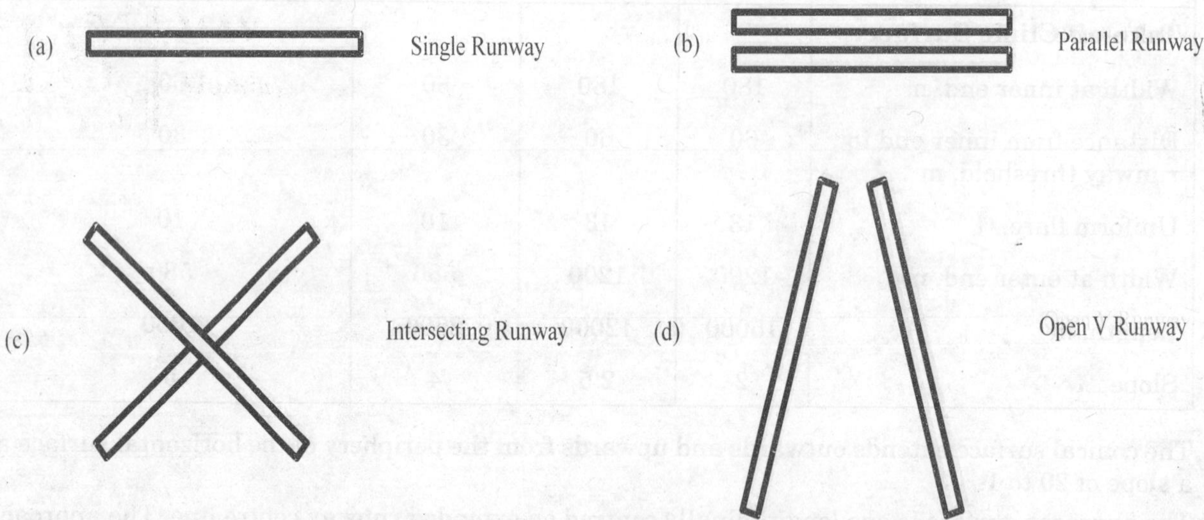

Fig. 43.4 : Perspective View of Surfaces Connected with Airport

The configurations are dictated by the prevailing wind directions. Designs (a) and (b) give the configurations parallel to the prevailing wind directions. Designs (c) and (d) are adopted when the winds are in more than one direction.

43.6.8. Runway and Taxiway dimensions

Various configurations of runways is given in Fig. 43.4.

Tables 43.4 and 43.5 give the dimensions of runway and taxiway recommended by ICAO.

Table 43.5 : ICAO Standards for Take-off Runways

Obstruction Clearance Surface	*ICAO Airport Code Destination*				
	A	B	C	D	E
Length of runway under standard conditions, m	2100 upwards	1500-2100	900-1500	750-900	600-750
Width of runway, m	45	45	30	23	18
Minimum distance between centre lines of parallel visual run-ways, m	210	210	150	120	120
Maximum longitudinal gradient i_1, %	1.25	1.25	1.5	2	2
Maximum radius of vertical curves, m	30000	30000	15000	5000	5000
Maximum traverse gradient $i_{(r)}$, %	1.5	1.5	1.5	2	2
Width of landing strip (precision instrument runway), m	300	300	300	--	--
Width of landing strip (visual run-way), m	150	150	150	80	60
Maximum gradients of unpaved portion of landing strip:					
longitudinal, %	1.5	1.75	2	2	2
transverse, %	2.5	2.5	2.5	3	3

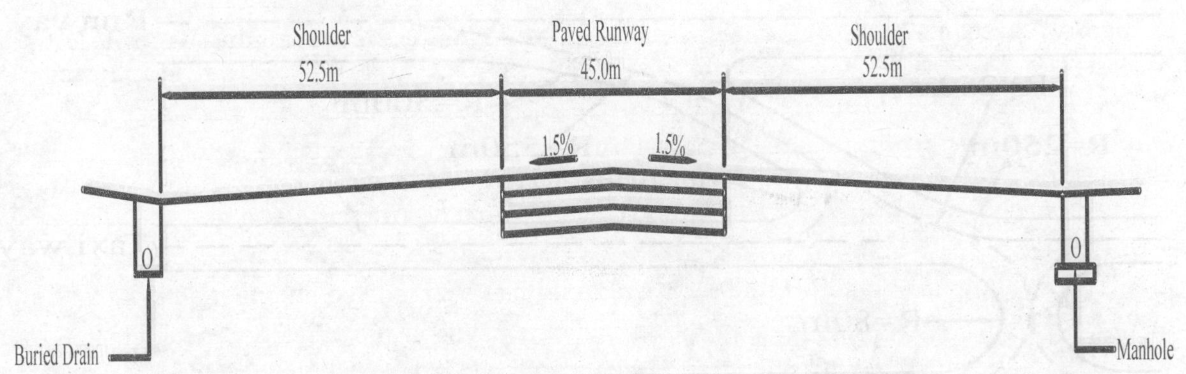

Fig. 43.5 : Runway Cross-section

The longitudinal gradient of runway affects the performance of the aircraft. Rising gradients demand extra power for take-off, and downward gradients need more braking distance. For large airports intended for commercial use, a maximum gradient of 1.5% is used. For utility airports, the gradients can be slightly increased to 2%.

Longitudinal grade changes should be avoided. When they are unavoidable, they should be serviced by suitably designed vertical curves. The minimum length of the vertical curve of each 1% grade change is 300 m for carrier airports and 90 m for utility airports. Transverse gradients should not exceed 1.5%.

Since aircraft travels at a much reduced speed on taxiways, their dimensional requirements are lower than for runways. ICAO standards for taxiway dimensions are given in Table 43.5

Table 43.6 : Taxiway Dimensions and Details

Items	***ICAO Airport Destination***				
	A	**B**	**C**	**D**	**E**
Width of taxiway (m)	23	23	15	10	7.5
Maximum transverse gradient taxiway (%)	1.5	1.5	1.5	2	2
Maximum longitudinal gradient of taxiway (%)	1.5	1.5	3	3	3
Width of taxiway margins (m)	10.5	10.5	7.5	- -	- -
Width of taxiway strip (taxiway + paved or un-paved margins) (m)	44	44	30	25	22

The exit taxiway from a runway can be either right angled or at some other angle. An angle of about 30° permits to clear the runway quickly at speeds of the above to 100 km per hour.

Such exit taxiways can be conveniently located at about 800 m, 1200 m, 1800 m, and 2000 m from the runway threshold such that a variety of aircraft can be served. Taxiway exit details are given in Figs. 43.6 and 43.7.

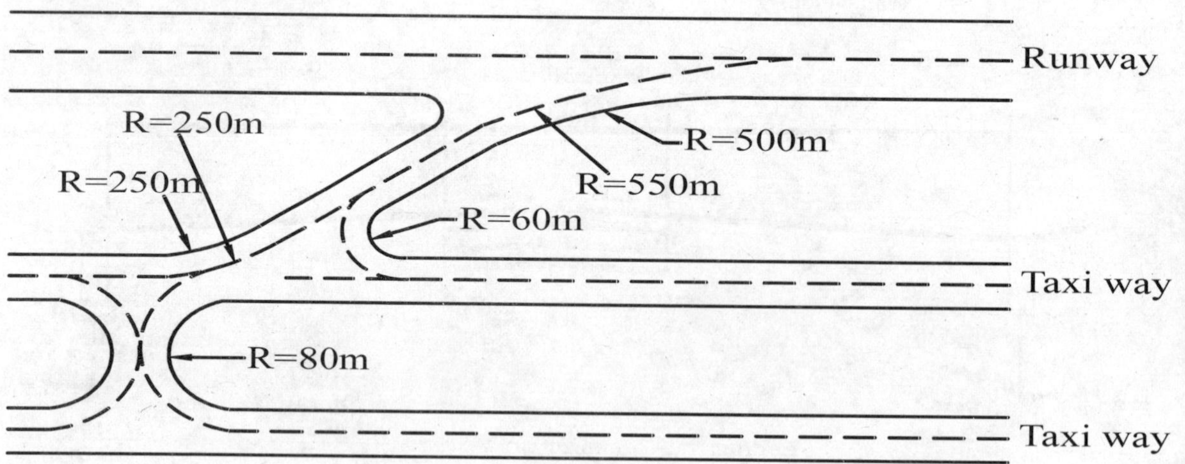

Fig. 43.6 : Taxiway Exit Details

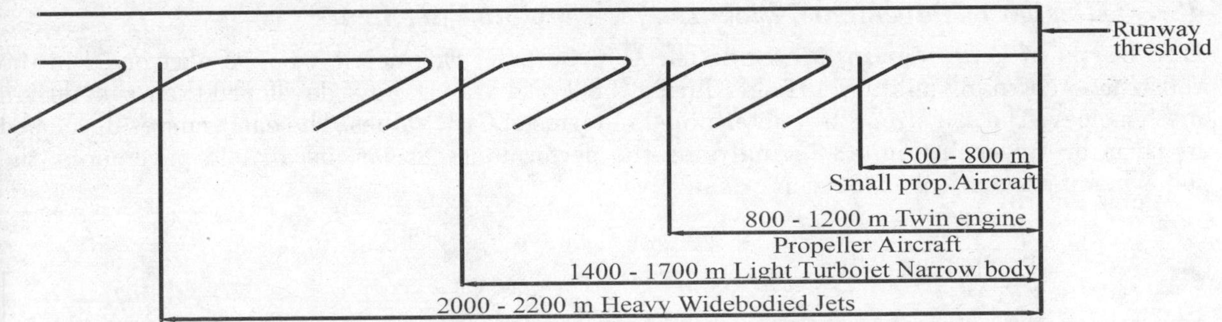

Fig. 43.7 : Taxiway Exit Details for Various Types of Aircraft.

For safety of operations during taxing, separation clearance is needed. Table 43.6 gives the clearances for various airport designations.

Table 43.7 : Separation Clearances

Items	*ICAO Airport Destination*				
	A	**B**	**C**	**D**	**E**
Minimum distance between taxiway and runway edges (m):					
• Precision instrument runway	150	150	150	--	--
• Non-precision instrument runway	75	73	73	36	29
Minimum distance between taxiway edges (m)	62	52	43	27	23
Minimum distance from edge of taxiway to fixed obstruction outside taxiway (m)	38	30	26	18	16

43.7. Airport Pavement

43.7.1. Runway and Taxiway dimensions

Airport pavements must meet the following requirements:

(1) They must have adequate strength to withstand the heavy loads imposed by modern aircrafts.

(2) They must take care of the repetition of loads over the design life of the pavement.

(3) Because of the high speeds at which aircrafts land and take-off, the pavement must take care of the interaction of vehicle response and pavement surface. For this purpose, the surface should be sufficiently smooth.

(4) The surface should be anti-skid.

(5) The pavement should be dust free and take care of jet blast.

(6) The maintenance needs of the pavement should be minimal.

(7) The pavement should withstand the detrimental effect caused by fuel and oil spillage.

(8) The cost of the pavement should be economical.

43.7.2. Types of Pavements

A bituminous surfaced pavement has a smooth surface and is the preferred choice for runways. For taxiways and aprons where joints in concrete pavement are not a disadvantage and the drippings of oil from the engines do not affect the pavement, a concrete pavement is generally preferred.

43.7.3. Design of Pavement Thickness for Flexible Pavements

U.S. Corps of Engineering Method: The U.S. Corps of Engineers, who were involved in the construction of many military airfields during the World War II, have developed their own design methodology. The soil strength is determined in terms of CBR values. The early curves developed by them are given in Fig. 43.8. Analysing the performance of various airfield pavements, the following mathematical expression was involved:

$$t = \sqrt{\frac{P}{8.1(CBR)} - \frac{A}{\pi}}$$

where t = pavement thickness, in inches

P = Single wheel load, in lb.

A = Tyre contact area, in sq.inch.

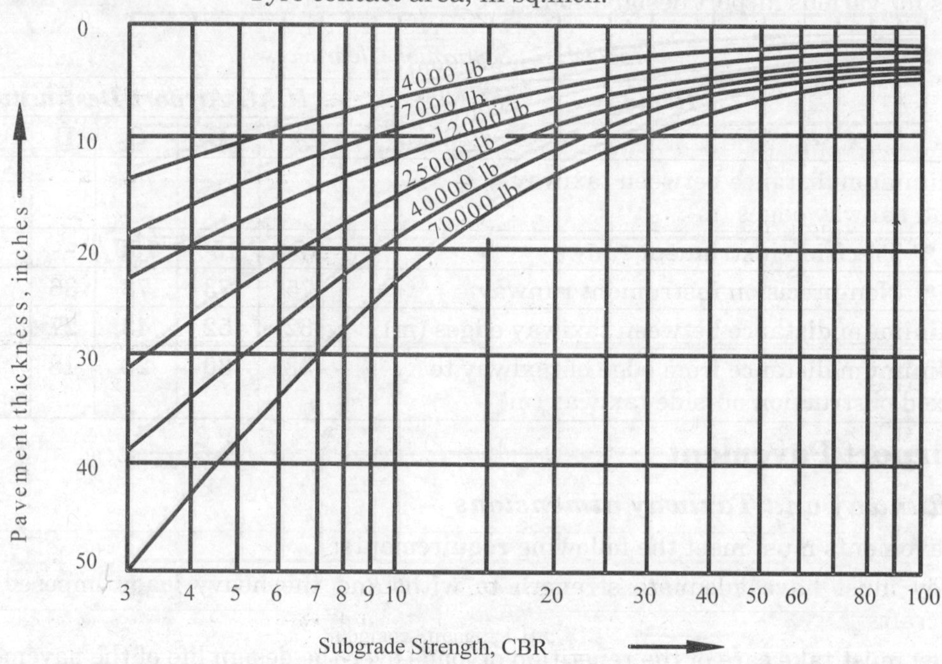

Fig. 43.8 : U.S. Corps of Engineers Curves for Airport Pavement Design for Different Wheel Loads A Set of Curves for 200 PSI Tyre Pressure is given in Fig. 43.9.

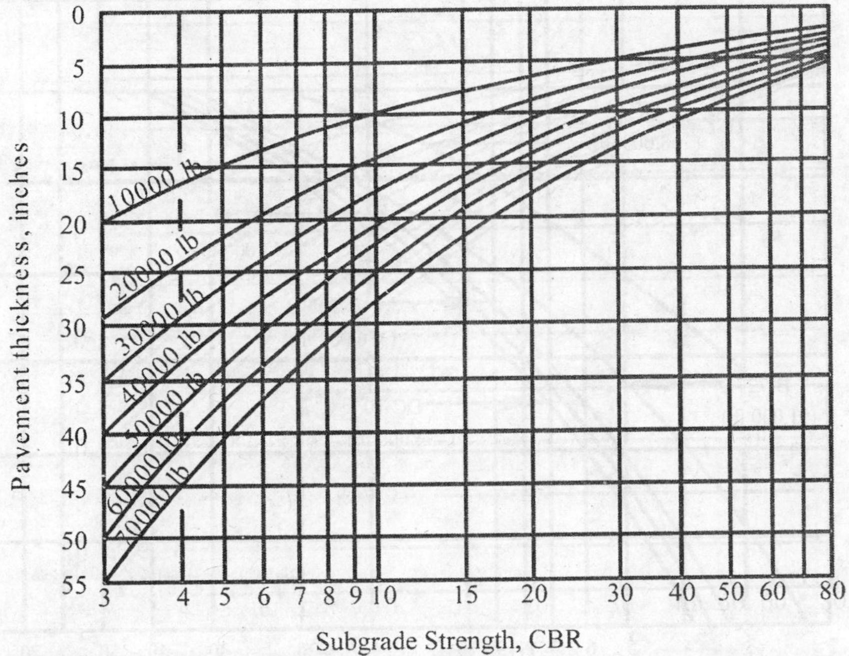

Fig. 43.9 : U.S. Corps of Engineers Curves for Airport Pavement Design for Different Wheel Loads for 200 PSI Tyre Pressure

Since different aircrafts had different wheel configurations, the concept of Equivalent Single Wheel Load (ESWL) was developed. ESWL for a particular wheel assembly is that single load which produces the same strain in the foundation as the multiple wheel loads. ESWL can be computed by using Boussinesq's theory.

If ESWL is considered, Equation above gets modified as under:

$$t = f\sqrt{\frac{ESWL}{8.1(CBR)} - \frac{A}{\pi}}$$

where

f = factor for adjusting design thickness for wheel load coverage

C = coverage (sufficient wheel passes to cover every point of traffic lane once)

ESWL = Equivalent Single Wheel Load

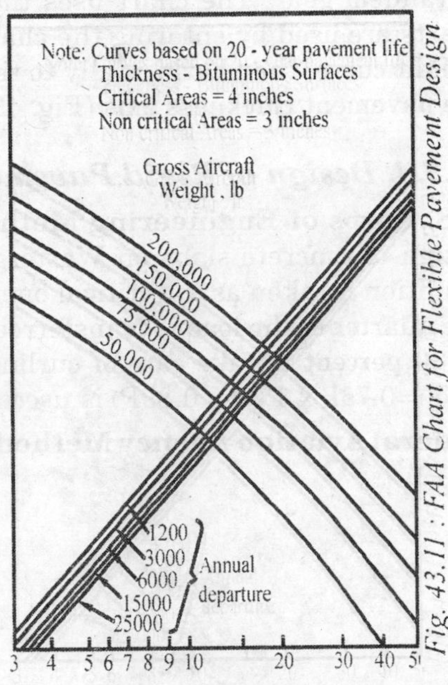

Fig. 43.11 : FAA Chart for Flexible Pavement Design

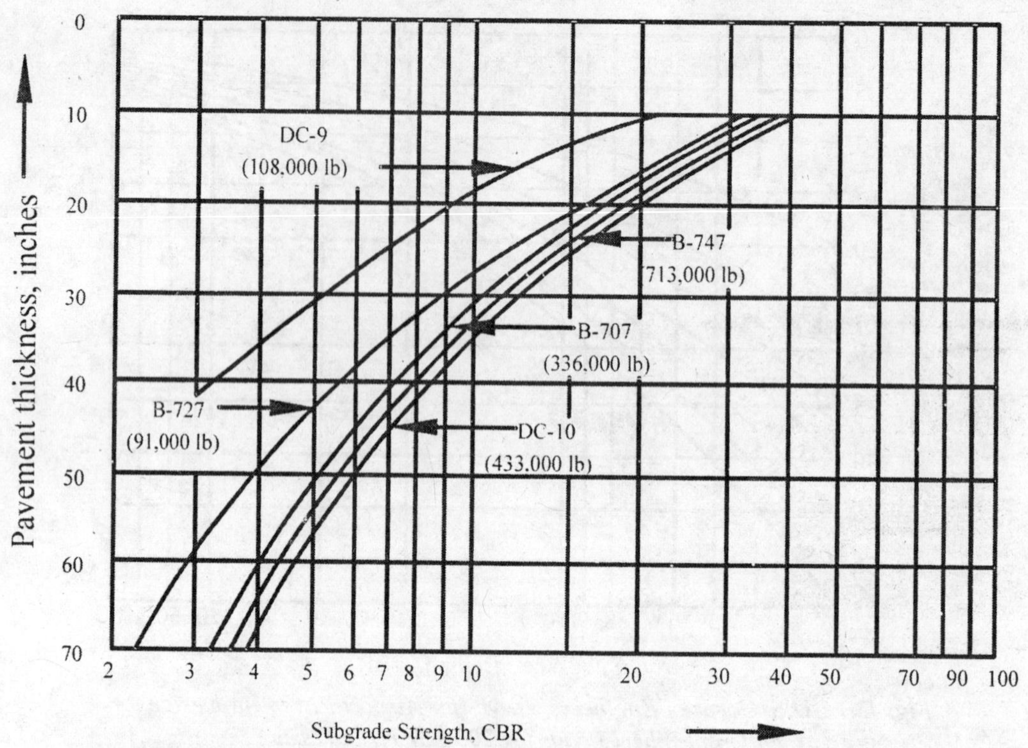

Fig. 43.10 : U.S. Corps of Engineers Thickness Design Chart for Selected Aircraft

Using this procedure, Corps of Engineers have developed design curves for typical commercial aircraft. The curves are given in Fig. 43.10.

Federal Aviation Agency Method: The FAA method is based upon the soil strength as determined by CBR. There are readymade curves for single wheel gear, dual wheel gear, and dual in tandem gear. The chart uses the Gross Aircraft Weight and annual departures. The design charts are used by entering the chart at the CBR axis, proceeding vertically to the gross take-off weight curve, then horizontally to the equivalent annual departure curve, and finally vertically to the pavement thickness axis (Fig. 43.11).

43.7.4. Design of Rigid Pavement for Airports

U.S. Corps of Engineering Method: The U.S. Corps of Engineers have evolved charts for the design of concrete slabs on Westergaard's formulae presented in Chapter 22. The edge loading condition is taken as the critical one. It is assumed that when the load is placed at an edge of slab, one quarter of the load is transferred to the adjacent slab. The remaining load (0.75P) is increased by 30 percent to take care of curling stresses, load repetitions and other contingencies. The net load (=0.75P x 1.30 = 0.89P) is used for designs.

Federal Aviation Agency Method: The typical FAA rigid pavement design chart is given in Fig. 43.12.

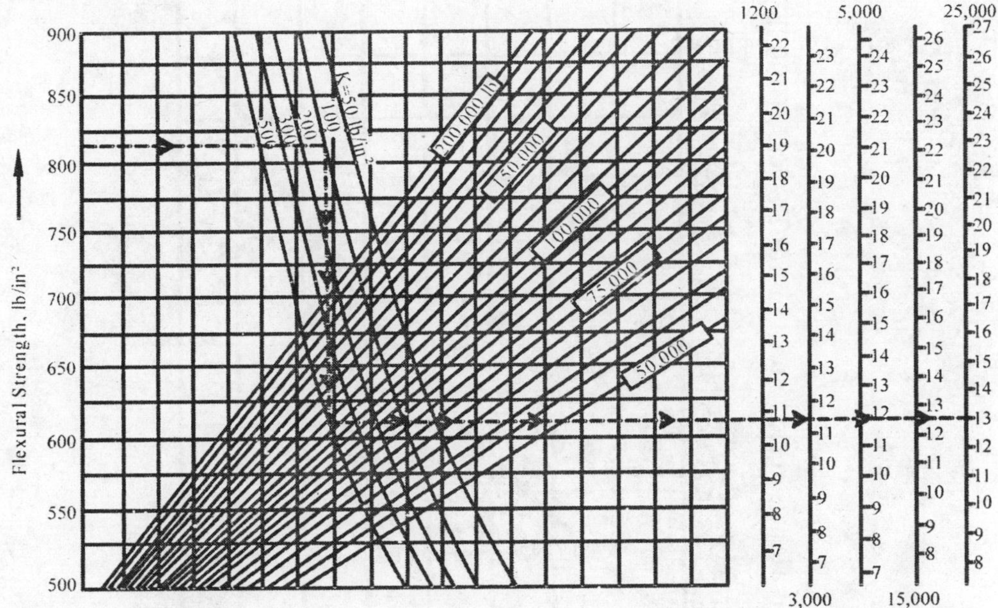

Fig. 43.12 : Rigid Pavement Design Chart for Critical Areas, Dual Wheel Gear

The chart given is for dual wheel gear aircraft. The chart uses inputs such as:

(1) Aircraft weight

(2) Annual departures

(3) Modulus of Subgrade Reaction

(4) Concrete Flexural strength.

General Features of Rigid Airport Pavements: Transverse contraction joints are provided at longer spacing than in highway. Typical range is 4 to 8 m. Transverse expansion joints are not generally used except at intersections. Dowels are used where load transfer occurs. A sub-base is generally provided to give a firm support. Reinforcement is desirable for controlling cracks.

LCN Method of Design: The Load Classification Number (LCN) is a method of aircraft evaluation of pavement design first used by the Air Ministry, UK. The method expresses the supporting capacity of a pavement in the terms of a number known as LCN. The wheel load of the aircraft can also be expressed in terms of LCN.

From the known values of the wheel loads and the contact areas of the aircraft in use when the LCN method was devised, a smooth curve was drawn (Fig. 43.13).

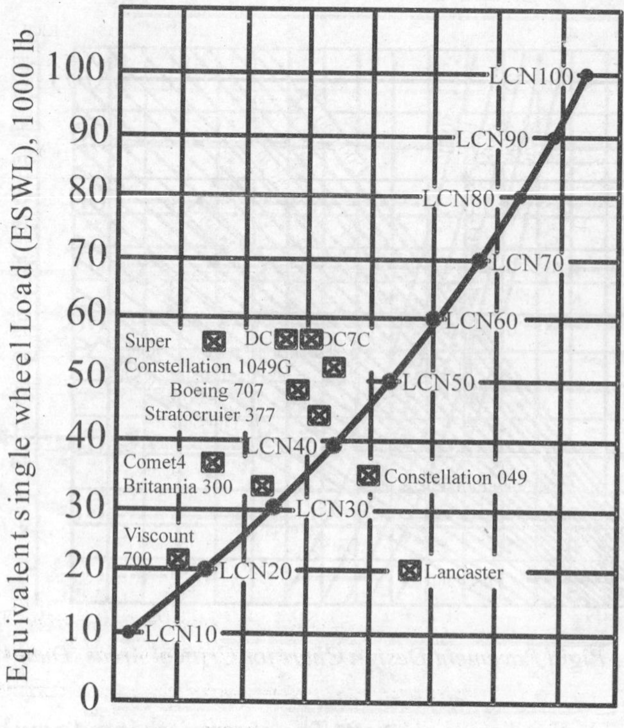

Fig. 43.13 : LCN Curve

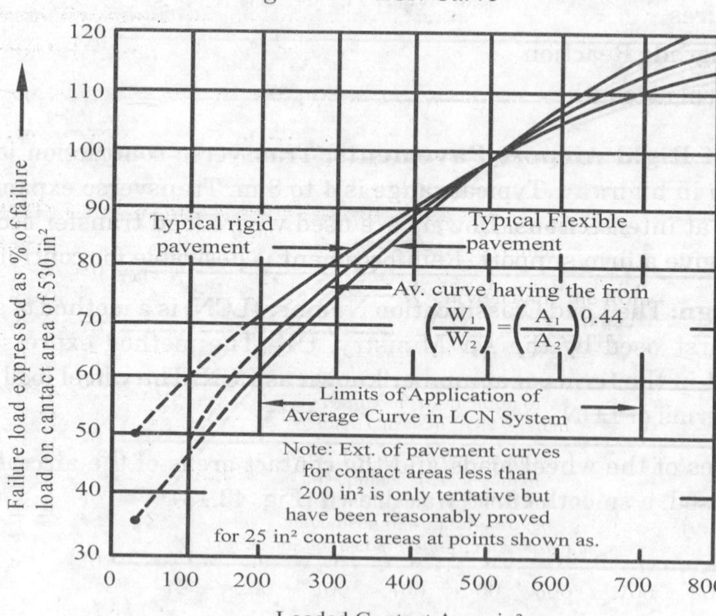

Fig. 43.14 : Failure Loaded/Contact Area Relationship

Various types of pavements, rigid and flexible, were tested with bearing plates having a range of contact areas (20 to 70 sq. in), which represented the aircraft then in use. Over this range of contact areas, both rigid and flexible pavements showed similar load deflection behaviour. From these studies, carried out till failure, the following relationship was established:

$$\frac{W_1}{W_2} = \left(\frac{A_1}{A_2}\right)^{0.44}$$

where W_1, W_2 = failure loads, lb

A_1, A_2 = load contact areas, sq.in

The curve that was developed is given in Fig. 43.14.

In this curve the ordinate (failure load) is expressed as a percentage of failure load on a contact area of 530 sq.in (plate diameter 26 in).

The LCN curve and failure load curve were then combined to produce a chart given in Fig. 43.15 using the following procedure.

(1) The tyre contact area lines were drawn from the relationship.

(2) Contact Area = $\dfrac{\text{Load}}{\text{Tyre pressure}}$

(3) One point on each LCN line comes directly from the standard load classification curve.

(4) Other points were calculated using the equation given above.

(5) Dotted lines are tentative extensions of the LCN system to accommodate contact areas less than 20 sq.in.

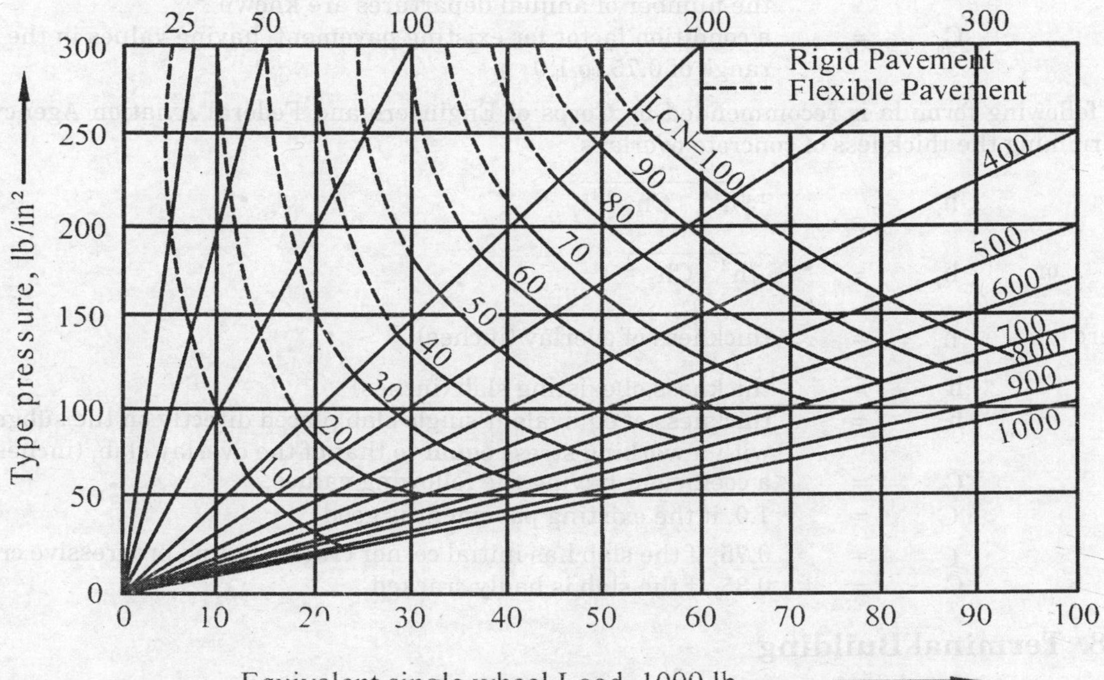

Fig. 43.15 : LCN, Load, Tyre Pressure, Contact Area Relationships for Flexible and Rigid Pavements

ACN-PCN System: The ICAO has developed an aircraft pavement classification system for reporting pavement strength.

The system, known as ACN-PCN, reports in pavement classification number (PCN) which indicates that an aircraft with an aircraft classification number (ACN) equal to or less than the PCN can operate on the pavement subject and will have a different ACN dependent upon limitations on the tyre pressure. A given aircraft will have a different ACN dependent upon the pavement on which it operates, whether flexible (F) or rigid (R) and the subgrade strength.

ICAO permits some overloading of the pavement by aircraft with ACNs slightly larger than the reported PCN.

43.7.5. Overlays

When an existing pavement deteriorates structurally and functionally, an overlay is provided. Overlays can be rigid overlays or flexible overlays.

The following equation is recommended by US Corps of Engineers for a flexible overlay on a rigid pavement.

$$t = 2.5\,(Fh_d - C_b h)$$

Where t = overlay (flexible) thickness, (inches)

h_d = thickness of cement concrete pavement assuming that the existing pavement did not exist, (inches)

h = thickness of existing pavement, (inches)

F = a factor having values in the range of 0.6 to 1.0, which can be determined from a chart if the Modulus of Subgrade Reaction and the number of annual departures are known.

C_b = a condition factor for existing pavement, having values in the range of 0.75 to 1.0

The following formula is recommended by Corps of Engineers and Federal Aviation Agency for determining the thickness of concrete overlays.

$$h_c = \sqrt[1.4]{h^{1.4} - Ch_e^{\;1.4}}$$

or

$$h_c = \sqrt{h^2 - Ch_e^{\;2}}$$

Where h_c = thickness of overlay (inches)

h_e = thickness of existing slab (inches)

h = thickness of equivalent single slab placed directly on the subgrade with a working stress equal to that of the overlay slab, (inches)

C = a coefficient having the following values:

C = 1.0, if the existing pavement is good

C = 0.75, if the slab has initial corner cracks, but no progressive cracks

C = 0.35, if the slab is badly cracked

43.8. Terminal Building

43.8.1. Functions

The terminal building receives passengers intending to travel, processes them and their baggage and transfers them to the aircraft. It also receives the passengers disembarking, handles their baggage and enables them to continue their journey to the city by taxies, buses or Metro lines.

43.8.2. Facilities required and design requirement

The facilities to be provided and their design requirements can be dealt with under the following heads:

(i) Passenger Needs
(ii) Airlines Needs
(iii) Airport Management Requirements
(iv) Community Requirements

Passenger Facilities Needed

(i) Passengers and their baggage should be handled within a short time. There should be adequate number of check-in counters.
(ii) There should be facilities for seating, lounges for passengers, drinking water, refreshments, shopping, toilets, bookstalls, postal services, telephone, banking (ATMs) and money exchange.
(iii) There should be good passenger information system, displaying the arrival/ departure schedules, a good public address system and pre-paid taxi counters.
(iv) Arrival and Departure zones should be separated to avoid conflict and confusion.
(v) Parking facilities should be adequate and should be close to the terminal building.
(vi) Baggage trolleys and wheel-chair facilities should be provided.
(vii) Baggage should be handled by means of conveyor belts

Airlines Requirements

(i) There should be space for ticketing, passenger check-in for the various airlines operating.
(ii) Immigration and Customs clearance facilities for international passengers should be available.
(iii) Security checking facilities should be available.

Airport Management Requirements

(i) Provision should be made for their office space.
(ii) Facilities for cleaning the airport spaces shall be available.

Community Requirements

(i) The passengers should be able to access the airport through their own cars, taxies, buses and Metro Lines.
(ii) The airport should be close to the city it serves so that a minimum time is lost in transit.

43.9. Air Traffic Control

Air traffic control aids are located in a separate space, accommodating the administrative staff, control tower, weather stations, and a radio communication centre.

43.10. Illumination

The lighting provided at the airport should be adequate. The edges of the runway and the taxiways and their centre should be clearly visible.

43.11. Radio Equipment

For controlling air traffic, use is made for radio and radar. The radio equipment is used for air surveillance, weather observation, monitoring take-offs and landings, guidance to aircraft to reach the terminal area and facilitating instrument aided landing.

43.12. Heliports

43.12.1. Helicopters and their role

Helicopter find extensive use in military operations, rescue operations, police patrolling and for civilian use for short hauls, such as Guwahati to Arunachal Pradesh. India has a public sector helicopter operation entity known as Pavan Hans. They can take-off or land in a nearly vertical position in heliports.

43.12.2. Characteristics of Helicopters

Helicopters of weight 3 to 5 tonnes can carry 1 to 3 passengers. Those of weight 5 to 10 tonnes can carry 10 to 30 passengers. Very large ones which carry upto 50 passengers weigh about 15 tonnes. Their speed is in the range of 150 to 200 km/hr. The rotors which are above the helicopter body have a diameter of 10 to 20 m. The overall length is 10 to 25 m and the height is 2.5 to 7.5.

43.12.3. Site selection for Heliports

The considerations for selecting a site for a Heliport are:-

(i) A clear approach and departure area free from obstructions with good visibility.

(ii) Avoidance of conflict with other air traffic.

(iii) Good potential for air traffic.

(iv) Good connectivity with other modes of transport.

(v) Avoidance of residential area so as not to cause noise.

43.12.4. Design Standards

The layout and dimensions are given in Figs. 43.16 and 43.17 and Table 43.7.

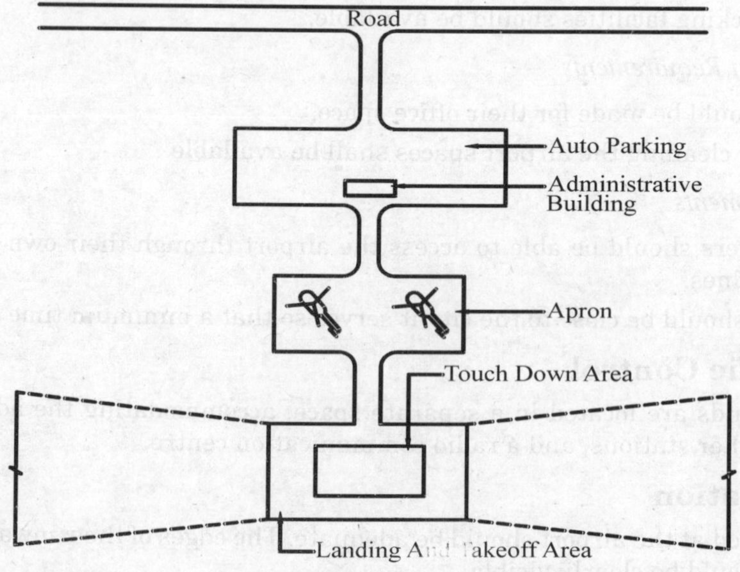

Fig. 43.16 : Heliport Layout

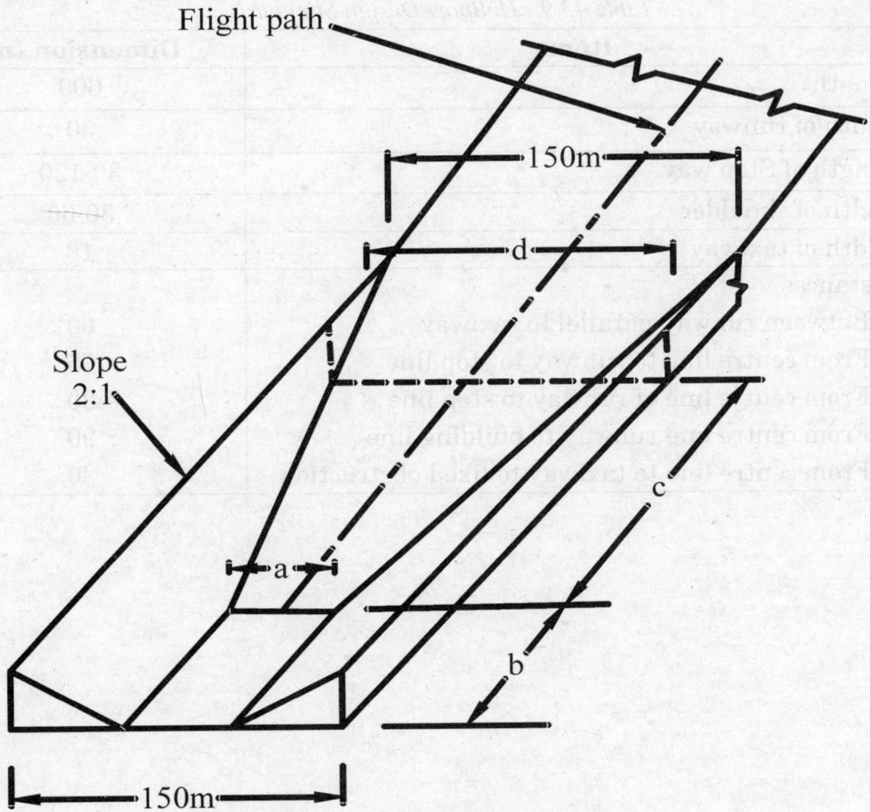

Fig. 43.17 : Heliport dimensions

Table 43.8 : Heliport Design Standards

Items	*Dimensions*			
	a	**b**	**c**	**d**
Private	1.5 A	1.5 A	90 m	60 m
Small Public	1.5 A	2.0 A	90 m	90 m
Large Public	1.5 A	2.0 A	120 m	90 m

[Note: A is the overall length of the helicopter. c is the recommended distance in metres prior to beginning of the curved path]

43.13. STOL

STOL means Short Take Off and Landing facility, where aircraft can operate on short runways of length less than 600 m. The climb is steep, at an angle of about 6° as against 3° needed by conventional aircraft. Table 43.8 gives the salient dimensions of STOL facility.

Table 43.9 : Heliport Design Standards

	Item	Dimension (m)
1.	Length	600
2.	Width of runway	30
3.	Length of Stop way	30-120
4.	Width of shoulder	30-60
5.	Width of taxiway	18
6.	Distances	
	Between runway parallel to taxiway	60
	From centre line to runway to stop line	75
	From centre line of runway to stop line	30
	From centre line runway to building line	90
	From centre line to taxiway to fixed obstruction	30

SECTION—5

CONTINUOUS FLOW SYSTEMS

44

Section

Pipe Line Transportation

44.1. Role of Pipelines in Transportation

Transport of fluids and gases are very efficiently handled by pipelines. The several advantages of pipelines transport are: -

(i) A continuous flow is ensured without interruptions and can be controlled as per actual requirements.

(ii) It is environmentally friendly.

(iii) There are no accidents as in the case of other modes.

(iv) The requirement of land is minimal.

(v) Maintenance cost is low.

(vi) No problem of return of empties (wagons and truck mounted tankers).

(vii) No losses during transportation

(viii) Very commercial.

44.2. Pipeline Transport in Selected Countries

In the U.S.A., the length of oil pipelines in 2010 was 276,844 Km. and the length of gas pipelines was 2,478,000 Km. (Ref. 1). In China, the length of oil pipelines in 2010 was 37,176 Km and the length of gas pipelines was 38,566 KM. In India, in 2010 the length of oil pipeline was 17,934 Km. and the length of gas pipelines was 11, 750 Km. In the oil producing countries, the pipelines are the principal mode of transport for this commodity.

In India, there are several pipelines operating conveying crude and oil products. Mention may be made of Nahorkatiya-Guwahati-Barauni pipeline, Ankleshwar-Koyali pipeline, Haldia-Barauni pipeline, Guwahati-Siliguri pipeline, Barauni-Kanpur pipeline, Koyali-Ahmedabad pipeline and Haldia-Mourigam pipeline, several pipelines conveying gas and the pipeline from Kudremukh to New Mangalore Port carrying iron ore.

44.3. Storage Facilities

For petroleum crude imported through oil tankers, storage depot are needed at the ports and at the points of transfer to railway wagons and oil tankers moving by road.

44.4. Types of Flow of Fluids

The flow of fluid particles move in parallel layers in one direction. In the turbulent flow, there are disturbance and eddies having both rotational and translation velocities.

There is a critical velocity at which the flow changes abruptly from laminar to turbulent. A dimensionless parameter known as Reynolds number is a reliable criterion for determining turbulence. The Reynolds number N_2 is given by:

$$N = \frac{Vdp}{u} \text{ seconds} \tag{44.1}$$

where V = mean velocity of flow, m/sec

d = diameter of pipe, m

u = absolute viscosity, Kg-sec/m^2

p = density = $\dfrac{\text{specific weight}}{u}$, Kg-sec^2/m^4

g = acceleration due to gravity, 9.81 m/sec^2

For round pipelines, the Reynolds number at critical velocity is taken as 2,000. If the number is less than 2,000 the flow is laminar, and if it is more than 3,000 the flow is turbulent. If the number is in the range 2,000-3,000, the flow is partly laminar and partly turbulent.

The loss of head occurs due to friction along the pipe.

44.5. Types of pipes used and construction

Generally cast iron or steel pipes are used for transport of liquids and gases. There may be a need to have booster stations at intervals to maintain the flow.

The pipes are generally laid below the ground. Iron pipes should be protected against corrosion by suitable coating. Where there are sudden changes in direction, the pipes should be anchored into solid concrete blocks.

REFERENCES

1. Encyclopaedia Britannica, Year Book 2014, Chicago, 2014.

45

Section

Aerial Ropeway

45.1. Use of Aerial Ropeways

Aerial Ropeways are used for transport of passengers and goods in areas where the provision of roads or railways would be too costly because of the hilly terrain conditions and the need to gain height is a short distance. They are economical of power consumption. They are either monocable or bicable. In the former, there is only one cable, whereas in the latter two parallel cables are provided, side by side of each. The system operates with a cable strung between towers. Single spans can be upto 300 m. They are provided in India where considerable tourist traffic exists. Examples are Mussorie, and the Hardwar-Mansa Devi Temple. In Switzerland, they are very popular.

45.2. Speed

The cars in the ropeway shall be driven at speeds which are safe. As per current standards in India, the following restrictions apply:-

(i) For attended cabins – 10 m/sec in long spans and 7.5 m/sec. over trestles.

(ii) For unattended cabins – 6 m/sec in long spans and 4 m/sec. over trestles

45.3. Considerations in Selection Layouts

(i) The alignment should be straight, and without curves. If any deviation is needed from the straight axles, it should not been more than 0.5 per cent.

(ii) There should be sufficient lateral and vertical clearance, with a minimum of 1.5 m on all sides.

(iii) The cars should not touch any portion of the trestle, and all trestles should be provided with guides.

45.4. Capacity of Car

The design of the various components should be done on the basis of 70 Kg per traveller in the case of cars having a capacity of 15 passengers. For cars carrying more than 15 passengers, the weight per passenger shall be 65 Kg. The area of the car shall be calculated on the basis of 0.222 sq. m per person upto 15 passengers and 0.15 sq. m. per person beyond 15 passengers. Where the car capacity is more than 15 passengers, a conductor should be present in the car.

45.5. Design of Cable Rope

The cable rope is made of stranded steel wires. The design should consider the moving load of the cars and wind loads.

The rollers moving on cables should have roller bearings, and the cables are lubricated to minimise the frictional forces.

45.6. Design of Trestles

The trestles supporting the cables shall be designed for the following loads:

(i) Dead load.

(ii) Load of moving cars and the oscillations generated by them.

(iii) Wind load.

(iv) Stresses caused by friction of moving ropes.

(v) Load of cables.

The factors of Safety shall be 3.

The trestles should have a firm foundation and be safe against displacement or subsidence. The trestles should be designed to resist torsion.

45.7. Stations

At the terminal stations, the cars run at fixed rails. The station should be designed keeping the following considerations:-

(i) Passengers should not cross the path of the cars.

(ii) The machinery and equipment at the station should be sufficiently away from the passenger movement area.

(iii) The stations should have covered roof to protect the passengers and equipment from snow and rain.

(iv) The operator of the cabins should be able to monitor the speed of travel, location of the car and detect any malfunctioning.

(v) The vehicle should be brought to rest at the station by means of a suitable device.

(vi) A good wireless communication facility shall be available between the terminal stations.

45.8. Maintenance

Periodic inspection and maintenance should be regularly carried out. Painting of steel members should be done at regular intervals. A Maintenance Manual specifying all the maintenance operations needed, and a register of maintenance carried out should be maintained.

46

Section

Belt Conveyors

46.1. Use

Belt conveyors are used for transporting materials on a continuous basis. The materials usually handled are: *(i)* Coal for thermal plants or at ports for handling import/export; *(ii)* Iron ore and other mineral raw materials at steel mills and at ports for handling import/export; and *(iii)* Stones at stone crushers. They are also used in airports for handling baggage of passengers, and also passengers at busy airports. The speed of passenger conveyors is 1.0 – 1.5 m/sec.

46.2. Components of the System

The three components are: *(i)* Belts; *(ii)* Idlers; *(iii)* Power system for driving.

46.2.1. Belts

Belts are made of rubber reinforced with cotton, rayon, steel, nylon (as in tyres). The width is in the range of 30 to 150 cm, as required by the capacity to be handled. Speed of movement is in the range of 100-250 metres per minute.

46.2.2. Idlers

Idlers are rollers over which the belt moves. They are of a diameter of 10 to 20 cm and are spaced at 0.75 – 2.0 m. The sketch of idlers and the belt is given in Fig. 46.1.

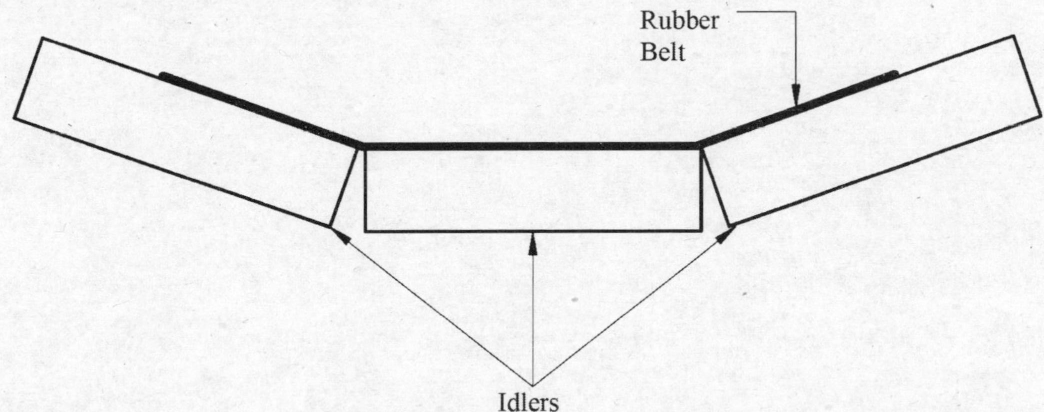

Fig. 46.1: Belt and Idlers in a Conveyor Belt System.

46.2.3. Escalators

Escalators having steps are provided for the movement (upwards or downwards) at airports, railway stations, pedestrian foot-over bridges and shopping malls. The rate of motion should be such that passengers can step into and disembark from the escalators.

Power requirement of belt conveyors

The normal arrangement is that the belt is driven at its ends by pulleys. The power requirement has three components:

(i) for operating an empty conveyors, P_o

(ii) for transporting load at a level stretch, P_1

(iii) for lifting the material from one elevation to another, P_2

$$P_o = C(L + L_O)QS \tag{46.1}$$

Where P_o = Power required in kg-m/min.

L = Horizontal length centre to centre between pulleys, m

L_O = Correction for friction losses, which are independent of belt length varying between 150-1000 m.

Q = Belt weight, weight of idlers, Kg/m of conveyor

S = Belt speed in m/min.

$$P_1 = \frac{1000\ \text{TC}\ (\text{L} + \text{L}_\text{O})\ \text{S}}{60S} \tag{46.2}$$

$$= \frac{50\ \text{TC}\ (\text{L} + \text{L}_\text{O})\ \text{S}}{3} \tag{46.3}$$

Where T = Weight of material, T/hour

$$\text{P}_2 = \frac{50\ \text{TH}}{3}\ \text{Kg-m/min.} \tag{46.4}$$

Where H = height to be lifted, m

Total $P = P_0 + P_1 + P_2$ (46.5)

The efficiency of the motor can be taken as 0.70 to 0.75.

SECTION—6

WATER TRANSPORT

47

Section

Water Transport : An Overview

47.1. Historical Perspective

Water transport has been an important factor in the development of civilisation and growth of nations. Rivers and seas have provided the primitive man with facilities for the movement of people and goods. Colonisation by countries like England, France, Portugal and Spain and discovery of new parts of the world (America, Australia etc.) have been possible only because of transport by sea. India, surrounded by seas, were quick to utilise water transport to send Indian goods to distant lands. Being blessed with good navigable rivers, India took advantage of inland water transport. India was colonised by England, France and Portugal by warriors and settlers who came from their faraway lands by ships. The Britishers built harbours and ports to facilitate export of raw materials and import of manufactured goods. They also built several canals along which inland water transport grew.

47.2. Advantages and Disadvantages of Water Transport

Advantages

(i) It is a cheap mode, because no expenditure is involved in constructing the paths like roads and railway lines. The only investments needed are in building ships and boats and providing terminal facilities. Water transport is also very energy efficient compared to other modes.

(ii) It is environment friendly and does not create pollution.

(iii) It is comparatively safe.

(iv) It is eminently suitable for defence

Disadvantages

(i) Its speed is lower than other modes like road transport, rail transport and air transport. Thus, whereas in earlier days, Indians used to travel to U.K in ships, they now travel only by air.

(ii) It is not a complete mode by itself as it has to depend on complementary modes like road transport, railways and pipelines. Thus, while petroleum crude comes to India in oil tankers, they have to be further transported by other modes to the refineries.

47.3. Types of Water Transport

Water transport is possible in several forms such as:

(i) Oceanic shipping

(ii) Coastal shipping

(iii) Inland Water Transport (IWT)

47.4. Water Transport in India Today

47.4.1. Oceanic Shipping

Oceanic shipping in India is heavily dependent on foreign ships. The share of Indian ships in 2010 in overseas trade is only 7.95 per cent (Ref. 1). Out of the Dead Weight Tonnage (DWT) of ships, 64 per cent are oil tankers, 29 per cent are bulk carriers and the rest 7 per cent are liner vessels. Indian Gross Tonnage (GT) is now 11.03 million tonnes and ranks sixteenth in the world. Thus there is obviously a need to increase the fleet.

47.4.2. Coastal Shipping

Coastal shipping in India is still in its infancy, despite the several advantages it enjoys. There are only 764 vessels accounting for about one million G.T.

47.4.3. Inland Water Transport

Inland Water Transport (IWT) is an underdeveloped mode of transport despite several advantages it has. This is mainly because some of the rivers are not perennial. The main movement takes place in the Brahmaputra River in Assam and Hoogly River in West Bengal. The provision of a lock in the Farakka Barrage has facilitated the IWT operations on the Ganges River. Kerala has a sizeable IWT operations covering passengers and goods.

47.4.4. National Waterways

The following waterways have been declared as National Waterways (NWWs)

(1) **NWW 1** : Ganga-Bhagirathi-Hoogly

(2) **NWW 2** : River Brahmaputra (Dhubri-Sadiya – 89 Km)

(3) **NWW 3** : West Coast Canal (Kottapuram-Kollam) along with Udyogamandal and Champakara Canals – 205 km in Kerala

(4) **NWW 4** : Kakinada-Puducherry canals along with Godavari and Krishna Rivers 1078 km

(5) **NWW 5** : East Coast Canal integrated with Brahmani and Mahanadi Rivers – 588 km.

47.4.5. Ports

India has a coast line of 7517 km served by 13 major ports and over 200 minor ports. The 13 major ports are: -

(1) Kolkatta

(2) Haldia

(3) Paradeep

(4) Vishakhapatnam

(5) Ennore

(6) Chennai

(7) Tuticorin

(8) Cochin
(9) New Mangalore
(10) Mormugao
(11) Mumbai
(12) JNPT
(13) Kandla.

The traffic handled at these major ports in 2012 was 560 million tonnes.

47.4.6. Shipyards

There are two major ship building yards, one at Vishakhapatnam and the other at Cochin.

REFERENCES

1. Twelfth Five Year Plan, Planning Commission, New Delhi, 2012.

48

Section

Planning for Water Transportation Facilities

48.1. Types of Facilities

The types of facilities needed for water transportation are:-

(1) Port
(2) Harbour
(3) Dock
(4) Jetty
(5) Quay
(6) Wharf
(7) Pier
(8) Transit Sheds
(9) Warehouse
(10) Cargo handling facilities

A port is a facility where transport of men and materials changes from a land based mode to a water based mode. A Free Trade Zone is an area adjacent to a port, but isolated, and which is provided with facilities for loading, unloading, processing goods without payment of customs duties. Kandla port is an example of Free Trade Zone in India.

A harbour is a safe haven for ships where they can be moored and sheltered from wind and sea waves.

A dock is an artificial enclosure for the reception of ships, loading and unloading of cargo and passengers. A wet dock is one which has impounded water at a uniform level, independent of external tide action and regulated by lock gates. The Princes and Victoria Docks in Mumbai is an example.

A dry dock is a dock to which a ship can be made to enter and from which water can be temporarily pumped away to make it completely dry to enable repairs to be carried out to the ship. The Merewether Dry Dock in Mumbai is an example.

A jetty is a structure which projects from the land to the sea. Marine Oil Terminal is a jetty for oil tankers in Mumbai.

A Quay is the surface upon which railway tracks, and crane tracks can be laid to handle cargo from ships.

A Wharf is a continuous structure, generally acting as a retaining wall, along a sea edge or along the banks of rivers, canal or any other waterway.

A pier is an isolated marine structure connected with the mainland. The Ballard Pier in Mumbai is an example.

Transit Sheds are covered structures meant for handling incoming and outgoing cargo for a short time and are provided behind quays.

Warehouse is a covered structure for storage of port cargo for a longer period. Bonded warehouses are those intended for storage of dutiable goods.

Cargo handling facilities include cranes, liquid cargo handling facilities, bulk cargo handling facilities and container handling facilities.

48.2. Types of Cargo Handling in Ports

The cargo handling in ports are of the following categories:

1. Dry bulk (food grains, mineral ores, coal, fertilisers)
2. Liquid bulk (petroleum crude and products, vegetable oil)
3. Containers
4. Parcels
5. Liquefied or Compressed Gas
6. General cargo

48.3. Considerations in Planning of Ports

The following considerations should be kept in view in the planning of ports:

1. Availability of cargo traffic to make the operations commercially viable
2. Availability of adequate land to accommodate all the facilities
3. Availability of good inland transport linkages like roads and railways
4. Good channel depth for accommodating the vessels that are likely to use the facility.

48.4. Traffic Forecast

The hinterland which generates the traffic that uses the facility such as the agricultural products, minerals, coal, industries, petroleum crude, gas reserves etc. dictates the amount of traffic likely to use the port. To have reasonably accurate forecast, a study of the past growth trends in the above categories of traffic and the expected future growth rate should be made. Trend analysis and Ecometric Models are generally used. The following expression gives the future traffic projection as per trend analysis.

$$T_n = T_0 (1 + r)^n \qquad (48.1)$$

where T_n = Traffic in the nth year

T_0 = Traffic in the 0th year

r : annual rate of growth (in decimals)

n : number of years

The following expression gives the future traffic projection as per econometric model which links traffic with economic indicators like population, GNP, Industrial output etc.

$$\text{Log}_e T_w = A + B\ \text{Log}_e\ (E.\ I) \qquad (48.2)$$

where T_n = Traffic in the nth year

A = Constant

B = Regression coefficient

EI = Economic Indicator like GNP, Population, Industrial Output etc.

n : number of years

48.5. Survey and Investigations

Apart from the surveys needed for traffic forecast, the following surveys are needed:

(i) Topographical Survey of the land area, with a total station

(ii) Hydrographic survey

- *(a)* to establish the ground profile using modern instruments such as echo soundings, and taking readings at 10 metre intervals in a rectangular grid;
- *(b)* to determine the high and low water lines.

(iii) Sub-soil investigations for the harbour area to obtain information of the nature of soil and rock, to be done by taking borings to a depth of 50 m or till hard rock is met with.

(iv) Sub-soil investigations in the land area to design the nature and depth of foundations for dock walls and buildings.

(v) Tidal studies to determine the high and low tide levels; velocity and direction of sea currents; direction, amplitude and height of sea waves.

(vi) Weather data, including wind velocity and direction and rainfall.

48.6. Data on Ship Dimensions and Weight

The data on the weight (Dead Weight Tonnage-DWT), length, width and draft should be available for designing the port facilities. The following table, Table 48.1 gives a rough guideline.

Table 48.1 : Current Dimensions of Ships

#	*Type of Ship*	*DWT*	*Length (m)*	*Width (m)*	*Draft (m)*
1.	General Cargo Ship	Upto 50,000	150-200	20-30	8-10
2.	Container Ship	Upto 50,000	150-275	20-35	8-12
3.	Bulk Carriers	60,000-1,00,000	200-250	30-40	12-15
4.	Tankers	Upto 50,000	Upto 450	Upto 65	Upto 30

49
Section

Harbours

49.1. Definition of Term

Harbour is a place where ships can find refuge from storms and other destructive elements of the sea. All ports must have harbours. Harbours can be of two types:

(i) Natural Harbours

(ii) Artificial Harbours

Natural Harbours are those that are located in places where the coast line is such that a part of the sea is protected by land. Fig 49.1 gives the sketch of a natural harbour.

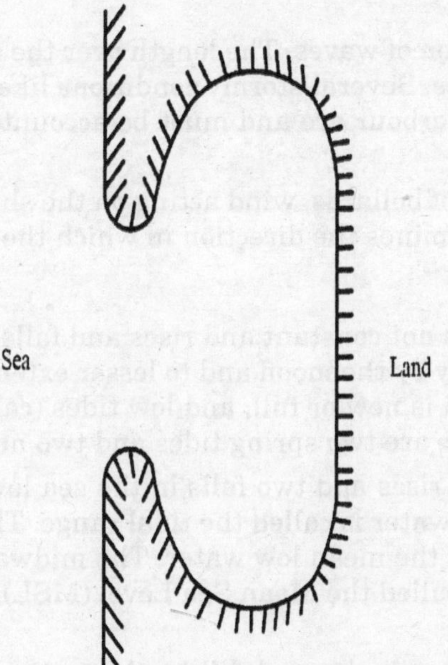

Fig. 49.1: Natural Harbour

Artificial harbours are those that need protective works like a breakwater to create a basin artificially. These are needed where the coastline is almost straight. Fig. 49.2 gives the layout of an artificial harbour created by constructing breakwaters.

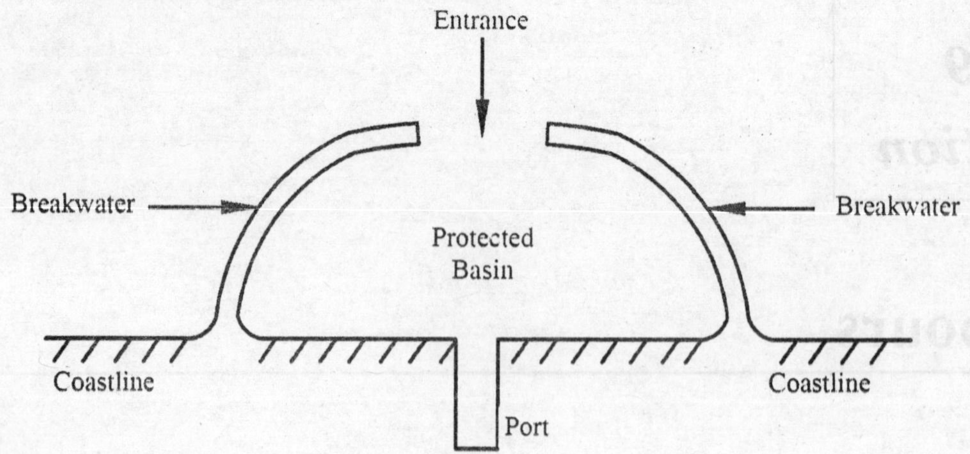

Fig. 49.2: Artificial Harbour

49.2. Factors Affecting Design of Harbours

The following three factors affect design of harbours:

(1) Wind

(2) Tides

(3) Waves

49.2.1. Wind

Wind is responsible to the formation of waves. The length over the see on which the storms prevail determines the height of the waves. Several stormy conditions like cyclones, tornado and typhoon cause considerable havoc in the harbour are and must be accounted for in the layout and design of harbours.

If ships are moored with the help of bollards, wind acting on the ships exerts a pulling force on the bollards. The wind direction determines the direction in which the waves travel.

49.2.2. Tides

The level of the water in the sea is not constant and rises and falls, mainly governed by the forces of attraction caused predominantly by the moon and to lesser extent by the sun. High tides (called spring tides) occur when the moon is new or full, and low tides (called neap tides) occur when the moon is in its quarters. Thus there are two spring tides and two neap tides in every lunar month.

On every lunar day there are two rises and two falls in the sea level. The difference between the high level and low level of the sea water is called the tidal range. The mean range is the difference between the mean high water and the mean low water. The midway level between the mean high water and the mean low water is called the Mean Sea Level (MSL). The tidal range is about 2.6 m in Mumbai.

Because of this, the berthed ships will rise and fall by this amount and it becomes necessary to provide an entrance lock so that the water level inside the basin is kept constant.

49.2.3. Waves

In the design of harbour works, the force exerted by the waves need to be considered. Fig. 49.3 gives a typical wave.

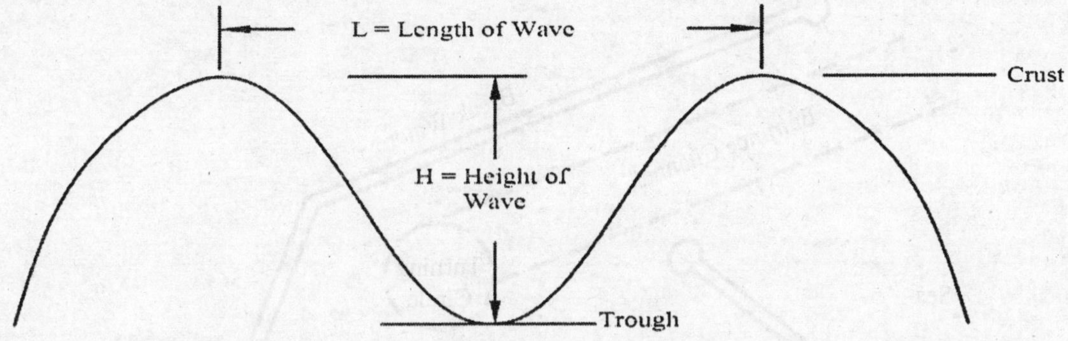

Fig. 49.3: A typical wave form

The speed wave v is given by

$$v = \frac{L}{T} = \frac{gT}{2\pi} \tag{49.1}$$

$$L = \frac{2\pi v^2}{g} = \frac{gT}{2\pi} \tag{49.2}$$

$$T = \sqrt{\frac{2\pi L}{g}} = \frac{2\pi v}{g} \tag{49.3}$$

In the above:

v	=	Speed of wave
L	=	Length of wave
T	=	Wave period
g	=	acceleration due to gravity (9.81 m/sec^2)

The height of the wave, H, is determined by the length upto which storms are limited, known as the fetch (F)

$$H = \sqrt[1.5]{F} \tag{49.4}$$

where H = Height in feet

F = Fetch in nautical miles

Waves exert enormous pressure against the marine structures.

49.3. Harbour Layout

A typical harbour has the following elements:

(1) **Entrance Channel, through which the ships enter the harbour.**

(2) **Turning Circle, where the ships can be manoeuvred into their moorings.**

(3) **Breakwater for protecting the harbour from waves.**

(4) **Docks where the ships are berthed.**

Fig. 49.4 gives a typical layout.

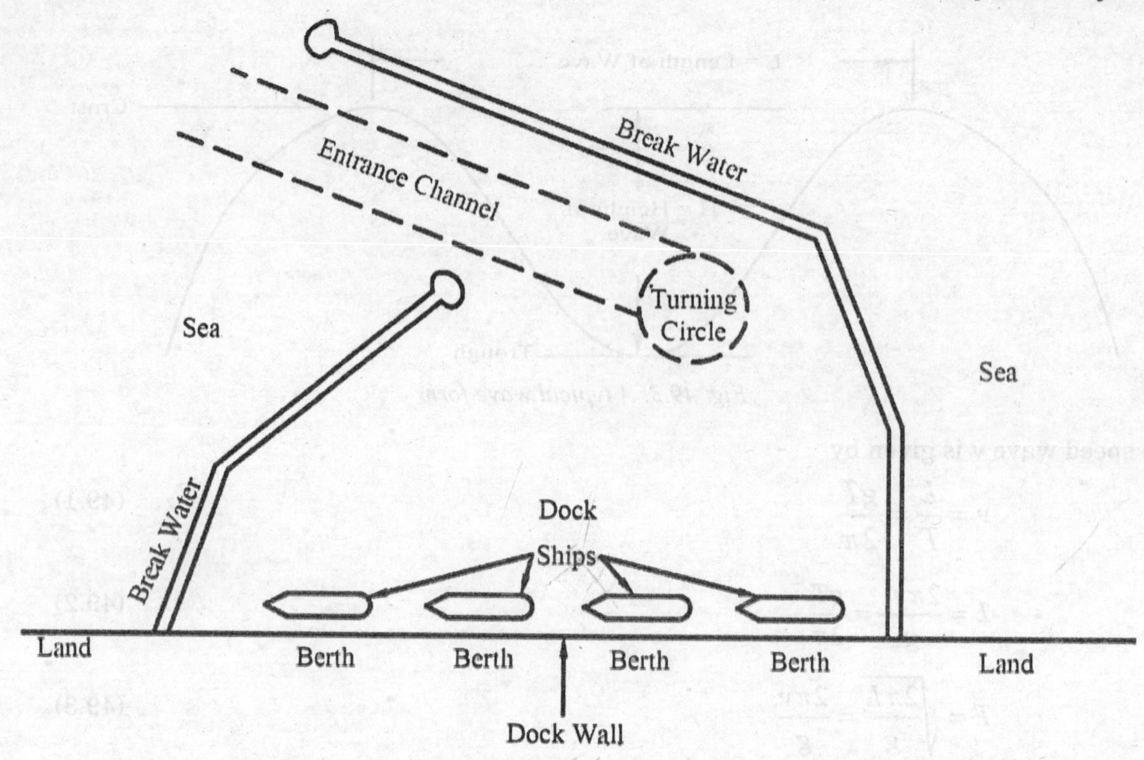

Fig. 49.4: Harbour Layout

49.4. Entrance Channel

From the open sea, the ships enter the harbour area through the entrance channel. The channel should be of adequate width to accommodate the movement of the ships of the expected sizes. The alignment should be straight with no curves. The entrance should be on the leeward side so that the ships entering or leaving are protected from wind and waves. The length varies from 100 m for small harbours to 300 m for large harbours.

The channel should have depth such that there is enough clearance (about 1.5 – 2.5 m) between the keel of the ship and the land underneath.

49.5. Turning Circle

The radius of the turning circle should be at least twice the length of the largest ship expected to use the facility.

49.6. Breakwater

The breakwater protects the harbour and the ships using it from the effect of sea waves. Breakwaters can be of rubble mound or vertical sided or composite as shown in Fig. 49.5, 49.6. Rubble mound breakwaters are the cheapest where adequate quantity of stone boulders is available from nearby quarries. The top width is generally about 3 m.

The end of a breakwater is bullet nosed as shown in Fig. 49.3.

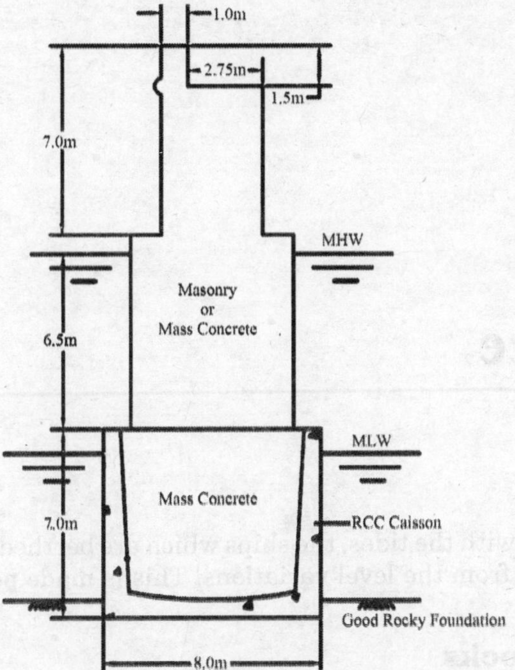

Fig. 49.5: Vertical Sided Breakwater

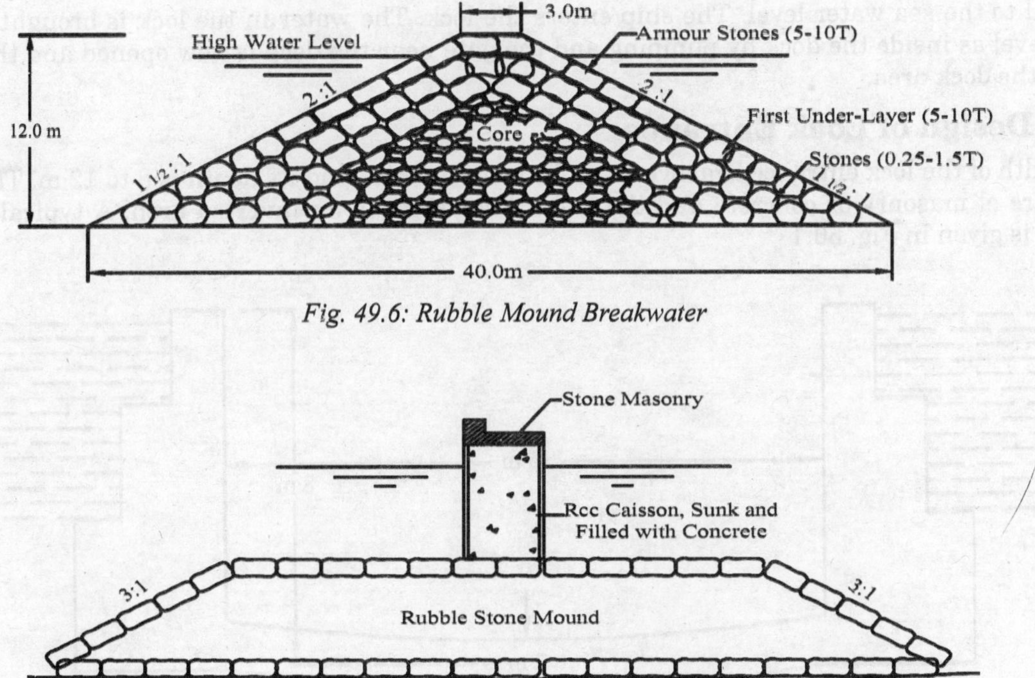

Fig. 49.6: Rubble Mound Breakwater

Fig. 49.7: Composite Breakwater

50
Chapter

Lock Entrance

50.1. Need for Locks

As the sea water level changes with the tides, the ships which are berthed for loading and unloading at a dock need to be protected from the level variations. This is made possible by the provision of locks.

50.2. Functioning of Locks

A lock is a concrete or masonry line channel with gates at its two ends. The length of lock entrance is governed by the length of the ship expected to use the port. To admit a ship into the dock area, the gate near to the dock is closed and the front gate is opened so that the water level in the lock is equal to the sea water level. The ship enters the lock. The water in the lock is brought to the same level as inside the dock by pumping and the gate near the dock is now opened and the ship enters the dock area.

50.3. Design of Lock Entrance

The width of the lock entrance is above 25 m for modern ships and its depth is 8 to 12 m. The side walls are of masonry or concrete and the bottom is generally an inverted arch. A typical cross-section is given in Fig. 50.1

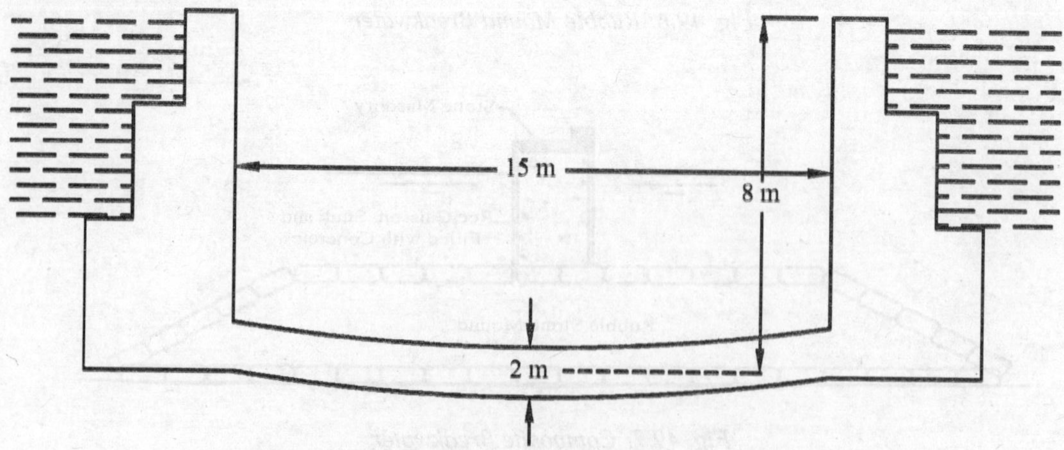

Fig 50.1: Typical Cross-section of the Lock Entrance

51

Chapter

Docks

51.1. Types of Docks

Three common types of docks are-

(i) Wet Docks

(ii) Dry Dock

(iii) Floating Dry Dock

51.2. Wet Docks

Wet Docks are sheltered and enclosed places where ships are received, berthed for loading and unloading. The water level is kept constant since any fluctuations can disrupt the operations. The berths are arranged in different layouts as given in Fig. 51.1.

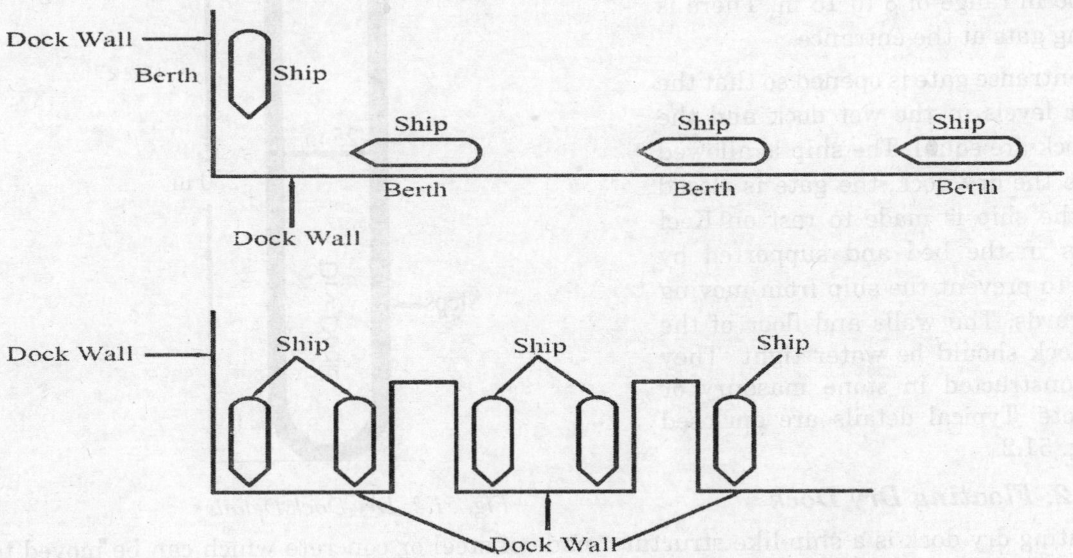

Fig. 51.1: Layout of Berths in a Dock

The length of ships varies from 150 to 350 m. An additional length of 25 m on either side is needed for manoeuvring. The draft of ships varies from 7 to 10 m, and for large tankers it may be as high as 30 m.

51.3. Design of Dock Walls

Dock walls are retaining walls of the gravity type. They are constructed in mass concrete or stone masonry. The front is vertical and is protected by fenders from the impact of ships. The forces for which dock walls are designed are:-

(i) Dead weight of the wall;

(ii) Earth pressure with live load surcharge, the live load varying from 1000 to 4000 kg/sq.m.;

(iii) Water pressure;

(iv) Seismic forces.

They should preferably be founded on rock. The quay is paved with hard-standing pavement like concrete pavement.

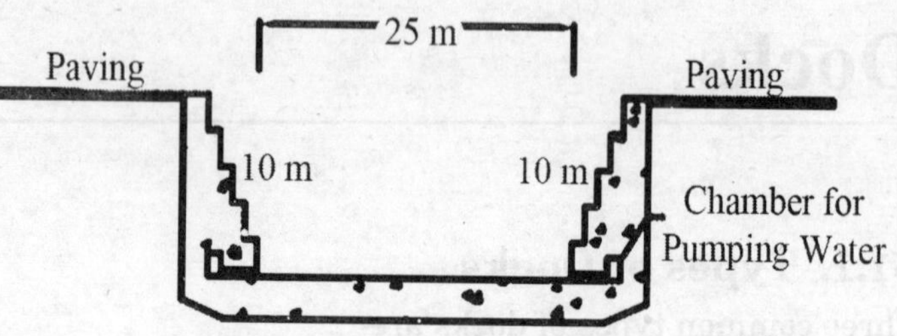

51.4. Dry Dock

51.4.1. A dry dock is a place where ships are repaired. The length depends upon the length of ships and can be in the range of 200 to 300 m. The width is in the range of 20 to 30 m. The depth can be in range of 8 to 15 m. There is sliding gate at the entrance.

The entrance gate is opened so that the water levels in the wet dock and the dry lock are equal. The ship is allowed inside the dry dock, the gate is closed and the ship is made to rest on Keel blocks in the bed and supported by stays to prevent the ship from moving sidewards. The walls and floor of the dry dock should be water tight. They are constructed in stone masonry or concrete. Typical details are enclosed in Fig. 51.2.

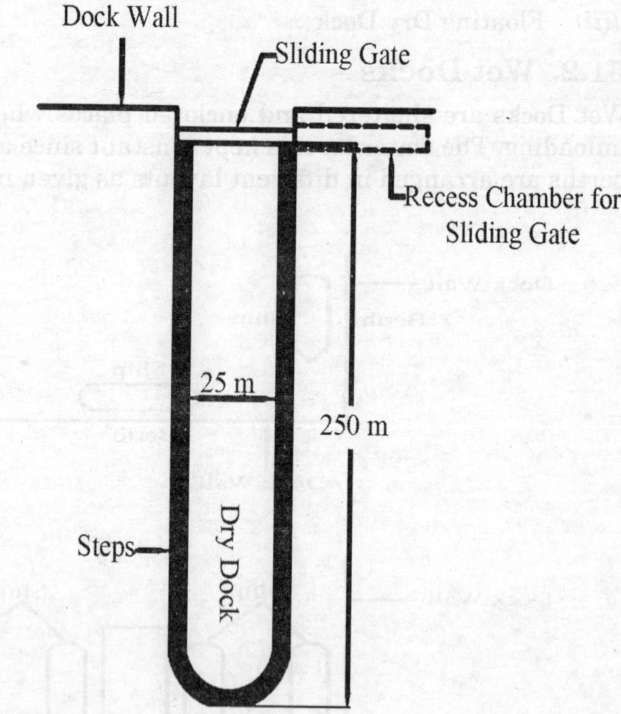

Fig. 51.2: Dry Dock Details

51.4.2. Floating Dry Dock

A floating dry dock is a ship-like structure made of steel or concrete which can be moved to any position. The ship to be repaired is brought inside the floating dry dock and the water inside is pumped out. The floating structure then rises up and the ship is made to rest on its keels. Shores are then put into position so that the ship does not move laterally. After the repair is done, water is made to enter the floating dry dock and the ship then moves out.

52

Chapter

Transit Sheds, Warehouses and Cargo Handling Facilities

52.1. Transit Sheds

When a ship arrives at a dock and is berthed, it has to be loaded or unloaded. The cargo has to be protected from rain and the sun temporarily. The cargo is stored in sheds called transit shed. They are provided at a distance of about 25 m from the dock wall. In this space in the quay, road and rail facilities and gantry cranes are provided. The transit shed can be single storied and two storied. It has a width of about 20 to 30 metres. Cargo handling facilities like lifts, forklifts are provided.

52.2. Warehouses

The warehouses are also covered sheds where the cargo is stored for a longer period. They are provided at a distance of about 25 m away from the transit sheds.

52.3. Typical Layout

A typical layout of transit sheds and warehouses is given in Fig. 52.1.

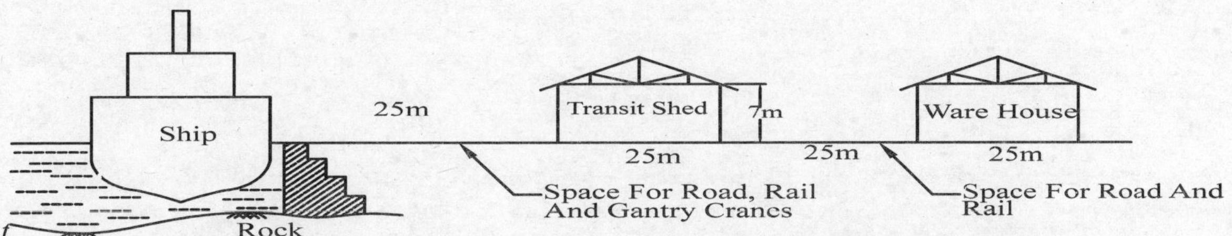

Fig. 52.1: Arrangement of Location of Transit Sheds and Warehouse in Quay

52.4. Cargo Handling Facilities

Cargo handling facilities include gantry cranes, forklifts and container handling cranes. Bulk cargo like coal, food grains, ores and fertilisers are handled by belt conveyors.

53

Chapter

Inland Water Transport

53.1. Draft of IWT (Inland Water Transport) Channels

IWT can take place in canals, rivers and backwaters of sea requires a minimum draft of the channels for operating the boats and barges. A minimum draft of 2 to 3 metres is needed. This requires dredging of rivers.

53.2. Land Facilities

Both passengers and cargo are handled by IWT. For this purpose, landing stages and jetties are needed. Small cranes can be of great help in handling cargo. For accommodating passengers in transit or cargo, sheds are provided to protect them from rain or sun.

SECTION—7

MULTI-MODAL TRANSPORT

54

Section

Multi-Modal Transport

54.1. Definition

Multi-Modal Transport is the carriage of goods (generally packed in containers or pallets) by two or more modes of transport from the place of acceptance of the goods to the place of delivery. It is sometimes also know as Inter-Modal Transport. In order to facilitate Multi-Modal Transport, among nations, the United Nations has adopted a convention on the system, and India has enacted the Multimodal Transport of Goods Act, 1993 to give legal backing (Ref. 1). Many other countries have also enacted laws enabling the multi-modal transport system.

54.2. Forms of Multi-Modal Transport

Multi-modal Transport can have several combinations of individual modes such as:

1. Rail and Road
2. Road, Rail and Overseas Shipping
3. Road, Rail and Coastal Shipping
4. Road, Rail and Inland Water Transport
5. Road and Air

Some of the specific forms of multi-modal transport are:

54.2.1. Roll-On, Roll Off (RORO)

This is a combination of road and rail, or road and sea. The truck moves from the place of origin loaded with goods and reaches a railhead, or a port, where it rolls on to a railway flat-car or a water-based vessel. The truck travels on the rail or water-based vessel to the destination rail-head or port, rolls-off and travels by road to the final destination. Thus, travel by road over long distances is avoided, saving fuel, avoiding emissions that pollute the air and improving safety. An example of RORO in India is the movement of trucks by Konkan Railway Corporation from a point near Mumbai to Kerala.

54.2.2. Piggy Back

In the piggy-back system, the road transport vehicle consists of a tractor unit and a trailer carrying a container. The tractor/trailer unit travels from the point of origin to a rail-head/port, and only the trailer unit is moved on to a railway flat car or a water-based vessel. At the end of the rail/water journey, the trailer rolls off and is attached to a tractor unit to continue its road journey to the final destination. The advantage of piggy-back over RORO is that the tractor unit does not have to spend idle hours during its rail/water leg.

54.2.3. Multi-Modal Transport

Oceanic shipping over long distances is slow and takes several days. If a land exists between two seas, the goods can be transported by land from one sea to the other. The Trans-Siberian land bridge is a classic example. Containers from Japan travel by sea to the Eastern Russian ports, they are then transferred to the Trans-Siberian Railway, they move to MOSCOW by rail and move by road or rail to an European port and then travel by sea to U.K.

54.2.4. Sea Train

This is similar to the RORO, except that the railway coaches/wagons move on an ocean carrier. This concept has been implemented for a long time on the English Channel crossings before the construction of the channel tunnel.

54.3. Containerisation

Containers, introduced in the mid-fifties, have revolutionised global trade and commerce. Under this system, the commodities are packed into a standard-size container which is transferred from one mode to another at the transhipment point, without emptying its contents. The entire container travels from the consigner to the consignee in various modes. The container is of standardised dimensions and is of such size and weight that mechanical equipment is needed for handling it. The International Standards Organisation (ISO) has standardised the sizes, and the containers are known as ISO containers. Some of the sizes are:

1. 8 ft. x 8 ft. x 10 ft.
2. 8 ft. x 8 ft. x 20 ft.
3. 8 ft. x 8 ft. x 30 ft.
4. 8 ft. x 8 ft. x 40 ft.

The term TEU means twenty foot equivalent units. The total container traffic handled in the world is over 150 million TEUs. Indian ports handle about 5 million TEUs whereas Singapore alone handles about 13 million TEUs a year.

The advantages of containerisation are:

1. Pilferage and damage to articles is avoided, and hence insurance costs are reduced.
2. Packaging costs are reduced.
3. Productivity of transport is increased and transport costs are reduced because rehandling of goods is avoided.
4. Dirty cargo like bones and hides, which are loathed by handling workmen because of smell can be packed in a container and conveniently handled.
5. The container is reusable and has a long life

54.4. Infrastructure Facilities

For handling multi-modal transport, some special infrastructural facilities are needed. These are Inland Container Depots (ICD) and Container Freight Stations (CFS).

54.4.1. Inland Container Depot (ICD)

These are facilities built near to the cargo generating hinterland. The ICDs serve as places where the mode is changed, for example, from road to rail and vice versa. Customs clearance facilities are also made available. The functions of ICDs cover stuffing, destuffing, locking, sealing, providing handling equipment, customs examination, and issuance of a combined transport by a carrier.

54.4.2. Container Freight Station (CFS)

CFSs are facilities where stuffing and destuffing and customs formalities are completed. The loaded containers are then moved to the nearest ICD.

In India, the Container Corporation of India (CONCOR), a subsidiary of the Indian Railways manages the ICDs and CFSs. The ICD at Tughlakabad near Delhi is one the biggest facilities run by CONCOR.

Dedicated container handling port facilities exist in some of the major ports of India. Recently, the private sector has been involved in developing some of the container terminals in ports. It is necessary to state that the multi-modal transport, including containerisation, has great potential for growth in the coming years. The infrastructure facilities like roads, railways and ports need to be upgraded.

REFERENCES

1. Multimodal Transport of Goods Act, 1993, Government of India, New Delhi, 1993
2. Raghuram, G., Multimodal Transport and Communication, Infrastructure Development and Financing, Macmillan India Ltd., 1999.

1. National Transport [illegible] 1980, Government of India, New Delhi, 1980.